MOLECULAR
BIOLOGY
INTELLIGENCE
UNIT 23

Nuclear Envelope Dynamics in Embryos and Somatic Cells

Philippe Collas, Ph.D.
Institute of Medical Biochemistry
University of Oslo
Oslo, Norway

LANDES BIOSCIENCE / EUREKAH.COM
GEORGETOWN, TEXAS
U.S.A

KLUWER ACADEMIC / PLENUM PUBLISHERS
NEW YORK, NEW YORK
U.S.A

Nuclear Envelope Dynamics in Embryos and Somatic Cells

Molecular Biology Intelligence Unit 23

Landes Bioscience / Eurekah.com
and
Kluwer Academic / Plenum Publishers

Kluwer Academic / Plenum Publishers, 233 Spring Street, New York, New York, U.S.A. 10013
http://www.wkap.nl/

Please address all inquiries to the Eurekah.com / Landes Bioscience:
Eurekah.com / Landes Bioscience, 810 South Church Street, Georgetown, Texas, U.S.A. 78626
Phone: 512/ 863 7762; FAX: 512/ 863 0081; www.Eurekah.com; www.landesbioscience.com
Landes tracking number: 1-58706-150-3

Nuclear Envelope Dynamics in Embryos and Somatic Cells edited by Philippe Collas/CRC, 184 pp. 6 x 9/ Landes/Kluwer dual imprint/ Landes series: Molecular Biology Intelligence Unit 23, ISBN 0-306-47439-5

While the authors, editors and publishers believe that drug selection and dosage and the specifications and usage of equipment and devices, as set forth in this book, are in accord with current recommendations and practice at the time of publication, they make no warranty, expressed or implied, with respect to material described in this book. In view of the ongoing research, equipment development, changes in governmental regulations and the rapid accumulation of information relating to the biomedical sciences, the reader is urged to carefully review and evaluate the information provided herein.

Library of Congress Cataloging-in-Publication Data

CIP information applied for but not received at time of publishing.

MOLECULAR BIOLOGY INTELLIGENCE UNIT

Recent Volumes in this Series

The Jak-Stat Pathway in Hematopoiesis and Disease
Edited by Alister C. Ward

Telomerases, Telomeres and Cancer
Edited by Guido Krupp and Reza Parwaresch

Reactivation of the Cell Cycle in Terminally Differentiated Cells
Edited by Marco Crescenzi

Ceramide Signaling
Edited by Anthony H. Futerman

Nuclear Envelope Dynamics in Embryos and Somatic Cells
Edited by Philippe Collas

The CDK-Activating Kinase (CAK)
Edited by Philipp Kaldis

Caspases—Their Role in Cell Death and Cell Survival
Edited by Marek Los and Henning Walczak

Quasispecies and RNA Virus Evolution: Principles and Consequences
Edited by Esteban Domingo, Christof Biebricher, Manfred Eigen and John Holland

Cell Cycle Checkpoints and Cancer
Edited by Mikhail V. Blagosklonny

RNA-Binding Antibiotics
Edited by Renée Schroeder and Mary G. Wallis

Lysosomal Pathways of Protein Degradation
Edited by J. Fred Dice

CONTENTS

EDITOR

Philippe Collas, Ph.D.
Institute of Medical Biochemistry
University of Oslo
Oslo, Norway
Chapter 9

CONTRIBUTORS

Teresa Barona
Biology Program
University Lusofona
Lisbon, Portugal
Chapter 9

Khaldon Bodoor
Department of Anatomy
  and Cell Biology
University of Florida
Gainesville, Florida, U.S.A.
Chapter 6

Gisele Bonne
Institut de Myologie
INSERM
Paris, France
Chapter 12

Brian Burke
Department of Anatomy
  and Cell Biology
University of Florida
Gainesville, Florida, U.S.A.
Chapter 6

Paul R. Clarke
Biomedical Research Centre
University of Dundee
Dundee, Scotland
Chapter 5

L. Clements
MRIC Biochemistry Group
North East Wales Institute
Wrexham, England
Chapter 11

Jan Ellenberg
Gene Expression and Cell Biology/
  Biophysics Programmes
European Molecular Biology Laboratory
Heidelberg, Germany
Chapter 2

Roland Foisner
Department of Biochemistry
  and Molecular Cell Biology
University of Vienna
Vienna, Austria
Chapter 3

Yosef Gruenbaum
Department of Genetics
The Hebrew University of Jerusalem
Jerusalem, Israel
Chapter 8

James M. Holaska
Department of Cell Biology
  and Anatomy
Johns Hopkins University School
  of Medicine
Baltimore, Maryland, U.S.A
Chapter 1

I. Holt
MRIC Biochemistry Group
North East Wales Institute
Wrexham, England
Chapter 11

Banafshe Larijani
Cell Biophysics Laboratory
Imperial Cancer Research Fund
London, England
Chapter 9

S. Manilal
MRIC Biochemistry Group
North East Wales Institute
Wrexham, England
Chapter 11

M. Mansharamani
Department of Cell Biology
 and Anatomy
Johns Hopkins University School
of Medicine
Baltimore, Maryland, U.S.A
Chapter 1

G.E. Morris
MRIC Biochemistry Group
North East Wales Institute
Wrexham, England
Chapter 11

Cecilia Östlund
Department of Medicine
Columbia University
New York, New York, U.S.A.
Chapter 4

Dominic L. Poccia
Department of Biology
Amherst College
Amherst, Massachusetts, U.S.A.
Chapter 9

C.A. Sewry
MRIC Biochemistry Group
North East Wales Institute
Wrexham, England
Chapter 11

George Simos
Laboratory of Biochemistry
University of Thessaly
Larissa, Greece
Chapter 7

N. thi Man
MRIC Biochemistry Group
North East Wales Institute
Wrexham, England
Chapter 11

D. Tunnah
MRIC Biochemistry Group
North East Wales Institute
Wrexham, England
Chapter 11

Yonatan B. Tzur
Department of Genetics
The Hebrew University of Jerusalem
Jerusalem, Israel
Chapter 8

Corinne Vigouroux
Laboratoire de Biologie Cellulaire
INSERM
Paris, France
Chapter 12

F.L. Wilkinson
MRIC Biochemistry Group
North East Wales Institute
Wrexham, England
Chapter 11

K.L. Wilson
Department of Cell Biology
 and Anatomy
Johns Hopkins University School
 of Medicine
Baltimore, Maryland, U.S.A
Chapter 1

Mariana F. Wolfner
Department of Molecular Biology
    and Genetics
Cornell University
Ithaca, New York, U.S.A.
Chapter 10

Howard J. Worman
Department of Medicine
Columbia University
New York, New York, U.S.A.
Chapter 4

Wei Wu
Department of Medicine
Columbia University
New York, New York, U.S.A.
Chapter 4

Chuanmao Zhang
Biomedical Research Centre
University of Dundee
Dundee, Scotland
Chapter 5

PREFACE

Roughly twenty-five years of studies of the nuclear envelope have revealed that it is more than just a bag of membranes enwrapping chromosomes. The nuclear envelope consists of several domains that interface the cell cytoplasm and the nucleus: the outer and inner nuclear membranes, connected by the pore membrane, the nuclear pore complexes and the filamentous nuclear lamina. Each domain is marked by specific sets of proteins that mediate interactions with cytoplasmic components (such as cytoskeletal proteins) or nuclear structures (such as chromosomes). The nuclear envelope is a highly dynamic structure that reversibly disassembles when cells divide. How these nuclear envelope domains and proteins are sorted at mitosis, and how they are targeted back onto chromosomes of the reforming nuclei in each daughter cell are two fascinating questions that have dominated the field for many years. Another item which in my mind makes the field of the nuclear envelope exciting is the range of organisms in which it has been studied: yeast, sea urchin, star fish, *C. elegans, Drosophila, Xenopus*, mammalian cells and more. Each model organism displays common features in the ways the nuclear envelope breaks down and reforms, but also pins differences in its organization and dynamics. Another source of enthusiasm is the variety of experimental systems that have been developed to investigate the dynamics of the nuclear envelope. These range from cell-free extracts (again, from eggs or cells of many organisms), to the use of synthetic beads (which a priori have nothing to do with a nucleus), genetic studies in *C. elegans* and recent elaborate 4-D imaging studies in living mammalian cells. All these provide unique angles to our view of nuclear envelope behavior. Finally, for many, the nuclear envelope has experienced a 'rebirth' after the identification of mutations in two of its components, the inner nuclear membrane protein emerin, and nuclear lamins A and C. Mutations in these proteins are the cause of several forms of dystrophies of skeletal and cardiac muscles and are life-threatening.

In twelve chapters, prominent experts in their field deliver the latest views on how molecules and pathways are orchestrated to build, or disassemble, the nuclear envelope. Each chapter is meant to lead the reader to a specific domain of the nuclear envelope or to a particular process, whether this takes place in an egg, an embryo or a somatic—healthy or diseased—cell.

Editing this book would have not been possible without the formidable contributions from all authors—many thanks to all of them, an initiative from Ron Landes and the technical support from Cynthia Dworaczyk.

I hope this volume will provide the reader with a better appreciation of the biology of the nuclear envelope. Have a good time reading it.

Philippe Collas

COLOR PLATE

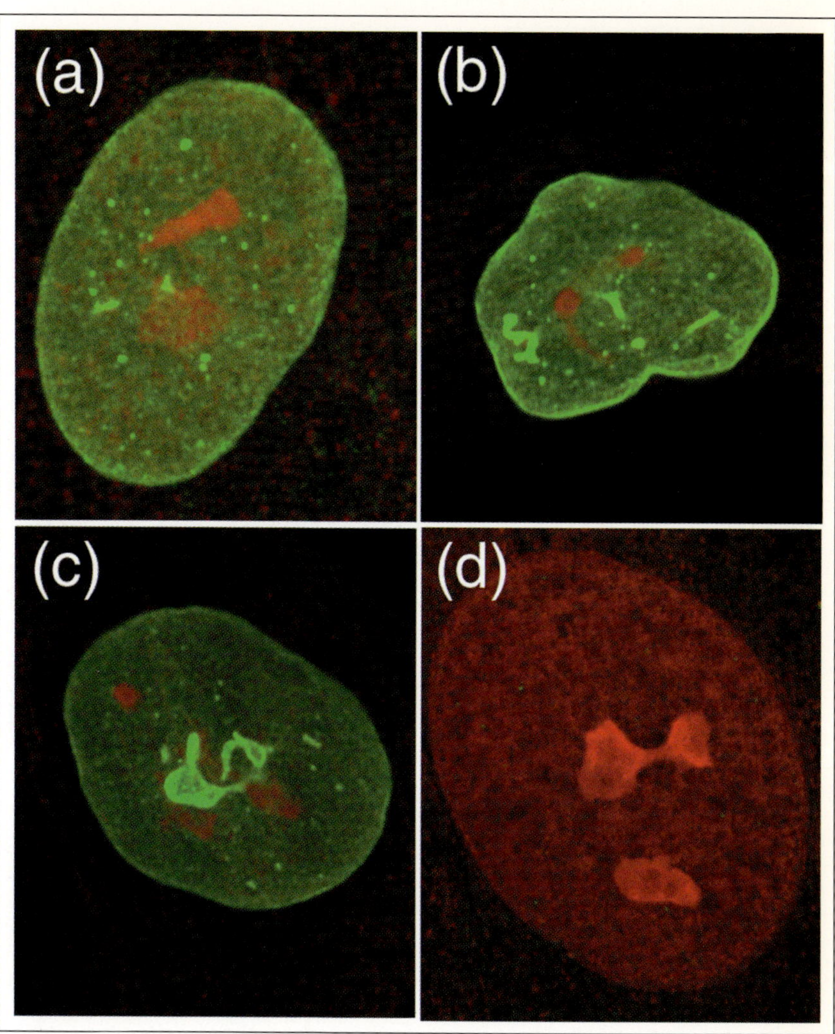

Figure C.1. Emerin distribution in cultured human skin fibroblasts as shown in Chapter 11, Figure 3. Emerin (green) was detected using monoclonal antibody MANEM5 and FITC-labelled secondary antibody with a BioRad MicroRadiance confocal microscope. Nuclei were counterstained with ethidium bromide (red) and the more intensely-stained bodies are nucleoli. (A)-(C) show different patterns observed in normal nuclei and (D) shows absence of emerin in an X-EDMD nucleus. Cells were fixed with a 50:50 acetone:methanol.

CHAPTER 1

Dynamics of the Vertebrate Nuclear Envelope

Malini Mansharamani, Katherine L. Wilson and James M. Holaska

Abstract

The cell nucleus is a complicated organelle that houses the genome of humans and other eukaryotic organisms. Chromosomes are enclosed by the nuclear envelope, and 'communicate' with the cytoplasm by the regulated movement of molecules across nuclear pore complexes. In multicellular animal eukaryotes ('metazoans'), a special set of nuclear membrane proteins and lamin filaments interact with chromatin to provide key structural and functional elements to the nucleus. Remarkably, these structures are reversibly disassembled during mitosis. This Chapter describes the structure and major constituent proteins of the metazoan nuclear envelope, our current understanding of nuclear envelope dynamics during mitosis, and pathways for the reversible breakdown and reassembly of the nuclear envelope and nuclear infrastructure. This field is moving quite fast. A better understanding of these fundamental aspects of nuclear envelope structure and dynamics will provide new insights into an emerging class of inherited human diseases, including Emery-Dreifuss muscular dystrophy, dilated cardiomyopathy, and lipodystrophy. Further work in this field may also suggest novel anti-viral therapies for HIV or herpesvirus, which specifically disrupt nuclear envelope structure during their life cycles.

Interphase Nuclear Envelope Structure

The nucleus of metazoan cells includes highly stable structures such as the chromosomes, the nuclear envelope and lamina plus highly mobile proteins responsible for RNA production and nuclear metabolism. This complex architecture is reversibly disassembled during mitosis. In this Chapter, we summarize interphase nuclear structure, and the events and mechanisms of nuclear envelope disassembly and reassembly during mitosis.

The nuclear envelope defines and encloses the cell nucleus. The envelope is composed of two concentric membranes (outer and inner) and nuclear pore complexes that are anchored by a network of filaments termed the nuclear lamina. The outer membrane is continuous with, and has the same protein composition as the rough endoplasmic reticulum (ER). The outer and inner nuclear membranes fuse periodically to form nuclear pores. Pores have a diameter of ~100 nm and are occupied by nuclear pore complexes. Nuclear pore complexes actively mediate the transport of macromolecules between the nucleus and the cytoplasm. They also provide aqueous channels through which ions and small proteins (<40-60 kD) can diffuse passively.[1] Unlike the nuclear outer membrane, the inner membrane has a unique protein composition and can thus be viewed as a highly specialized subdomain of the ER. Proteins unique to the nuclear inner membrane include the lamin B receptor (LBR),[2] several isoforms each of the lamina associated polypeptides (LAPs) 1 and 2,[3-5] emerin,[6,7] MAN1,[8,9] nurim[10] and the RING Finger Binding Protein (RFBP).[11] Many of these proteins can bind directly to nuclear lamins, which are abundant near the inner membrane. We will refer to the lamin

Nuclear Envelope Dynamics in Embryos and Somatic Cells
edited by Philippe Collas. ©2002 Eurekah.com and Kluwer Academic/Plenum Publishers.

filaments and lamin-binding proteins collectively as the nuclear lamina. The lamina comprises a major element of nuclear architecture and nuclear function.[12,13]

Lamin filaments are composed of nuclear-specific type V intermediate filament proteins named lamins (reviewed in 14). Lamins have a small N-terminal 'head' domain followed by an α-helical coiled-coil 'rod' and a C-terminal globular 'tail'. The rod sequence is highly conserved with other intermediate filament proteins, except for a lamin-specific extension in the second α-helical segment of the rod domain,[14] whereas the head and tail domains of lamins are more divergent from cytoplasmic intermediate filaments.[15] The coiled-coil rod mediates the formation of parallel dimers, which pair into anti-parallel tetramers. The tetramers polymerize head-to-tail into polymeric filaments.[16] There are two classes of lamins, A-type and B-type, based on their biochemical properties and sequence homology. B-type lamins are found in all cell types, including embryonic cells.[17] In contrast, A-type lamins are expressed predominantly in differentiated cells and are therefore proposed to contribute to cell-type specific functions.[18] Vertebrates have three lamin genes. Two genes code for B-type lamins; *LMNB1* encodes lamin B1[19] and *LMNB2* encodes lamins B2 and B3.[14] The third gene, *LMNA*, encodes four isoforms of A-type lamins, through differential splicing: lamins A, AΔ10, C1 and C2.[20-22] Interestingly, lamins C2 and B3 are both expressed uniquely in germ cells,[23, 24] suggesting roles in the meiosis-specific reorganization of nuclear and chromosomal structure.[25] As discussed in a later Chapter, mutations in A-type lamins are now linked to at least five hereditary diseases that affect a variety of specific tissues.

Without lamins, nuclear structure is severely impaired. In cell-free extracts of *Xenopus* eggs, which will assemble nuclei around added chromatin, the addition of dominant negative mutant lamin proteins prevents nuclear envelope reassembly after mitosis.[26] When lamins are immunodepleted from extracts, the resulting nuclei are fragile and cannot replicate their DNA[27,28] indicating that nuclear lamins are required for DNA replication. *LMNA* knockout mice are born normal, but by three weeks after birth, develop a severe form of muscular dystrophy.[29] These mice die by eight weeks. Similarly, RNAi-mediated depletion of the only lamin in *C. elegans* (B-type) causes embryonic lethality.[30] There is currently no data for the phenotype of B-lamin knockout in any organism with more than one lamin gene. Nevertheless, the phenotypes of lamin-null *C. elegans* and the *LMNA* knockout mice strongly suggest that B-type lamins are essential for life, whereas A-type lamins are tailored to the functions of specific cell types and tissues. A-type lamins are also essential for long-term viability of individuals.

Inner Nuclear Membrane Proteins

Many different proteins located at the inner nuclear membrane are known to bind lamins, and these interactions are important for attaching lamin filaments to the inner membrane. The first lamin-binding membrane protein to be discovered was LBR, the 'lamin B receptor'. LBR is a 58 kD membrane protein, which has a ~25 kD nucleoplasmic domain followed by eight transmembrane domains.[2] The nucleoplasmic domain of LBR binds directly to lamin B[31,32] and also interacts with a chromatin protein named HP1, which is required for repressive chromatin structure in *Drosophila*.[33,34] The next proteins to be discovered were named Lamina associated polypeptides (LAPs)1 and 2. The *LAP1* gene is proposed to encode three LAP1 protein isoforms (A, B and C) by alternative splicing.[35] The A and B isoforms of LAP1 interacted with all lamins tested, including lamins A, C and B1.[36] LAP1C is the most abundant, and is currently the only isoform for which the full cDNA sequence is published.[35] LAP1C binds strongly to lamin B.[37] More is known about the *LAP2* gene, which encodes six isoforms in mammals by alternative splicing.[4,5] LAP2β is the largest membrane-bound LAP2 isoform, and directly binds lamin B. LAP2β also has a growing number of additional partners including BAF (Barrier-to-Autointegration Factor; a small DNA binding protein),[38,39] DNA[40] and GCL, a transcriptional repressor.[41] LAP2β is known to play roles in DNA replication competence and nuclear reassembly[42,43] by mechanisms that are not yet understood. Interestingly, LAP2β itself can function as a transcriptional repressor.[41] The other widely expressed isoform of LAP2,

LAP2α, does not have a transmembrane domain and is distributed throughout the nucleoplasm, where it forms stable complexes with A-type lamins and an unidentified chromatin partner.[5] In theory, all LAP2 isoforms are capable of binding to chromatin through the DNA- and BAF-binding domains present at their shared N-terminal constant region.[39,40]

The BAF-binding domain of LAP2 is conserved in several other nuclear envelope proteins, including emerin and MAN1, in a 40-residue region called the LEM domain.[9] The atomic structure of the LEM-domain has been solved for LAP2 and emerin.[40,44,45] The LEM-domain mediates binding to BAF.[38] BAF is a 10 kD protein that forms a stable 20 kD homodimer, and binds to double-stranded DNA non-specifically.[46-48] BAF is highly conserved in metazoans,[46] but is absent (along with lamins and all other nuclear envelope proteins discussed here) in yeast and plants. Like the B-type lamins, BAF is essential for the viability of dividing cells, suggesting fundamental roles in nuclear structure and function.[48]

Emerin directly binds both A- and B-type lamins as determined by in vitro binding assays and coimmunoprecipitations, but may have higher affinity for A-type lamins,[49] and specifically lamin C.[50] The nuclear localization of emerin depends on lamins, since deletion of the only lamin in *C. elegans* causes the loss of emerin protein from the nuclear envelope.[51] Another inner nuclear membrane protein, named RFBP (RING Finger Binding Protein) binds the SWI2/SNF2 related RUSH transcription factors.[11] RFBP has nine transmembrane domains; it has not yet been tested for binding to lamins. Important areas for future work include the identification of binding partners for RFBP and other newly identified nuclear membrane proteins, including the LEM-domain protein MAN1[9] and nurim.[10]

The interactions of inner nuclear membrane proteins with lamins are thought to have functional implications for the nucleus. Importantly, a growing number of transcription factors are localized to the lamina. Oct-1, a ubiquitously expressed transcription factor, co-localizes with lamin B.[52] Retinoblastoma (Rb), a transcriptional repressor with major roles in growth control, co-localizes with lamin A in its functionally repressive form.[53] Finally, a transcriptional repressor named GCL (germ-cell-less) binds directly to the β-specific region of LAP2β, and GCL and LAP2β together are as effective as Rb in repressing transcription.[41] Therefore, the lamina provides not only mechanical strength to the nucleus, but may also help localize or stabilize protein complexes essential for gene regulation, as discussed here, and DNA replication as discussed earlier.

Nuclear Envelope Disassembly

During mitosis in multicellular eukaryotes, prophase is marked by the disassembly of the nuclear envelope. The nuclear envelope then begins to re-form even while the chromosomes segregate during late anaphase and telophase. Nuclear disassembly is a regulated process, in which the chromosomes condense, the pore complexes disassemble, the lamina filaments undergo a slow depolymerization, and nuclear membranes and inner membrane proteins are dispersed into the ER.[54-56] Disassembly is driven by site-specific phosphorylation of key target proteins by the mitotic cyclin-dependent kinase cdc2 (also known as p34^{cdc2} or MPF, maturation promoting factor)[57] or protein kinase C.[58] The major events and mechanisms of nuclear disassembly are discussed below. It is important to note that lamin disassembly, chromatin condensation and membrane dispersal are all independent events.[59] Functional nuclear pore complexes are required for nuclear lamina disassembly, to allow the entry of cell-cycle regulatory proteins including cdc25[60] and cyclin B,[61] which are critical for activating mitotic kinase activity inside the nucleus.[62,63]

Nuclear Pore Complex Disassembly

Nuclear pore complexes (NPCs) are the first structures to disassemble during mitosis.[64,65] Terasaki and colleagues[65] proposed an elegant model in which the disappearance of NPCs leaves behind open unstabilised pores (holes) in the nuclear envelope; these holes allow the mitotically-phosphorylated inner membrane proteins to diffuse freely to the outer membrane and hence into the ER network. The NPC is a supramolecular structure with an estimated maximum mass of 124×10^6 Da.[1] Each NPC is composed of about 40 distinct proteins (nucleoporins or Nups), each of which is present in 8 copies or multiples of 8 copies. Identified vertebrate Nups have been reviewed elsewhere.[66] Similar to other nuclear envelope components, NPC disassembly appears to be driven by mitotic phosphorylation. Interestingly, the disassembled Nups remain associated as soluble subcomplexes that disperse throughout the cytoplasm during mitosis.[67] The nucleoporins Nup97 and Nup200 are directly phosphorylated by the cdc2/cyclin B kinase, and exist in complexes of masses ~1000 kD and 450 kD, respectively, in mitotic *Xenopus* egg cytosol.[68] Other mitotically hyperphosphorylated nucleoporins include Nup153 (a component of the intranuclear NPC basket), and Nup214 and Nup358 (which are found on cytoplasmic NPC filaments). It is worth noting that some nucleoporins may also be phosphorylated during interphase, potentially for the purpose of regulating NPC function.[69] Gp210, which is one of only two known integral membrane nucleoporins, is not phosphorylated during interphase but is specifically phosphorylated at Ser 1880 during mitosis, by cdc2/cyclinB.[67,70] It is proposed that this phosphorylation disrupts binding between the exposed gp210 tail and an unknown partner, and might disrupt the anchoring of NPCs to the pore membrane. More work, particularly on gp210 and the other membrane nucleoporin POM121, is needed to understand the mechanisms of NPC disassembly.

Membrane Disassembly

The mechanism of nuclear membrane disassembly has been a matter of some confusion until recently. In fractionated egg extracts from *Xenopus laevis*, heterogenous 80-300 nm vesicles were seen to bind chromatin and fuse to form the nuclear envelope.[71,72] It was proposed that these vesicles arose during mitosis by a mechanism similar to the formation of ER transport vesicles, and that the nuclear membranes therefore disassembled by vesiculation.[73] Upon the inactivation of mitotic kinases, these nuclear vesicles would be permitted to fuse and reassemble the envelope.[74,75] However, cell fractionation procedures might have converted tubular membranes into vesicles. The question of nuclear membrane disassembly has been clearly answered for mammalian cells using live-cell imaging studies, in which the dynamic properties of LBR and POM121, an integral membrane protein of the NPC, were studied during mitosis using fluorescence recovery after photobleaching (FRAP).[56,76] These experiments showed that GFP-labeled LBR and POM121 were stably localized at the inner membrane and NPC during interphase, but were dispersed into a relatively intact ER network during mitosis.

It has been suggested that all nuclear proteins that are mitotically phosphorylated may contribute to nuclear envelope structure in some way. LBR contains phosphorylation sites for PKA[31] and the mitotic cyclin-dependent kinase cdc2/cyclin B.[62,77] Consistent with mutual structural roles, the chromatin partner for LBR, a protein named HP1, is also phosphorylated in a cell cycle dependent manner.[78] The β and γ isoforms of LAP2 are phosphorylated on multiple residues during interphase, suggesting that the interphase functions of LAP2 are regulated by several different kinases.[79] In addition, LAP2 isoforms become differentially phosphorylated at mitosis.[36] Phosphorylation at mitotic-specific residues causes LAP2β to dissociate from lamin B in vitro.[36] Emerin is also differentially phosphorylated during mitosis, and like other nuclear membrane proteins becomes dispersed throughout the ER.[80]

Until recently, phosphorylation was viewed only in the context of mitosis. However because of growing evidence for regulated phosphorylation during interphase, investigators now need to map both mitotic and interphase sites of phosphorylation. It will also be useful to consider additional modifications such as glycosylation, which is a common modification of several nucleoporins.[81]

Lamina Disassembly

Gerace and colleagues[82] localized lamins by immunofluorescence and electron microscopy during both interphase and mitosis. They showed that lamins are prominent at the nuclear periphery during interphase, yet become dispersed throughout the cell during mitosis. Their subsequent discovery that nuclear lamin proteins were reversibly depolymerized during mitosis, and that depolymerization correlated with hyperphosphorylation, provided key insights into the mechanism of nuclear envelope breakdown.[54] Furthermore, the A-and B-type lamins had distinct behaviors during mitosis in many cell types; the B-type lamins remained associated with membranes during mitosis, whereas lamins A and C were completely solubilized.[54,83] Both A- and B-type lamins are post-translationally modified by prenylation at their C-termini, which confers greater affinity for membranes.[84] Prenylated A-type lamins are recognized by Narf, a newly-identified protein inside the nucleus.[85] However, the prenyl modification is then cleaved from A-type lamins during proteolytic processing of the C-terminus, to yield mature lamins A and C.[86] Both A- and B-type lamins are mitotically phosphorylated on conserved serine residues, located at each end of the coiled-coil rod domain.[57,87,88] Phosphorylation appears to change the conformation of lamin dimers such that the dimers or tetramers are released from the lamin polymer. Further work on lamins will be facilitated by knowing the structure of the tail domain of lamin A, which was recently solved independently by the Shoelson and Zinn-Justin laboratories.

Nuclear Assembly

The metaphase-anaphase transition during mitosis is triggered by the proteolytic degradation of cyclins and cohesins, which inactivates the mitotic kinases and sister-chromatid 'glue' proteins, respectively.[89-91] Once the mitotic kinase is inactivated, phosphatases rapidly de-phosphorylate the dispersed nuclear membrane proteins, lamins, and nucleoporins.[92] Nuclear envelope components begin to re-assemble while the chromosomes are segregating during anaphase and telophase. Typically, each set of daughter chromosomes is completely re-enclosed within a nascent nuclear envelope by late telophase.[74] In the ensuing second phase of re-assembly, which is more poorly understood, the nascent nucleus re-imports a multitude of dispersed soluble nuclear proteins, including lamins, and must re-assemble its interior infrastructure, decondense the chromatin, and expand to reform a functional interphase nucleus.

The mechanisms of nuclear envelope formation have been studied primarily using fractionated, reconstituted extracts from *Xenopus* eggs, *Drosophila* embryos, and sea urchin eggs which contain stockpiles of mitotically disassembled nuclear components.[74, 93] Recent advances in fluorescent imaging and the ability to express specific nuclear envelope proteins fused to the Green Fluorescent Protein (GFP), have allowed investigators to follow nuclear assembly in living cells in real time. These advances, combined with the recent use of *C. elegans* as an experimental organism,[64] are increasing our understanding of nuclear envelope assembly.[66,75]

The mechanisms of nuclear envelope assembly will be discussed chronologically, starting with the proposed mechanisms for targeting (or sorting) membranes that contain inner nuclear membrane proteins (e.g., LAP2β, emerin, MAN1) to the chromatin during late anaphase and telophase. We will then discuss the assembly of nuclear pores and NPCs, which are essential to re-establish nuclear transport activity, and the role of nuclear transport in nuclear growth. Finally, we will discuss nuclear lamina assembly, which remains an important open question.

Nuclear Membrane Targeting and Fusion

The nuclear envelope can be viewed as a highly-specialized subdomain of the ER. To re-assemble the nuclear envelope, ER membranes that carry the inner nuclear membrane proteins must (a) bind the chromatin surface, (b) fuse together to enclose the chromosomes, and (c) flatten and fuse periodically to form pores and NPCs.[74,94,95] As discussed earlier, the nuclear membranes (and inner membrane proteins) mix with the ER membrane proteins during mitosis. It is still not understood which membrane proteins have structural roles in nuclear envelope assembly, primarily due to our inability to specifically deplete integral membrane proteins from isolated membranes and extracts. For this reason genetic systems such as *C. elegans* and *Drosophila* will be critical for the functional analysis of nuclear membrane proteins.

Nuclear Membrane Protein Targeting to Chromatin

Despite being dispersed throughout the ER, most nuclear-specific membrane proteins re-accumulate at the chromatin surface within minutes after the metaphase-anaphase transition, as determined by fluorescence imaging.[43,56,75] The kinetics of nuclear membrane protein recruitment (or sorting) to the reforming nuclear envelope is of great interest to the field, and has been followed using various fluorescently-labeled nuclear envelope proteins in both fixed and living cells.[56,96,97] ER membranes gain access to chromatin during late anaphase and telophase, and membrane-chromatin contacts are likely to be stabilized by the binding of nuclear membrane proteins to their appropriate ligands on chromatin. Chromatin contacts gradually increase in number as additional inner membrane proteins reach the chromatin surface by diffusing along the ER membrane.[75,97] Access of dispersed nuclear proteins to the reforming nuclear envelope may be enhanced by the ongoing fusion and fission activities of ER tubules with the outer nuclear membrane.

DNA itself is recognized directly by both LAP2[39,40] and lamins.[98-100] Lamins also bind directly and specifically to histones H2A and H2B,[98,101] as well as DNA.[100] In addition, two non-histone chromatin-associated proteins, named heterochromatin binding protein 1 (HP-1) and Barrier to autointegration factor (BAF), interact with one or many nuclear membrane proteins. HP-1 localizes to chromatin during metaphase and binds LBR.[33,34,102] The other chromatin-associated protein, BAF, is proposed to interact with all LEM domain proteins, including LAP2, emerin, MAN1, and LEM-3 and otefin.[12,64,103] BAF has been shown bio-chemically to bind directly to both LAP2[38-40] and emerin.[104] New evidence suggests that emerin and BAF interact in vivo, and that BAF is required for the recruitment of emerin, LAP2β and lamin A (but remarkably not lamin B) to reforming nuclear envelopes.[104,105] BAF and HP-1 are sub-localized to different regions of the chromatin for a brief time (~4 minutes) during late anaphase and telophase. BAF and emerin co-localize at the so-called 'core' region, which comprises the surfaces of the massed telophase chromatin that are closest to, and opposite to, the spindle pole.[105] In contrast, HP-1 localizes to centromeres, whereas another isoform of HP1, HP1γ, localizes to the chromosome arms.[78] These transient spatial distinctions are important because emerin co-localizes with BAF at the 'core',[105] and LBR co-localizes separately with HP-1,[33] suggesting that these interactions may regulate specific steps in nuclear assembly. A few minutes later, all of these proteins spread out and become uniformly distributed around the nuclear envelope. An emerin mutant that cannot bind BAF also cannot re-assemble into reforming nuclear envelopes,[105] suggesting that BAF-emerin co-localization at the 'core' is critical to recruit emerin during nuclear envelope assembly, and possibly also to assemble lamin A-dependent structures inside the nucleus. Preliminary experiments in tissue culture cells suggest that HP1-LBR interactions are also important for nuclear assembly.[102]

Nuclear Membrane Fusion

Nuclear membrane fusion at the chromatin surface is likely to involve a mechanism by which ER tubules fuse, termed 'homotypic fusion'.[106] 'Lateral' fusion between adjacent nuclear envelope cisternae encloses the chromatin. Further **fusion events enlarge the nucleus.**[107] The fusion of nuclear membranes also requires GTP hydrolysis.[108-110] The GTPase responsible for this hydrolysis has not been identified, although it was shown that one particular GTP binding protein, ARF, is not required.[111,112] There is growing evidence that Ran, a GTPase that regulates the directionality of nuclear transport, may also mediate membrane fusion events. However, more work is needed to determine if Ran's role in membrane fusion is direct or indirect.[113,114]

Nuclear Pore Formation

The assembly of nuclear pore complexes (NPCs) is essential for the growth phase of nuclear assembly, because many different structural and regulatory proteins must be re-imported and assembled in the nucleus. One major class of proteins that are transported through NPCs are the A-type nuclear lamins.[74] Whereas B-type lamins are membrane- associated during mitosis in vertebrate cells, A-type lamins are soluble and must be imported into nascent nuclei prior to their polymerization at the nuclear envelope.[74,115] Interestingly, NPCs are also needed to re-assemble nuclear membrane proteins and lipids. Because the chromatin is completely enclosed by membranes, the pore membrane domain is essential for nuclear membrane proteins to gain diffusional access to the inner membrane. The pore membrane domain also allows lipids to move to the inner membrane, which might be essential for the nuclear envelope to expand back to its full, interphase size.

Individual nuclear pore complexes form very rapidly and are seen within seven minutes in reconstituted *Xenopus* egg extracts.[116] Pore formation begins as soon as membranes attach and flatten onto the chromatin surface.[72,116] However, pore formation does not require chromatin, since NPCs can assemble into stacked ER-like membranes termed 'annulate lamellae' in the absence of chromatin.[117,118] Initiation of pore formation is an interesting question, because eukaryotic evolution depended on having a mechanism to allow the genome to communicate with the cytoplasm. The fusion of all biological membranes is triggered by proteins, which help lipid bilayers overcome their mutual surface charge repulsion.[119,120] 'Porogenic' fusion between the inner and outer nuclear membranes is proposed to require protein domains that extend into the lumenal space of the nuclear envelope. Vertebrates have two known integral membrane nucleoporins, named POM121[121] and gp210.[122,123] Based on the membrane topology of gp210, which has a small, exposed C-terminal domain and a massive lumenal domain, gp210 was hypothesized to be the fusogenic protein.[124,125] Recent evidence is consistent with gp210 having a fusogenic role, and also suggests that the exposed C-terminal tail of gp210 may have a role in dilating small, nascent pores (1-5 nm diameter) immediately after the membrane fusion event.[126] FRAP experiments show that during interphase, POM121 is an extremely stable component of the NPC, with a half-time of turnover of more than 20 hours.[76] Unexpectedly, gp210 is relatively mobile during interphase, with an estimated half-time for movement of 6 hours (Bodoor and Burke, personal communication). These different mobilities may explain why POM121, but not gp210, re-accumulates rapidly during nuclear assembly,[127] but do not yet reveal the mechanism of pore membrane fusion. Determining the 'porogenic' membrane fusion mechanism is essential for understanding how functional nuclei form at the end of mitosis.

Assembly of the NPC

Pore formation and NPC assembly are challenging unsolved problems in cell biology. Functional NPCs are required for the first morphologically detectable step in nuclear envelope growth, termed 'smoothing'.[116] An ordered self-assembly pathway for NPC formation has been proposed based on the visualization of structures termed membrane 'dimples', 'stabilizing pores', and 'star-rings' in *Xenopus* nuclear assembly reactions.[128] 'Dimples' are indentations in the outer membrane, and are inferred to be intermediates in the membrane fusion events that generate the pores. 'Stabilizing pores' have irregular shapes, but are typically ~35-45 nm in diameter, and are thought to reveal how the NPC appears at a very early stage of assembly. The next two proposed intermediates in NPC assembly, termed star-rings and thin-rings, contain additional structures (cytoplasmic ring and underlying components) and can exhibit the characteristic eight-fold symmetry of mature NPCs.[128] Similar NPC-related structures have been seen in vivo in *Drosophila* embryonic nuclei.[129] Among the final steps of NPC formation are the assembly of filaments that emanate from the nucleoplasmic ring.[66,129] Little is known about the assembly of NPC substructures located within the nucleus. Even less is known about the formation of interior filaments that attach to NPCs, except that these filaments may consist of the Tpr protein in association with nucleoporins Nup98 and Nup153.[130-132]

Biochemical intermediates in NPC formation have been characterized using annulate lamellae, which are NPC-rich stacks of membrane cisternae.[117] NPCs in annulate lamellae have similar biochemical composition and structure as NPCs formed in vivo. Forbes and colleagues used annulate lamellae formation in *Xenopus* egg extracts to study NPC assembly in vitro.[118,133-136] They found that reagents that block the homotypic fusion of membranes into cisternae (GTPγS and NEM) also block NPC assembly. They further discovered that BAPTA, a calcium buffering agent,[137] profoundly inhibited pore formation at the earliest stages.[133] BAPTA can also arrest NPC formation at the 'star-ring' stage.[128] A number of soluble nucleoporins are O-glycosylated, and removal of these nucleoporins from cytosol, by WGA-Sepharose depletion, can either produce malformed NPCs[118,128] or completely block pore formation.[128,138] The removal of O-GlcNAc-modified nucleoporins has distinct effects on both early and later stages of NPC assembly.[118,139]

Enlargement of the Nucleus

Because lamins play such fundamental roles in nuclear structure and shape, their re-assembly may be central to the formation of functional nuclei.[74,115] In reconstituted *Xenopus* egg extracts, chromosome decondensation is inhibited by blocking the polymerization of lamin B.[26] In mammalian cells, most A-type lamins do not integrate into filaments until the middle of G1.[115,140] These differences suggest that A- and B-type lamin filaments may assemble distinctly and deliberately, providing a plausible mechanism to both drive and control the increase in nuclear volume over time.

One of the biggest open questions for lamin polymerization is what role(s), if any, are played by the growing number of lamin-binding inner membrane proteins. Nuclei assembled in the presence of exogenous LAP2β fragments fail to expand, even though the nuclear envelope has NPCs and appears normal.[42,43] This block of nuclear growth might be due to LAP2 being required for lamin assembly. For example, LAP2β may promote lamin B polymerization at the nuclear envelope, and the soluble isoform, LAP2α may promote filament formation by A-type lamins in the nuclear interior. Alternatively, LAP2 proteins might bind lamins only as a localization mechanism, and contribute to nuclear expansion by regulating chromatin structure or the transcription of genes required for expansion.

Assembly of Non-Lamin Intra-Nuclear Structures

While the structure and assembly of nuclear lamins are still poorly understood, even less is known about other interior structures of the nucleus. These structures include an extensive network of intra-nuclear filaments that attach to the NPC and extend throughout the nucleus.[130,131,141] One component of these NPC-linked filaments is Tpr, a coiled-coil protein of 270 kDa that is localized at NPC baskets and can form parallel homodimers in solution.[142] NPC-linked filaments are proposed to facilitate the intra-nuclear movement of cargo destined for nuclear export.[130,131] A mobile nucleoporin, Nup98, which associates with export receptors, co-localizes extensively with NPC-linked filaments.[130] The three-dimensional assembly and function of the NPC-linked filaments and lamin filaments are important challenges for future work.

Concluding Remarks

Further study of the structure, assembly and dynamics of the nucleus will be important to understand the functions of this complex organelle, which is home to the human genome. A basic understanding of nuclear envelope structure may lead to rational therapies for an emerging class of human diseases, including Emery-Dreifuss muscular dystrophy, dilated cardiomyopathy, limb girdle muscular dystrophy and familial partial lipodystrophy, which are caused by defects in nuclear lamins and lamin-binding proteins.[13,50,143] In addition, an understanding of nuclear envelope structure and dynamics may also lead to improved anti-viral therapy in the case of HIV[144] and herpesvirus,[145] both of which disrupt nuclear envelope structure as a required part of their life-cycle.

References

1. Allen TD, Cronshaw JM, Bagley S et al. The nuclear pore complex: Mediator of translocation between nucleus and cytoplasm. J Cell Sci 2000; 113:1651-1659.
2. Worman HJ, Yuan J, Blobel G et al. A lamin B receptor in the nuclear envelope. Proc Natl Acad Sci USA 1988; 85:8531-8534.
3. Senior A, Gerace L. Integral membrane proteins specific to the inner nuclear membrane and associated with the nuclear lamina. J Cell Biol 1988; 107:2029-2036.
4. Berger R, Theodor L, Shoham J et al. The characterization and localization of the mouse thymopoietin/lamina-associated polypeptide 2 gene and its alternatively spliced products. Genome Res 1996; 6:361-370.
5. Dechat T, Vlcek S, Foisner R. Review: Lamina-associated polypeptide 2 isoforms and related proteins in cell cycle-dependent nuclear structure dynamics. J Struct Biol 2000;129:335-345.
6. Bione S, Maestrini E, Rivella S et al. Identification of a novel X-linked gene responsible for Emery-Dreifuss muscular dystrophy. Nat Genet 1994; 8:323-327.
7. Manilal S, Sewry CA, Man N et al. Diagnosis of X-linked Emery-Dreifuss muscular dystrophy by protein analysis of leukocytes and skin with monoclonal antibodies. Neuromuscul Disord 1997; 7:63-66.
8. Paulin-Levasseur M, Blake DL, Julien M et al. The MAN antigens are non-lamin constituents of the nuclear lamina in vertebrate cells. Chromosoma 1996; 104:367-379.
9. Lin F, Blake DL, Callebaut I et al. MAN1, an inner nuclear membrane protein that shares the LEM domain with lamina-associated polypeptide 2 and emerin. J Biol Chem 2000; 275:4840-4847.
10. Rolls MM, Stein PA, Taylor SS et al. A visual screen of a GFP-fusion library identifies a new type of nuclear envelope membrane protein. J Cell Biol 1999; 146:29-44.
11. Mansharamani M, Hewetson A, Chilton BS. Cloning and characterization of an atypical Type IV P-type ATPase that binds to the RING motif of RUSH transcription factors. J Biol Chem 2001; 276:3641-3649.
12. Cohen M, Lee KK, Wilson KL et al. Transcriptional repression, apoptosis, human disease and the functional evolution of the nuclear lamina. Trends Biochem Sci 2001; 26:41-47.
13. Wilson KL. The nuclear envelope, muscular dystrophy and gene expression. Trends Cell Biol 2000; 10:125-129.
14. Stuurman N, Heins S, Aebi U. Nuclear lamins: Their structure, assembly, and interactions. J Struct Biol 1998; 122:42-66.
15. Franke WW. Nuclear lamins and cytoplasmic intermediate filament proteins: A growing multigene family. Cell 1987; 48:3-4.
16. Stuurman N, Sasse B, Fisher PA. Intermediate filament protein polymerization: molecular analysis of Drosophila nuclear lamin head-to-tail binding. J Struct Biol 1996; 117:1-15.
17. Stewart C, Burke B. Teratocarcinoma stem cells and early mouse embryos contain only a single major lamin polypeptide closely resembling lamin B. Cell 1987; 51:383-392.
18. Broers JL, Machiels BM, Kuijpers HJ et al. A- and B-type lamins are differentially expressed in normal human tissues. Histochem Cell Biol 1997; 107:505-517.
19. Lin F, Worman HJ. Structural organization of the human gene (LMNB1) encoding nuclear lamin B1. Genomics 1995; 27:230-236.
20. Fisher DZ, Chaudhary N, Blobel G. cDNA sequencing of nuclear lamins A and C reveals primary and secondary structural homology to intermediate filament proteins. Proc Natl Acad Sci USA 1986; 83:6450-6454.
21. Machiels BM, Zorenc AH, Endert JM et al. An alternative splicing product of the lamin A/C gene lacks exon 10. J Biol Chem 1996; 271:9249-9253.
22. McKeon FD, Kirschner MW, Caput D. Homologies in both primary and secondary structure between nuclear envelope and intermediate filament proteins. Nature 1986; 319:463-468.
23. Furukawa K, Hotta Y. cDNA cloning of a germ cell specific lamin B3 from mouse spermatocytes and analysis of its function by ectopic expression in somatic cells. Embo J 1993; 12:97-106.
24. Furukawa K, Inagaki H, Hotta Y. Identification and cloning of an mRNA coding for a germ cell-specific A- type lamin in mice. Exp Cell Res 1994; 212:426-430.
25. Alsheimer M, von Glasenapp E, Hock R et al. Architecture of the nuclear periphery of rat pachytene spermatocytes: distribution of nuclear envelope proteins in relation to synaptonemal complex attachment sites. Mol Biol Cell 1999; 10:1235-1245.
26. Lopez-Soler RI, Moir RD, Spann TP et al. A role for nuclear lamins in nuclear envelope assembly. J Cell Biol 2001; 154:61-70.
27. Meier J, Campbell KH, Ford CC et al. The role of lamin LIII in nuclear assembly and DNA replication, in cell-free extracts of *Xenopus* eggs. J Cell Sci 1991; 98:271-279.

28. Newport JW, Wilson KL, Dunphy WG. A lamin-independent pathway for nuclear envelope assembly. J Cell Biol 1990; 111:2247-2259.
29. Sullivan T, Escalante-Alcalde D, Bhatt H et al. Loss of A-type lamin expression compromises nuclear envelope integrity leading to muscular dystrophy. J Cell Biol 1999; 147:913-920.
30. Liu J, Ben-Shahar TR, Riemer D et al. Essential roles for *Caenorhabditis elegans* lamin gene in nuclear organization, cell cycle progression, and spatial organization of nuclear pore complexes. Mol Biol Cell 2000; 11:3937-3947.
31. Ye Q, Worman HJ. Primary structure analysis and lamin B and DNA binding of human LBR, an integral protein of the nuclear envelope inner membrane. J Biol Chem 1994; 269:11306-11311.
32. Smith S, Blobel G. Colocalization of vertebrate lamin B and lamin B receptor (LBR) in nuclear envelopes and in LBR-induced membrane stacks of the yeast *Saccharomyces cerevisiae*. Proc Natl Acad Sci USA 1994; 91:10124-10128.
33. Ye Q, Worman HJ. Interaction between an integral protein of the nuclear envelope inner membrane and human chromodomain proteins homologous to *Drosophila* HP1. J Biol Chem 1996; 271:14653-14656.
34. Ye Q, Callebaut I, Pezhman A et al. Domain-specific interactions of human HP1-type chromodomain proteins and inner nuclear membrane protein LBR. J Biol Chem 1997; 272:14983-14989.
35. Martin L, Crimaudo C, Gerace L. cDNA cloning and characterization of lamina-associated polypeptide 1C (LAP1C), an integral protein of the inner nuclear membrane. J Biol Chem 1995; 270:8822-8828.
36. Foisner R, Gerace L. Integral membrane proteins of the nuclear envelope interact with lamins and chromosomes, and binding is modulated by mitotic phosphorylation. Cell 1993; 73:1267-1279.
37. Maison C, Pyrpasopoulou A, Theodoropoulos PA et al. The inner nuclear membrane protein LAP1 forms a native complex with B-type lamins and partitions with spindle-associated mitotic vesicles. Embo J 1997; 16:4839-4850.
38. Furukawa K. LAP2 binding protein 1 (L2BP1/BAF) is a candidate mediator of LAP2- chromatin interaction. J Cell Sci 1999; 112:2485-2492.
39. Shumaker DK, Lee KK, Tanhehco YC et al. LAP2 binds to BAF-DNA complexes: requirement for the LEM domain and modulation by variable regions. Embo J 2001; 20:1754-1764.
40. Cai M, Huang Y, Ghirlando R et al. Solution structure of the constant region of nuclear envelope protein LAP2 reveals two LEM-domain structures: One binds BAF and the other binds DNA. Embo J 2001; 20:4399-4407.
41. Nili E, Cojocaru GS, Kalma Y et al. Nuclear membrane protein LAP2beta mediates transcriptional repression alone and together with its binding partner GCL (germ-cell-less). J Cell Sci 2001; 114:3297-3307.
42. Gant TM, Harris CA, Wilson KL. Roles of LAP2 proteins in nuclear assembly and DNA replication: truncated LAP2beta proteins alter lamina assembly, envelope formation, nuclear size, and DNA replication efficiency in *Xenopus laevis* extracts. J Cell Biol 1999; 144:1083-1096.
43. Yang L, Guan T, Gerace L. Lamin-binding fragment of LAP2 inhibits increase in nuclear volume during the cell cycle and progression into S phase. J Cell Biol 1997; 139:1077-1087.
44. Laguri C, Gilquin B, Wolff N, Romi-Lebrun R, Courchay K, Callebaut I et al. Structural characterization of the LEM motif common to three human inner nuclear membrane proteins. Structure (Camb) 2001; 9:503-511.
45. Wolff N, Gilquin B, Courchay K, et al, Zinn-Justin S. Structural analysis of emerin, an inner nuclear membrane protein mutated in X-linked Emery-Dreifuss muscular dystrophy. FEBS Lett 2001; 501:171-176.
46. Lee MS, Craigie R. A previously unidentified host protein protects retroviral DNA from autointegration. Proc Natl Acad Sci USA 1998; 95:1528-1533.
47. Chen H, Engelman A. The barrier-to-autointegration protein is a host factor for HIV type 1 integration. Proc Natl Acad Sci USA 1998; 95:15270-15274.
48. Zheng R, Ghirlando R, Lee MS et al. Barrier-to-autointegration factor (BAF) bridges DNA in a discrete, higher-order nucleoprotein complex. Proc Natl Acad Sci USA 2000; 97:8997-9002.
49. Ellis JA, Yates JR, Kendrick-Jones J et al. Changes at P183 of emerin weaken its protein-protein interactions resulting in X-linked Emery-Dreifuss muscular dystrophy. Hum Genet 1999; 104:262-268.
50. Hutchison CJ, Alvarez-Reyes M, Vaughan OA. Lamins in disease: why do ubiquitously expressed nuclear envelope proteins give rise to tissue-specific disease phenotypes? J Cell Sci 2001; 114:9-19.
51. Gruenbaum Y, Lee KK, Liu J et al. The expression, lamin-dependent localization and RNAi depletion phenotype for emerin in *C. elegans*. J Cell Sci 2002; 115:923-929.

52. Imai S, Nishibayashi S, Takao K et al. Dissociation of Oct-1 from the nuclear peripheral structure induces the cellular aging-associated collagenase gene expression. Mol Biol Cell 1997; 8:2407-2419.
53. Mancini MA, Shan B, Nickerson JA et al. The retinoblastoma gene product is a cell cycle-dependent, nuclear matrix-associated protein. Proc Natl Acad Sci USA 1994; 91:418-422.
54. Gerace L, Blobel G. The nuclear envelope lamina is reversibly depolymerized during mitosis. Cell 1980; 19:277-287.
55. Yang L, Guan T, Gerace L. Integral membrane proteins of the nuclear envelope are dispersed throughout the endoplasmic reticulum during mitosis. J Cell Biol 1997; 137:1199-1210.
56. Ellenberg J, Siggia ED, Moreira JE et al. Nuclear membrane dynamics and reassembly in living cells: targeting of an inner nuclear membrane protein in interphase and mitosis. J Cell Biol 1997; 138:1193-1206.
57. Peter M, Nakagawa J, Doree M et al. In vitro disassembly of the nuclear lamina and M phase-specific phosphorylation of lamins by cdc2 kinase. Cell 1990; 61:591-602.
58. Collas P. Sequential PKC- and Cdc2-mediated phosphorylation events elicit zebrafish nuclear envelope disassembly. J Cell Sci 1999; 112:977-987.
59. Newport J, Spann T. Disassembly of the nucleus in mitotic extracts: membrane vesicularization, lamin disassembly, and chromosome condensation are independent processes. Cell 1987; 48:219-230.
60. Takizawa CG, Morgan DO. Control of mitosis by changes in the subcellular location of cyclin-B1- Cdk1 and Cdc25C. Curr Opin Cell Biol 2000; 12:658-665.
61. Pines J, Hunter T. Human cyclins A and B1 are differentially located in the cell and undergo cell cycle-dependent nuclear transport. J Cell Biol 1991; 115:1-17.
62. Courvalin JC, Segil N, Blobel G et al. The lamin B receptor of the inner nuclear membrane undergoes mitosis-specific phosphorylation and is a substrate for p34cdc2-type protein kinase. J Biol Chem 1992; 267:19035-19038.
63. Collas P. Nuclear envelope disassembly in mitotic extract requires functional nuclear pores and a nuclear lamina. J Cell Sci 1998; 111:1293-1303.
64. Lee KK, Gruenbaum Y, Spann P et al. *C. elegans* nuclear envelope proteins emerin, MAN1, lamin, and nucleoporins reveal unique timing of nuclear envelope breakdown during mitosis. Mol Biol Cell 2000; 11:3089-3099.
65. Terasaki M, Campagnola P, Rolls MM et al. A new model for nuclear envelope breakdown. Mol Biol Cell 2001; 12:503-510.
66. Vasu SK, Forbes DJ. Nuclear pores and nuclear assembly. Curr Opin Cell Biol 2001; 13:363-375.
67. Favreau C, Worman HJ, Wozniak RW et al. Cell cycle-dependent phosphorylation of nucleoporins and nuclear pore membrane protein Gp210. Biochemistry 1996; 35:8035-8044.
68. Macaulay C, Meier E, Forbes DJ. Differential mitotic phosphorylation of proteins of the nuclear pore complex. J Biol Chem 1995; 270:254-262.
69. Feldherr CM, Akin D. The location of the transport gate in the nuclear pore complex. J Cell Sci 1997; 110:3065-3070.
70. Nurse P. Universal control mechanism regulating onset of M-phase. Nature 1990; 344:503-508.
71. Lohka MJ, Masui Y. Formation in vitro of sperm pronuclei and mitotic chromosomes induced by amphibian ooplasmic components. Science 1983; 220:719-721.
72. Lohka MJ, Masui Y. Roles of cytosol and cytoplasmic particles in nuclear envelope assembly and sperm pronuclear formation in cell-free preparations from amphibian eggs. J Cell Biol 1984; 98:1222-1230.
73. Warren G. Membrane partitioning during cell division. Annu Rev Biochem 1993; 62:323-348.
74. Gant TM, Wilson KL. Nuclear assembly. Ann Rev Cell Develop Biol 1997; 13:669-695.
75. Collas P, Courvalin JC. Sorting nuclear membrane proteins at mitosis. Trends Cell Biol 2000; 10:5-8.
76. Daigle N, Beaudouin J, Hartnell L et al. Nuclear pore complexes form immobile networks and have a very low turnover in live mammalian cells. J Cell Biol 2001; 154:71-84.
77. Nikolakaki E, Meier J, Simos G et al. Mitotic phosphorylation of the lamin B receptor by a serine/arginine kinase and p34(cdc2). J Biol Chem 1997; 272:6208-6213.
78. Minc E, Allory Y, Worman HJ et al. Localization and phosphorylation of HP1 proteins during the cell cycle in mammalian cells. Chromosoma 1999; 108:220-234.
79. Dreger M, Otto H, Neubauer G et al. Identification of phosphorylation sites in native lamina-associated polypeptide 2 beta. Biochemistry 1999; 38:9426-9434.
80. Manilal S, Nguyen TM, Morris GE. Colocalization of emerin and lamins in interphase nuclei and changes during mitosis. Biochem Biophys Res Commun 1998; 249:643-647.
81. Miller MW, Caracciolo MR, Berlin WK et al. Phosphorylation and glycosylation of nucleoporins. Arch Biochem Biophys 1999; 367:51-60.

82. Gerace L, Blum A, Blobel G. Immunocytochemical localization of the major polypeptides of the nuclear pore complex-lamina fraction. Interphase and mitotic distribution. J Cell Biol 1978; 79:546-566.

83. Stick R, Angres B, Lehner CF et al. The fates of chicken nuclear lamin proteins during mitosis: evidence for a reversible redistribution of lamin B2 between inner nuclear membrane and elements of the endoplasmic reticulum. J Cell Biol 1988; 107:397-406.

84. Vorburger K, Kitten GT, Nigg EA. Modification of nuclear lamin proteins by a mevalonic acid derivative occurs in reticulocyte lysates and requires the cysteine residue of the C-terminal CXXM motif. Embo J 1989; 8:4007-4013.

85. Barton RM, Worman HJ. Prenylated prelamin A interacts with Narf, a novel nuclear protein. J Biol Chem 1999; 274:30008-30018.

86. Weber K, Plessmann U, Traub P. Maturation of nuclear lamin A involves a specific carboxy-terminal trimming, which removes the polyisoprenylation site from the precursor; implications for the structure of the nuclear lamina. FEBS Lett 1989; 257:411-414.

87. Heald R, McKeon F. Mutations of phosphorylation sites in lamin A that prevent nuclear lamina disassembly in mitosis. Cell 1990; 61:579-589.

88. Ward GE, Kirschner MW. Identification of cell cycle-regulated phosphorylation sites on nuclear lamin C. Cell 1990; 61:561-577.

89. Hixon ML, Gualberto A. The control of mitosis. Front Biosci 2000; 5:D50-57.

90. Heck MM. Condensins, cohesins, and chromosome architecture: How to make and break a mitotic chromosome. Cell 1997; 91:5-8.

91. Kirsch-Volders M, Cundari E, Verdoodt B. Towards a unifying model for the metaphase/anaphase transition. Mutagenesis 1998; 13:321-325.

92. Lohka MJ. Analysis of nuclear envelope assembly using extracts of *Xenopus* eggs. Meth Cell Biol 1998; 53:417-452.

93. Newport JW, Forbes DJ. The nucleus: Structure, function, and dynamics. Annu Rev Biochem 1987; 56:535-565.

94. Gerace L, Burke B. Functional organization of the nuclear envelope. Annu Rev Cell Biol 1988; 4:335-374.

95. Wiese C, Wilson KL. Nuclear membrane dynamics. Curr Opin Cell Biol 1993; 5:387-394.

96. Dabauvalle MC, Muller E, Ewald A et al. Distribution of emerin during the cell cycle. Eur J Cell Biol 1999; 78:749-756.

97. Haraguchi T, Koujin T, Hayakawa T et al. Live fluorescence imaging reveals early recruitment of emerin, LBR, RanBP2, and Nup153 to reforming functional nuclear envelopes. J Cell Sci 2000; 113:779-794.

98. Taniura H, Glass C, Gerace L. A chromatin binding site in the tail domain of nuclear lamins that interacts with core histones. J Cell Biol 1995; 131:33-44.

99. Burke B. On the cell-free association of lamins A and C with metaphase chromosomes. Exp Cell Res 1990; 186:169-176.

100. Glass JR, Gerace L. Lamins A and C bind and assemble at the surface of mitotic chromosomes. J Cell Biol 1990; 111:1047-1057.

101. Goldberg M, Harel A, Brandeis M et al. The tail domain of lamin Dm0 binds histones H2A and H2B. Proc Natl Acad Sci USA 1999; 96:2852-2857.

102. Kourmouli N, Theodoropoulos PA, Dialynas G et al. Dynamic associations of heterochromatin protein 1 with the nuclear envelope. EMBO J 2000; 19:6558-6568.

103. Goldberg M, Lu H, Stuurman N et al. Interactions among *Drosophila* nuclear envelope proteins lamin, otefin, and YA. Mol Cell Biol 1998; 18:4315-4323.

104. Lee KK, Haraguchi T, Lee RS, Koujin T et al. Distinct functional domains in emerin bind lamin A and DNA-bridging protein BAF. J Cell Sci 2001; 114:4567-4573.

105. Haraguchi T, Koujin T, Segura-Totten M et al. BAF is required for emerin assembly into the reforming nuclear envelope. J Cell Sci 2001; 114:4575-4585.

106. Rose MD. Nuclear fusion in the yeast *Saccharomyces cerevisiae*. Annu Rev Cell Dev Biol 1996; 12:663-695.

107. Hetzer M, Meyer HH, Walther TC et al. Distinct AAA-ATPase p97 complexes function in discrete steps of nuclear assembly. Nature Cell Biol 2001; 3:1086-1091.

108. Vigers GP, Lohka MJ. A distinct vesicle population targets membranes and pore complexes to the nuclear envelope in *Xenopus* eggs. J Cell Biol 1991; 112:545-556.

109. Boman AL, Delannoy MR, Wilson KL. GTP hydrolysis is required for vesicle fusion during nuclear envelope assembly in vitro. J Cell Biol 1992; 116:281-294.

110. Newport J, Dunphy W. Characterization of the membrane binding and fusion events during nuclear envelope assembly using purified components. J Cell Biol 1992; 116:295-306.

111. Boman AL, Taylor TC, Melancon P et al. A role for ADP-ribosylation factor in nuclear vesicle dynamics. Nature 1992; 358:512-514.
112. Gant TM, Wilson KL. ARF is not required for nuclear vesicle fusion or mitotic membrane disassembly in vitro: evidence for a non-ARF GTPase in fusion. Eur J Cell Biol 1997; 74:10-19.
113. Hetzer M, Bilbao-Cortes D, Walther TC et al. GTP hydrolysis by Ran is required for nuclear envelope assembly. Mol Cell 2000; 5:1013-1024.
114. Zhang C, Clarke PR. Roles of Ran-GTP and Ran-GDP in precursor vesicle recruitment and fusion during nuclear envelope assembly in a human cell-free system. Curr Biol 2001; 11:208-212.
115. Moir RD, Yoon M, Khuon S et al. Nuclear Lamins A and B1: different pathways of assembly during nuclear envelope formation in living cells. J Cell Biol 2000; 151:1155-1168.
116. Wiese C, Goldberg MW, Allen TD et al. Nuclear envelope assembly in *Xenopus* extracts visualized by scanning EM reveals a transport-dependent 'envelope smoothing' event. J Cell Sci 1997; 110:1489-1502.
117. Dabauvalle MC, Loos K, Merkert H et al. Spontaneous assembly of pore complex-containing membranes ("annulate lamellae") in *Xenopus* egg extract in the absence of chromatin. J Cell Biol 1991; 112:1073-1082.
118. Meier E, Miller BR, Forbes DJ. Nuclear pore complex assembly studied with a biochemical assay for annulate lamellae formation. J Cell Biol 1995; 129:1459-1472.
119. Robinson LJ, Martin TF. Docking and fusion in neurosecretion. Curr Opin Cell Biol 1998; 10:483-492.
120. Skehel JJ, Wiley DC. Receptor binding and membrane fusion in virus entry: The influenza hemagglutinin. Ann Rev Biochem 2000; 69:531-569.
121. Hallberg E, Wozniak RW, Blobel G. An integral membrane protein of the pore membrane domain of the nuclear envelope contains a nucleoporin-like region. J Cell Biol 1993; 122:513-521.
122. Filson AJ, Lewis A, Blobel G et al. Monoclonal antibodies prepared against the major *Drosophila* nuclear matrix-pore-complex-lamina glycoprotein bind specifically to the nuclear envelope in situ. J Biol Chem 1985;260:3164-72.
123. Gerace L, Ottaviano Y, Kondor-Koch C. Identification of a major polypeptide of the nuclear pore complex. J Cell Biol 1982; 95:826-837.
124. Wozniak RW, Bartnik E, Blobel G. Primary structure analysis of an integral membrane glycoprotein of the nuclear pore. J Cell Biol 1989; 108:2083-2092.
125. Greber UF, Senior A, Gerace L. A major glycoprotein of the nuclear pore complex is a membrane-spanning polypeptide with a large lumenal domain and a small cytoplasmic tail. Embo J 1990; 9:1495-1502.
126. Drummond SP, Wilson, KL. Interference with the cytoplasmic tail of gp210 disrupts "close apposition" of nuclear membranes and blocks nuclear pore dilation. J Cell Biol 2002; 158:53-62.
127. Bodoor K, Shaikh S, Salina D et al. Sequential recruitment of NPC proteins to the nuclear periphery at the end of mitosis. J Cell Sci 1999; 112:2253-2264.
128. Goldberg MW, Wiese C, Allen TD et al. Dimples, pores, star-rings, and thin rings on growing nuclear envelopes: evidence for structural intermediates in nuclear pore complex assembly. J Cell Sci 1997; 110:409-420.
129. Kiseleva E, Goldberg MW, Cronshaw J et al. The nuclear pore complex: structure, function, and dynamics. Crit Rev Eukaryot Gene Expr 2000; 10:101-112.
130. Fontoura BM, Dales S, Blobel G et al. The nucleoporin Nup98 associates with the intra-nuclear filamentous network TPR. Proc Natl Acad Sci USA 2001; 98:3208-3213.
131. Paddy MR. The Tpr protein: linking structure and function in the nuclear interior? Am J Hum Genet 1998; 63:305-310.
132. Cordes VC, Reidenbach S, Kohler A et al. Intranuclear filaments containing a nuclear pore complex protein. J Cell Biol 1993; 123:1333-1344.
133. Macaulay C, Forbes DJ. Assembly of the nuclear pore—Biochemically distinct steps revealed with NEM, GTPγS, and BAPTA. J Cell Biol 1996; 132:5-20.
134. Miller BR, Forbes DJ. Purification of the vertebrate nuclear pore complex by biochemical criteria. Traffic 2000; 1:941-951.
135. Miller BR, Powers M, Park M et al. Identification of a new vertebrate nucleoporin, Nup188, with the use of a novel organelle trap assay. Mol Biol Cell 2000; 11:3381-3396.
136. Finlay DR, Forbes DJ. Reconstitution of biochemically altered nuclear pores: Transport can be eliminated and restored. Cell 1990; 60:17-29.
137. Sullivan KM, Busa WB, Wilson KL. Calcium mobilization is required for nuclear vesicle fusion in vitro: implications for membrane traffic and IP3 receptor function. Cell 1993; 73:1411-1422.
138. Dabauvalle MC, Loos K, Scheer U. Identification of a soluble precursor complex essential for nuclear pore assembly in vitro. Chromosoma 1990; 100:56-66.

139. Powers MA, Macaulay C, Masiarz FR et al. Reconstituted nuclei depleted of a vertebrate GLFG nuclear pore protein, p97, import but are defective in nuclear growth and replication. J Cell Biol 1995; 128:721-736.
140. Broers JLV, Taylor TC, Melancon P et al. Dynamics of the nuclear lamina as monitored by GFP-tagged A-type lamins. J Cell Sci 1999; 112:3463-3475.
141. Cordes VC, Reidenbach S, Rackwitz HR et al. Identification of protein p270/Tpr as a constitutive component of the nuclear pore complex-attached intranuclear filaments. J Cell Biol 1997; 136:515-529.
142. Hase ME, Kuznetsov NV, Cordes VC. Amino acid substitutions of coiled-coil Tpr abrogate anchorage to the nuclear pore complex but not parallel, in-register homodimerization. Mol Biol Cell 2001; 12:2433-2452.
143. Wilson KL, Zastrow MS, Lee KK. Lamins and disease: insights into nuclear infrastructure. Cell 2001; 104:647-650.
144. de Noronha CMC, Sherman MP, Lin HW et al. Dynamic disruptions in nuclear envelope architecture and integrity induced by HIV-1 Vpr. Science 2001; 294:1105-1108.
145. Scott ES, O'Hare P. Fate of the inner nuclear membrane protein lamin B receptor and nuclear lamins in herpes simplex virus type 1 infection. J Virol 2001; 75:8818-8830.

CHAPTER 2

Dynamics of Nuclear Envelope Proteins During the Cell Cycle in Mammalian Cells

Jan Ellenberg

Abstract

Breakdown and reformation of the nuclear envelope (NE) during cell division is one of the most dramatic structural and functional changes in higher eukaryotic cells. NE breakdown (NEBD) marks a highly regulated switch in chromosome confinement by membranes in interphase to microtubules in M-phase. The boundary of interphase nuclei has a rigid and highly interconnected architecture made up of a concentric double membrane with embedded nuclear pores, underlying intermediate filaments and the connected chromosome territories. Upon entering mitosis, cells completely and rapidly dismantle the connections between these structures to allow chromosomes to condense and be captured by the mitotic spindle which then accurately partitions them to daughter cells. Once segregation is accomplished, the complex interphase architecture is quickly re-established to enable essential functions such as transcription and replication to start anew. Several excellent recent reviews have touched upon this subject from several angles.[1-6] In this Chapter, I intend to present a global picture of the dynamics of nuclear envelope proteins during mitosis in mammalian cells and also touch upon other cellular structures important for nuclear envelope remodeling including chromosomes and the mitotic spindle.

Why Should Nuclear Envelope Proteins Be Dynamic?

The NE forms a selective boundary around the chromosomes and acts as a peripheral scaffold to spatially organize chromatin. As a consequence, most NE proteins have structural functions in organizing the interphase nuclear architecture. For structural proteins the intuitive assumption is that their behavior is rather static. However, both in non-dividing and dividing cells there are aspects of NE function that require dynamic exchange of its proteins. Before we review these, it is useful to remind ourselves that the NE has a unique topology. Its two membranes, inner nuclear membrane (INM) and outer nuclear membrane (ONM) are connected at several thousand nuclear pores via a short stretch of lipid bilayer sometimes referred to as the pore membrane (POM) (Fig. 1). The outer membrane is continuous with the endoplasmic reticulum (ER) and is indistinguishable from the ER in terms of its protein composition including attached and translating ribosomes. Viewed from the cytoplasm, the NE is simply a specialized subcompartment of the ER, a large spherical ER cisterna studded by nuclear pores and wrapped around lamins and chromosomes (Fig. 1).

What then are the situations in which NE proteins have to be dynamic? The first need arises when cells replicate their set of chromosomes which causes nuclear volume and NE surface to grow significantly. This expansion requires the targeting of proteins to the NE to equip it with new molecules. A good example for this is that the number of nuclear pores doubles during this time.[7] Secondly, nuclear architecture needs to be remodeled in response to external stimuli. It

Nuclear Envelope Dynamics in Embryos and Somatic Cells

edited by Philippe Collas. ©2002 Eurekah.com and Kluwer Academic/Plenum Publishers.

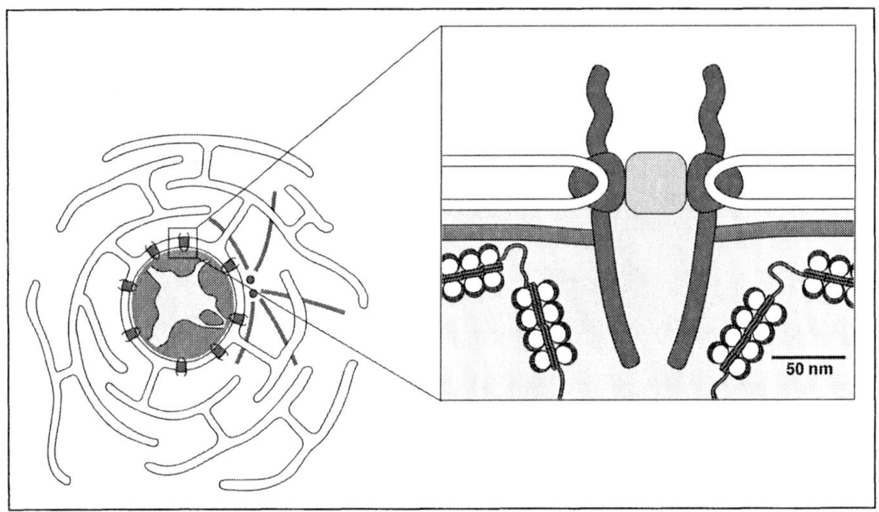

Figure 1. Schematic view of the organization of the interphase nuclear envelope.
(Left) Nuclear membranes can be seen clearly as a subcompartment of the ER studded by nuclear pores and closely apposed to the nuclear lamina and peripheral chromatin. Also shown are the exclusively cytoplasmic microtubules and centrosomes.
(Right) The four major structural components of the NE drawn to their approximate molecular scale. The cytoplasm (white) is separated from the nucleoplasm (gray) by the nuclear membranes consisting of outer nuclear membrane (ONM) facing the cytoplasm, pore membrane embedded in the nuclear pore complex (NPC), and inner nuclear membrane (INM) facing the nuclear lamina. Peripheral chromatin is shown schematically as a 30 nm fiber composed of DNA wrapped around nucleosomes. Substructures of the nuclear pore complex shown are the central spoke ring complex (embedded in the membranes), as well as cytoplasmic and nuclear fibrils and the central plug. Scale bar: 50 nm.

is becoming increasingly clear that chromosome attachment to the nuclear envelope can influence replication timing and transcription activity.[8-10] When cells activate peripherally located genes or replicate them, these attachments must be remodeled in a dynamic fashion. NE protein dynamics become essential when a cell divides. The stable structure of the NE poses a formidable barrier to mitosis in metazoan cells which have exclusively cytoplasmic microtubules. These cells undergo an open mitosis, disassembling their NE at the transition to M-phase so that the mitotic spindle can access and attach to the chromosomes. Conversely, after sister chromatids have been successfully separated, new NEs have to be reformed quickly to allow a new cycle of metabolic activity.

What is the Nuclear Envelope Made of?

Over the recent years we have obtained an almost comprehensive list of the proteins present in the NE in vertebrates especially with the advances made by recent proteomics studies.[11,12] Based both on the identified proteins and on morphological considerations it makes sense to subdivide the NE into four main structures (Fig. 1), each of which are reviewed in more detail in other Chapters of this volume. The first of these, the nuclear lamina consists of lamins, proteins of the intermediate filament family that are divided into two classes, namely B type (ubiquitous) and A/C type lamins (found only in differentiated cells).[13,14] These rod-shaped proteins form a peripheral branched polymer of 10-nm filaments which provides structural support to the NE.[15] The second NE structure is the inner nuclear membrane (INM). It contains a unique set of membrane proteins and protein families which reside only at low levels in the ER and the secretory pathway. Most of these more than 10 proteins function as adaptors linking the INM to the lamina and/or chromosomes[2] and some authors have now extended the definition of the lamina to encompass also the lamina associated proteins.[16,17] The third structure is the nuclear pore complex (NPC), a 125-MDa large protein assembly that forms an aqueous channel through the NE, thereby joining the inner and outer NM. Mammalian cells contain one to several thousand of these channels per nucleus. Each NPC is made of nucleoporins (Nups), a class of more than 20 soluble and only two integral membrane proteins. The NPC mediates all nucleocytoplasmic traffic[18] but may also be involved in nuclear organization in general.[19] Some nucleoporins interact both with the lamina and chromosomes[19,20] and direct connections between the lamina and the nuclear face of the NPC can be visualized by electron microscopy.[21] The last structure of the NE, which is classically not counted among NE components is the peripheral chromatin which contains several proteins that interact with the lamina and/or the INM. In interphase these four units of NE architecture are connected by a multitude of protein-protein interactions and the NE appears as a complex, highly cross-linked structural protein network (Fig. 1).[16]

Studying Nuclear Envelope Protein Dynamics

True insight into NE protein dynamics has mostly come from studying these proteins in their natural environment in living cells. In mammalian cells this has been achieved through the analysis of fluorescently labeled derivatives of NE proteins. Fusion to green fluorescent protein (GFP)[22] and subsequent stable or transient expression has been the method of choice in many cases, especially for the many transmembrane proteins, for which recombinant expression, labeling with chemical fluorophores and reintroduction into live cells is not feasible. Once the NE protein of interest has been labeled successfully (and without impairing its function!) several techniques can be used to characterize its dynamics. In this Chapter, I will describe results mostly from two approaches. The first is time-lapse fluorescence imaging. Here a fluorescence microscope, either confocal or wide field, is used to take images of the protein distribution in live cells and document changes of localization over time. Time-lapse imaging, if performed quantitatively, can document the fluxes of a given protein within the cell with high spatial resolution and even in three dimensions.[23] The second method is fluorescence recovery after photobleaching or FRAP.[24] In FRAP a portion of the fluorescently labeled protein is bleached irreversibly with a high intensity laser beam. After the bleach, the exchange of the bleached molecules with the surrounding unbleached molecules is then measured by monitoring the recovery of fluorescence in the bleached area. If the bleached molecules do not exchange during the time of the experiment, fluorescence does not recover and the patterns bleached by the laser can be used to mark regular geometries inside cells. This approach is referred to as pattern bleaching and has been very useful to characterize surface dynamics of the NE as we will see below.

Dynamics in Interphase

INM Proteins are Targeted by Selective Retention

INM proteins are defined by their specific localization to the nucleoplasmic face of the nuclear membranes. Since the NE is an ER subcompartment, it is interesting to ask how these proteins are confined to just the INM and largely excluded from the ER. Initial experiments focused on identifying "sorting signals" in INM proteins, analogous to the short consensus sequences that govern localization of membrane proteins in the secretory pathway.[25,26] However, the sequences identified turned out to be binding motifs to nuclear proteins rather than classical signals for transport adaptors. We now know that most INM proteins contain sequence motifs in their nucleoplasmic domains that mediate interactions to lamins, chromatin or other INM proteins in an often redundant fashion. The ability of INM proteins to bind to nuclear partners turns out to be sufficient to account for their specific localization by a mechanism based on selective retention (Fig. 2A). INM proteins start their life in the ER where they are inserted into the membrane. In the ER, their binding domains are exposed to the cytoplasm and do not encounter nuclear proteins. As a result, INM proteins can freely diffuse within the ER and also have access to the INM through the membrane connection between ONM and INM at the periphery of each NPC (Figs. 1 and 2A). Importantly, this access is driven by diffusion and is thus independent of signals and not directional. The only restriction to diffusion through the POM appears to be the size of the cytoplasmic domain; it can inhibit localization when it becomes too bulky to pass through the peripheral channels of the NPC.[27] Once an INM protein has reached the inner face of the NE, its now nucleoplasmic binding domain encounters nuclear interaction partners to which it attaches, preventing its diffusion back into the ER. This selective retention of INM proteins but not of general ER proteins in the INM elegantly explains their retention and concentration in the INM ER-subdomain (Fig. 2A). Selective retention makes two clear predictions for the dynamics of integral INM proteins: (i) INM proteins can be targeted in interphase (as opposed to just after mitosis) and (ii) the mobility of these proteins should be reduced upon localization to the NE. Indeed, interphase targeting was demonstrated by following the localization of newly synthesized GFP-tagged lamin B receptor (LBR) after microinjection of an expression plasmid in interphase cells. Initially fluorescence was equally distributed between ER and NE, but after a few hours it was five times more concentrated in the NE.[28] The reduced mobility in the INM has been confirmed by FRAP of three INM proteins, GFP-tagged emerin, LBR and MAN1 [28-30]. The fact that localization of emerin to the INM depends in part upon lamin A provides further evidence for this mechanism.[31,32] We will revisit selective retention again when discussing nuclear membrane dynamics in mitosis where switching on and off the retaining interactions is responsible for loss and reestablishment of the INM domain of the ER (Fig. 2B, C).

The Interphase Lamina: A Stable but Elastic Polymer

Several recent studies have examined the properties of GFP tagged A and B type lamins.[33-35] Time lapse sequences on interphase cells demonstrate that the lamina can undergo dynamic deformations, such as folds and indentations that typically occur during cellular movements or nuclear rotations (Fig. 3A). To assay how stable fluorescent lamins were incorporated into the lamin polymer, FRAP was used to determine if bleached lamin molecules could be replaced by new fluorescent lamins. Both for A and B type lamins, recovery was found to be extremely slow and complete recovery could not be observed in experiments ranging from 10 minutes [35] to more than 40 hours.[34] This indicated a very low dissociation rate of lamins from the polymer in interphase. On the other hand overexpressed lamins can be incorporated into the lamina of interphase cells in less than 20 h probably reflecting the capability of excess lamin monomers to be absorbed into the lamina in addition to, but not replacing the already polymerized filaments. The elasticity of the lamina was directly addressed by taking advantage of the very slow recovery of GFP-tagged B type lamins in pattern bleaching experiments. Here bleaching by a laser beam is used to create geometrical patterns such as stripes and grids on the surface of the smooth peripheral lamina surface, which can then be tracked during cellular movements (Fig. 3B). These experiments clearly demonstrated that the lamina behaves as a two dimensional polymer that can undergo elastic deformations during cellular movement but relaxes back into the original geometrical arrangement when movement ceases.[34,36] The stable and elastic properties of the lamin polymer have confirmed in vivo what could be predicted from its ultrastructural mesh-like appearance[15] and its resistance to biochemical extractions since the 70's.[37]

NPCs Form Networks and Have a Stable Core

So far only three studies have started to characterize the dynamics of NPCs in intact mammalian cells.[34,38,39] The NPC is a remarkable protein complex in many ways. It is very large (125 MDa), consists of more than 30 different proteins in vertebrates each of which occurs in probably 8-24 copies, reflecting the eightfold rotational symmetry of the complex.[12] The core of the NPC forms a flat hollow cylinder with dimensions of ~120 nm in width and ~40 nm in length, and an inner channel diameter of ~40 nm whose walls are embedded in the POM (Fig. 1). This cylinder surrounds the so-called central plug, proteinaceous material located in the

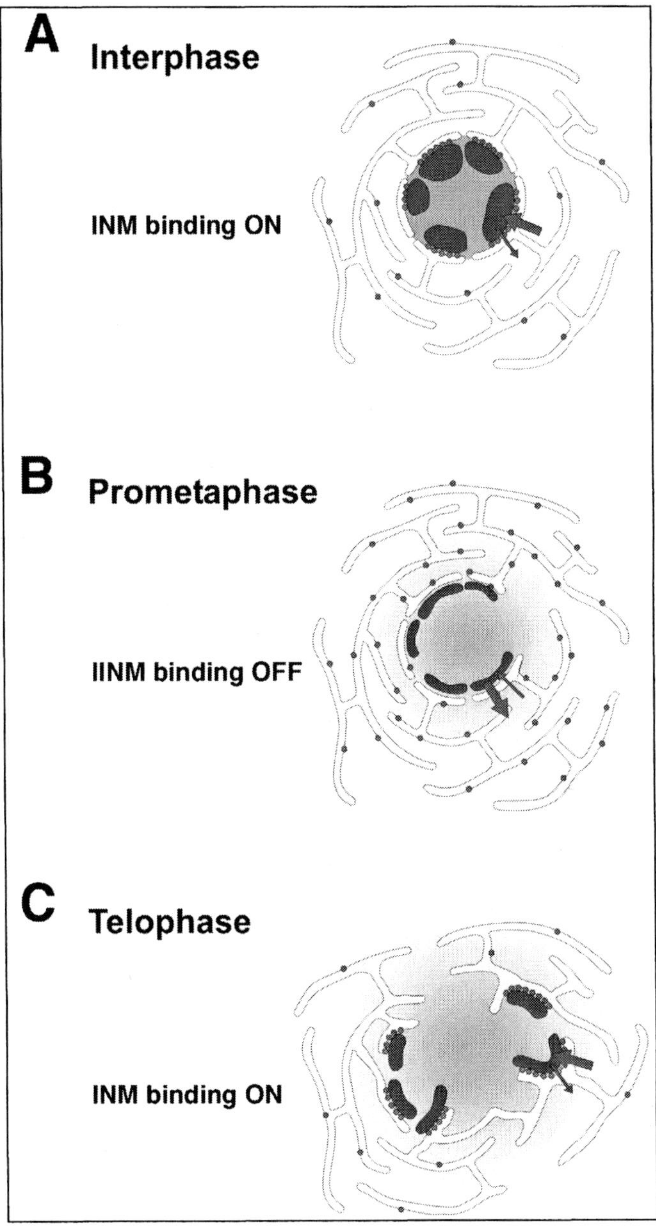

Figure 2. Selective retention in interphase and mitosis.
Schematic illustrating how INM proteins can be localized to the ER and INM-subdomain in interphase and mitosis. ER/nuclear membranes contain a typical chromatin binding INM protein (dots) and are in close proximity to chromatin. In interphase binding is enabled (arrows), the INM protein can exchange between ER and INM by diffusion and is retained in the INM by binding to chromatin. In prometaphase binding is disabled (arrows) by phosphorylation and the INM protein dissociates from chromatin and equilibrates with the ER by diffusion. In telophase binding is switched back on (arrows) by dephosphorylation and INM proteins diffusing in ER cisternae that come in contact with chromatin are retained and thus reform the INM subdomain by attaching this face of the ER cisterna.

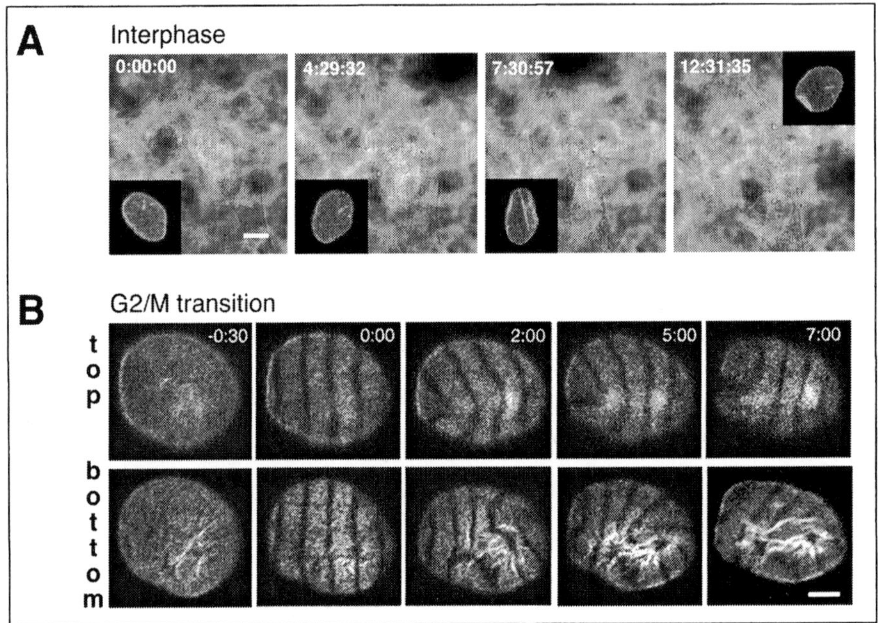

Figure 3. Dynamic properties of the peripheral lamin polymer.
(A) 3D confocal time-lapse sequence of a PtK$_2$ cell expressing GFP-lamin B1[34] in interphase. DIC and fluorescence images are overlaid. Insets show a top projection of GFP fluorescence only. Note nuclear rotations and reversible deformation of the nuclear lamina. Time: hh:mm:ss, bar:10 μm.
(B) Elastic deformations of the prophase lamina. Confocal time lapse sequence of a NRK expressing GFP-lamin B1. Vertical stripes were bleached across the whole nuclear surface in prophase as geometrical landmarks. Note the pronounced stretching occurring on the top and the contraction on the bottom surface. Time: m:ss, t = 0 corresponds to nuclear permeabilization; bar: 5 μm.

middle of the aqueous channel. From the rims of the cylinder emanate eight cytoplasmic and nuclear filaments the latter being joined by a distal ring to form the nuclear basket (Fig. 1). Using five GFP-tagged nucleoporins Nup98, Nup 153, POM121, and Nup107/Nup 133, these studies again employed time-lapse fluorescence microscopy and FRAP to assay the dynamics of nucleoporins in interphase. The core of the NPC represented by the transmembrane protein POM121 and Nup107/Nup133 was found to form an extremely stable complex that did not exchange any of the three Nups over many hours in interphase. Strikingly, Nup153 and Nup98 which are both localized to the nuclear face of the NPC were found to associate only transiently with the NPC.[34,39]

Using markers of the NPC core, the mobility of the whole NPC itself in the plane of the NE was also examined. The notion that NPCs might be mobile was prompted by earlier studies in yeast, which reported movement of NPC across the surface of nuclei after karyogamy of haploid cells.[40,41] In contrast to yeast, mammalian NPCs were found to be completely immobile in the surface of the NE unless it was deformed by folds and indentations. Under those circumstances NPC movements correlated precisely with those of the underlying lamina.[34] These in vivo experiments support ultrastructural data that proposed a direct link between the NPC and the lamina meshwork.[21]

Chromosomes Do not Move Much in Interphase

Our insight into the dynamics of the chromatin class of NE envelope proteins is unfortunately very limited at the moment. Only for one of them, the DNA crosslinker barrier to autointegration factor (BAF), do we have any data from living cells, which mostly addresses the localization of a GFP fusion during nuclear assembly.[42] No FRAP data on the lifetime of chromatin-NE interactions are available at the moment, although for BAF and heterochromatin protein 1 it appears as if they might be more dynamic than those reported for the INM, NPC and lamina (J.E., unpublished observations). However, we know more about the dynamics of chromosomes themselves in interphase mammalian nuclei from several approaches. In one approach developed by Daniele Zink and coworkers, chromatin domains are labeled with pulses of microinjected fluorescent nucleotides during replication and can then be traced over several cell cycles.[43-45] A second approach pioneered by Andrew Belmont and coworkers employs a system of multimeric repeats of lac operators integrated into the genome of cell lines. These arrays can then be labeled by expression of lac repressor-GFP fusion proteins.[46] Using global DNA labeling with intercalating dyes, FRAP has also been used to address chromatin dynamics [47]. The consensus from all of these studies is that chromatin typically does not undergo long range movement over several hours in interphase but is restricted to local constrained motion. However this rather static picture can change if transcription is activated, which can lead to decondensation and movement to the interior of the affected locus.[48,49] Another phase of repositioning seems to be replication of a locus, which again can be associated with movement towards the interior.[50] In summary, we can assume that the position of peripheral chromatin is rather static during interphase, consistent with the exceptional stability of the NE protein network. It will be important to find out in the future how long chromatin-NE adaptors stay bound to chromosomes and if these interactions are specifically regulated during transcription activation or replication of peripheral chromatin.

Overall the interphase dynamics of all NE proteins studied so far have reinforced the view of a protein network that is very stable, made up of long lived interactions that serve to maintain the structure of the interphase nucleus.

Dynamics in Mitosis

The interphase NE which so efficiently separates nuclear from cytoplasmic processes complicates life of metazoan cells when it is time to divide. To successfully complete mitosis, the microtubules of the spindle apparatus which are exclusively cytoplasmic must come in contact with chromosomes which are shielded by the NE protein network. To achieve this, mammalian cells break down their NE completely in prometaphase and undergo an "open" mitosis, releasing chromosomes into the cytoplasm to accomplish segregation. The process of NE breakdown (NEBD) and reformation involves the disassembly and dispersal of all four structural units of the NE. Once mitosis is completed, the dispersed NE proteins are then used again to assemble new nuclei in the next cell generation. As expected from the complex interphase architecture, NE breakdown and assembly are complicated processes that require the coordinated action of many cellular activities such as mitotic phosphorylation/ dephosphorylation, nucleocytoplasmic transport, membrane fusion as well as the action of microtubule motorproteins. Currently, a consensus model of NE dynamics is emerging that can explain all the changes in NE structure and dynamics that have been documented during cell division.

INM Proteins: Switching Retention Off and Back On

The Old Model: Mitotic Phosphorylation of NE Proteins and Vesiculation of Nuclear Nembranes

Many biochemical studies have shown that NE proteins are subject to phosphorylation in M-phase by MPF, the complex of cyclin B and p34^{cdc2} in mammalian cells. Phosphorylation depolymerizes and disperses lamins[34,51-53] and some nucleoporins.[54,55] Several INM proteins have also been shown to be targeted by cdc2 (ref. 2) but the consequence of their modification is much less clear. We currently assume that it abolishes their ability to interact with lamins and/or chromatin, which would allow the INM to detach from chromosomes. The fate of nuclear membrane proteins during M-phase has been an issue of some contention in the recent literature.[1] Nevertheless, most textbooks present a seemingly simple model according to which the NE vesiculates after the lamin polymer has been depolymerized through mitotic phosphorylation.[56] It is useful to take a brief look at how we arrived at this model. In the early '80s nuclear assembly and breakdown was reconstituted in amphibian oocyte extracts[57,58] a system that subsequently lead to a wealth of biochemical data from many other laboratories. Since the procedure of this assay results in fragmented membrane homogenates, such "vesicles" were assumed to be the natural starting material to assemble new nuclear membranes. Additional support for mitotic NE vesicles came from a contemporary EM study showing ER vesiculation in dividing rat thyroid cells.[59] Based on these two lines of evidence, NE vesiculation was quickly accepted as the mechanism that would do in cells what homogenization did in nuclear reconstitution assays: produce precursor membrane fragments for nuclear assembly. Another attractive feature postulated by this model was that many small precursor membrane fragments can be partitioned efficiently by a stochastic mechanism such as diffusion between the two daughter cells.

The Modern (and Traditional!) View: ER Absorption by Switching Off Retention

However, if one steps back even further in time and looks at the pioneering electron microscopic work done on mitotic plant and animal cells in the '60s[60-62] it is clear that assays in extracts are not ideally suited to evaluate the dynamic morphological changes nuclear membranes undergo in mitosis. The first EM observations of mitotic cells already documented that mitotic nuclear membranes became indistinguishable from tubules and cisternae of the ER when cells entered M-phase and that nuclear membranes assembled after mitosis seemed to derive from the ER. This view has been confirmed strongly in recent studies in intact mammalian cells that revisited the fate of the NE in mitosis and demonstrated that the ER serves as the reservoir for nuclear membrane proteins in M-phase.[28,34,36,63] That the ER network, rather than membrane vesicles, is the precursor for NE assembly is also suggested by recent dynamic in vitro studies on NE assembly, which show that also in *Xenopus* egg extracts, network formation from vesicles is an intermediate step prior to NE assembly.[64-66]

How then do INM proteins move back into the ER in prometaphase and how is the INM subdomain of the ER reestablished? If we remind ourselves how INM proteins are targeted in interphase, and take into account the disruptive force of mitotic phosphorylation on protein-protein interactions the answer becomes immediately clear. In interphase nuclear membrane proteins diffuse between the ER and the INM but are trapped in the latter by selective binding interactions when they meet lamins and chromatin (Fig. 2A). When these interactions are switched off by mitotic phosphorylation in prophase, INM proteins will equilibrate with the ER, since they are no longer retained and set free to diffuse back into the ER (Fig. 2B). Simple diffusion can equilibrate the INM pool with the ER efficiently and rapidly through many connections between INM and ONM and the continuity between ONM and ER. Exactly such an equilibration process from nuclear rim to the ER network can be observed in vivo for several INM proteins at different times in prophase[36] (J. Beaudouin and J.E., unpublished observations) leading to a uniform dispersed distribution of INM proteins in the intact mitotic ER.[28] The reverse mechanism, i.e., switching the retaining binding interactions back on by dephosphorylation at the end of mitosis, elegantly explains how the INM subdomain can be reformed. Degradation of cyclin B after metaphase inactivates MPF kinase and allows dephosphorylation to reactivate the interactions between INM proteins and their chromatin binding partners. In anaphase, when more and more attachment sites for membranes are becoming available through the combined effect of dephosphorylation and chromosome decondensation, NE assembly can proceed by coating of the chromosome surface with ER cisternae. The cisternae contain INM proteins which bind to chromatin as soon as they are in close proximity (Fig. 2C). Thus, nuclear membrane proteins are immediately concentrated at the membrane chromatin interface, again by diffusion from the ER and selective retention on chromatin, which drives an increases in the membrane surface around the chromosome template. Precisely this process can be observed in living cells by following GFP-labeled INM and ER proteins (Fig. 4).[28,42,67] However, even with the ER network as a precursor for nuclear membranes, membrane fusion will be necessary to enclose the chromosomes by a sealed NE.[68] Recent studies have begun to shed light on the molecular machinery in NE fusion processes[65,66] and it will be very interesting to investigate the dynamics of this process in intact cells.

Lamina: Tearing of a Polymer, Dispersion and Re-Import of Monomers

The same pioneering ultrastructural studies that reported the merging of NE and ER in mitosis, also noted that centrosomes were closely associated with the NE and often buried in an invagination in prophase.[60,61] More recent biochemical and genetic studies of microtubule motors have shown that cytoplasmic dynein is required to attach centrosomes to the nucleus in *C. elegans* and *Drosophila*.[69-71] In addition, dynein localizes to the NE of mammalian cells in prophase.[72,73] Although the molecular basis of the dynein-NE interaction is still unclear, we have gained some insight into its functions such as centrosome separation and nuclear movement.[74] Two recent studies have now also linked NEBD to the action of dynein and the mitotic spindle. Using quantitative live cell imaging and electron microscopy, these studies showed that spindle microtubules facilitate NEBD by literally tearing the lamin polymer open. This is apparently accomplished by immobilizing dynein on the outer surface of the nuclear envelope, which is then drawn towards the centrosomes of the forming mitotic spindle by dynein's minus end directed motion. Pulling on the nucleus by the mitotic spindle results in massive distortion of nuclear shape, which could be documented by pattern photobleaching of the nuclear lamina (Fig. 3B). Most prominently deep invaginations are formed close to the centrosomes while the lamina is stretched further away from the asters.[36,73] The NE remained intact during these deformations until holes appeared in the lamina at the sites of maximum stretching, suggesting a tearing mechanism. The opening of this physical discontinuity in the NE allows even large cytoplasmic molecules to freely enter the nucleus. This then triggers the gradual disassembly of the lamina, a process that is only completed in metaphase, when even the lamina fragments that have been drawn to the centrosomes by dynein are completely solubilized. These observations nicely demonstrate that formation of the mitotic spindle and NEBD are two mitotic processes which are highly coordinated. By doing this the mammalian cells could have evolved an additional mechanism to control the transition of chromosome organization by nuclear membranes to microtubules.

Although it is clear that the lamina plays an essential role in maintaining nuclear integrity and shape in somatic cells and is probably a key structure resisting transition into mitosis, it seems to play only a minor role in the early stages of nuclear assembly. According to most studies, the majority of both A and B type lamins are re-imported into post-mitotic nuclei that have already assembled a fully sealed nuclear membrane containing functional nuclear pores,[34,35,75] although some studies have suggested an earlier association.[33] Interestingly, recent work has shown that the assembly of B type lamins is regulated by protein phosphatase 1. This protein binds to the integral membrane protein A-kinase anchoring protein (AKAP)149 and then dephosphorylates lamins at sites of contact between ER and chromosomes.[76] Without the interactions of PP1 and AKAP149 lamins do not assemble, but cells still complete mitosis. Thus it appears that the assembly of a functional nuclear lamina is secondary to the assembly of nuclear membranes and dispensable for nuclear assembly.

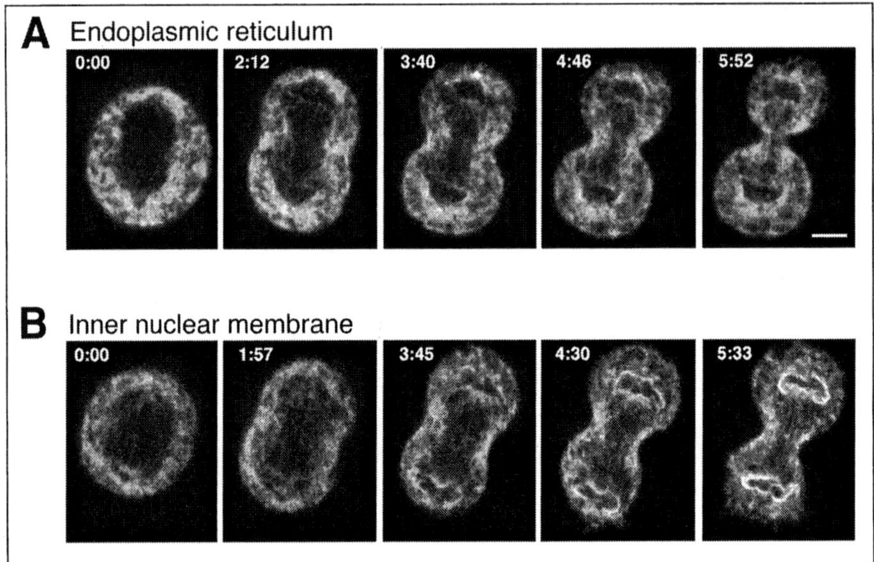

Figure 4. Reestablishment of the INM subdomain from the ER.
2D confocal time-lapse sequences of a NRK cells expressing the ER membrane protein SRβ-CFP [90](A) and the INM protein LBR-GFP [28](B). Note how ER cisternae and tubules surround the chromatin area in anaphase and how the INM protein, but not the ER protein becomes enriched in membranes in contact with chromosomes. Time: mm:ss, t = 0 corresponds to the metaphase to anaphase transition; bar: 10 μm.

Pore Complex Disassembly and Assembly: Many Open Questions

The nuclear pore is a topologically unique structure. It forms an aqueous channel that spans and connects a double membrane. We know very little about the mechanism of disassembling or reassembling the NPC, apart from the fact that some nucleoporins undergo mitotic phosphorylation.[54,55] It is completely unclear how or in which order this large complex is disassembled. In mammalian cells we only know that dispersal of a core nucleoporin such as POM121 only starts after the NE is permeabilized by tearing of the lamina.[36] However there is evidence from very different cell systems such as starfish oocytes and *Drosophila* embryos that point to a key role for NPC disassembly in triggering nuclear permeabilization[77,78] and it will be very interesting to investigate this process in more detail in mammalian cells. Once disassembly is accomplished, the NPC is not broken down to individual polypeptides but rather into Nup subcomplexes that are stable in mitosis and probably form the building blocks from which the NPC can be assembled anew after mitosis.[38,79] While the majority of nucleoporins show a dispersed cytoplasmic distribution in mitotic cells, the transmembrane nucleoporins are absorbed by the ER similar to INM proteins.[34,63] Some nucleoporin (subcomplexes) however show striking localizations in mitosis. The Nup133/107 complex binds to kinetochores from prophase to anaphase[38] while Nup358 (RanBP2) can be seen to localize to the spindle apparatus in mitotic cells.[80] So far however, the mitotic function—if any—of these nucleoporins is unclear. The reassembly of the NPC after mitosis is also mysterious. Two principally different ways of NPC assembly can be envisioned and available data are supporting aspects of both mechanisms. In the first mechanism the soluble core structure of the NPC would be assembled on the surface of chromosomes and then connect to ER cisternae that attach to chromosomes and the side of the core NPC. This model does not require a fusion event between the INM and ONM and is supported by the very early appearance of some nucleoporins on the chromosome surface during anaphase.[34,38,81] Alternatively, NPCs could be inserted into large intact double membranes by a specific intralumenal fusion event. This model is supported by studies on artificial nuclei in the presence of inhibitors as well as in Drosophila embryos where different stages of NPC assembly on the surface of intact membranes can be distinguished by electron microscopy.[77,82-84] It will be a main challenge of future work to shed more light on this mechanism and identify key molecules involved in this process.

Chromosomes: A Complex Template for Nuclear Assembly

We understand even less about the mitotic dynamics of peripheral chromatin proteins linked to the NE than we know about the mechanism of NPC disassembly and reassembly. From the limited data available, it appears that chromosomes retain at least some of their NE adaptor proteins in mitosis. In intact cells, only the behavior of a GFP fusion to barrier to autointegration factor (BAF) has been described. The GFP tagged protein appeared soluble in mitosis and assembled on chromatin concomitantly with one of its binding partners the INM protein emerin.[42] However this data is in conflict with previous localization of BAF to mitotic chromosomes[85] and it remains to be tested if the GFP fusion employed is DNA binding competent, as other GFP-BAF fusions show different behavior (J.E. unpublished observations). Another important group of peripheral chromatin proteins, the heterochromatin protein 1 family has also been localized to chromosomes in mitotic cells using antibodies.[86] Similar to BAF, a second study reported a different localization[87] and more experiments are required to clarify the picture. It is interesting to note that both the HP1 family as well as the third peripheral chromatin protein lamina associated protein 2α (LAP2α) localize to specific subchromosomal domains such as centromeres and telomeres.[88] In anaphase this creates a patchwork like template for nuclear membrane assembly and probably explains the differential localization patterns found for different INM proteins at this time.[67,89]

Concluding Remarks

Our understanding of NE dynamics during the cell cycle has increased dramatically over the recent years. Although areas such as NPC assembly and the precise role of the heterochromatin proteins remain poorly studied, we have arrived at several important mechanistic conclusions. In mammalian cells it is clear now that the ER functions as the mitotic reservoir for all nuclear membrane proteins tested so far and this has had fundamental implications to interpret nuclear membrane protein dispersal and the reformation of the INM ER-subdomain after mitosis. For the latter it seems clear that the binding interactions between INM proteins and chromatin are the driving force of nuclear reformation and probably important in determining the nuclear architecture of the next cell generation. Most likely we still have to discover many chromatin bound factors involved in this process. At the G2/M transition we have seen that mechanical forces exerted by the mitotic spindle on the stable NE protein network facilitate NEBD and complement the biochemical machinery that disrupts protein-protein interactions by phosphorylation. Functional dynamics of the NE promises to be an exciting subject for future research in the coming years.

Acknowledgements

The author would like to thank Péter Lénárt for preparing Figures 1 and 3, Joël Beaudouin for preparing Figure 2B and Nathalie Daigle for preparing Figure 4.

References

1. Collas P, Courvalin J C. Sorting nuclear membrane proteins at mitosis. Trends Cell Biol 2000; 10:5-8.
2. Worman H, Courvalin JC. The inner nuclear membrane. J Membr Biol 2000; 177:1-11.
3. Buendia B, Courvalin JC, Collas P. Dynamics of the nuclear envelope at mitosis and during apoptosis. Cell Mol Life Sci 2001; 58:1781-1789.
4. Aitchison JD, Rout MP. A tense time for the nuclear envelope. Cell 2002; 108:301-304.
5. Gonczy P. Nuclear envelope: Torn apart at mitosis. Curr Biol 2002; 12:R242-244.
6. Lippincott-Schwartz J. Cell biology: Ripping up the nuclear envelope. Nature 2002; 416:31-32.

7. Maul GG, Maul HM, Scogna JE et al. Time sequence of nuclear pore formation in phytohemagglutinin-stimulated lymphocytes and in HeLa cells during the cell cycle. J Cell Biol 1972; 55:433-447.
8. Andrulis ED, Neiman AM, Zappulla DC et al. Perinuclear localization of chromatin facilitates transcriptional silencing. Nature 1998; 394:592-595.
9. Gasser SM. Positions of potential: nuclear organization and gene expression. Cell 2001; 104:639-642.
10. Feuerbach F, Galy V, Trelles-Sticken E et al. Nuclear architecture and spatial positioning help establish transcriptional states of telomeres in yeast. Nat Cell Biol 2002; 4:214-221.
11. Dreger M, Bengtsson L, Schoneberg T et al. Nuclear envelope proteomics: Novel integral membrane proteins of the inner nuclear membrane. Proc Natl Acad Sci USA 2001; 98:11943-11948.
12. Rout MP, Aitchison JD, Suprapto A et al. The yeast nuclear pore complex: Composition, architecture, and transport mechanism. J Cell Biol 2000; 148:635-651.
13. Stuurman N, Heins SAebi U. Nuclear lamins: Their structure, assembly, and interactions. J Struct Biol 1998; 122:42-66.
14. Gruenbaum Y, Wilson KL, Harel A et al. Review: nuclear lamins—Structural proteins with fundamental functions. J Struct Biol 2000; 129:313-323.
15. Aebi U, Cohn J, Buhle L et al. The nuclear lamina is a meshwork of intermediate-type filaments. Nature 1986; 323:560-564.
16. Wilson KL. The nuclear envelope, muscular dystrophy and gene expression. Trends Cell Biol 2000; 10:125-129.
17. Cohen M, Lee KK, Wilson KL et al. Transcriptional repression, apoptosis, human disease and the functional evolution of the nuclear lamina. Trends Biochem Sci 2001; 26:41-47.
18. Gorlich D, Kutay U. Transport between the cell nucleus and the cytoplasm. Annu Rev Cell Dev Biol 1999; 15:607-660.
19. Galy V, Olivo-Marin JC, Scherthan H et al. Nuclear pore complexes in the organization of silent telomeric chromatin. Nature 2000; 403:108-112.
20. Smythe C, Jenkins HE, Hutchison CJ. Incorporation of the nuclear pore basket protein nup153 into nuclear pore structures is dependent upon lamina assembly: Evidence from cell-free extracts of *Xenopus* eggs. Embo J 2000; 19:3918-3931.
21. Goldberg MW, Allen TD. The nuclear pore complex and lamina: Three-dimensional structures and interactions determined by field emission in-lens scanning electron microscopy. J Mol Biol 1996; 257:848-865.
22. Tsien RY. The green fluorescent protein. Annu Rev Biochem 1998; 67:509-544.
23. Gerlich D, Beaudouin J, Gebhard M et al. Four-dimensional imaging and quantitative reconstruction to analyse complex spatiotemporal processes in live cells. Nat Cell Biol 2001; 3:852-855.
24. Lippincott-Schwartz J, Snapp EKenworthy A. Studying protein dynamics in living cells. Nat Rev Mol Cell Biol 2001; 2:444-456.
25. Soullam B, Worman HJ. The amino-terminal domain of the lamin B receptor is a nuclear envelope targeting signal. J Cell Biol 1993; 120:1093-1100.
26. Smith S, Blobel G. The first membrane spanning region of the lamin B receptor is sufficient for sorting to the inner nuclear membrane. J Cell Biol 1993; 120:631-637.
27. Soullam B, Worman HJ. Signals and structural features involved in integral membrane protein targeting to the inner nuclear membrane. J Cell Biol 1995; 130:15-27.
28. Ellenberg J, Siggia ED, Moreira JE et al. Nuclear membrane dynamics and reassembly in living cells: Targeting of an inner nuclear membrane protein in interphase and mitosis. J Cell Biol 1997; 138:1193-1206.
29. Wu W, Lin F, Worman HJ. Intracellular trafficking of MAN1, an integral protein of the nuclear envelope inner membrane. J Cell Sci 2002; 115:1361-1371.
30. Ostlund C, Ellenberg J, Hallberg E et al. Intracellular trafficking of emerin, the Emery-Dreifuss muscular dystrophy protein. J Cell Sci 1999; 112:1709-1719.
31. Vaughan A, Alvarez-Reyes M, Bridger JM et al. Both emerin and lamin C depend on lamin A for localization at the nuclear envelope. J Cell Sci 2001; 114:2577-2590.
32. Gruenbaum Y, Lee KK, Liu J et al. The expression, lamin-dependent localization and RNAi depletion phenotype for emerin in *C. elegans*. J Cell Sci 2002; 115:923-929.
33. Moir R D, Yoon M, Khuon S et al. Nuclear lamins A and B1: Different pathways of assembly during nuclear envelope formation in living cells. J Cell Biol 2000; 151:1155-1168.
34. Daigle N, Beaudouin J, Hartnell L et al. Nuclear pore complexes form immobile networks and have a very low turnover in live mammalian cells. J Cell Biol 2001; 154:71-84.
35. Broers JL, Machiels BM, van Eys G J et al. Dynamics of the nuclear lamina as monitored by GFP-tagged A-type lamins. J Cell Sci 1999; 112:3463-3475.

36. Beaudouin J, Gerlich D, Daigle N et al. Nuclear envelope breakdown proceeds by microtubule-induced tearing of the lamina. Cell 2002; 108:83-96.
37. Aaronson RP, Blobel G. Isolation of nuclear pore complexes in association with a lamina. Proc Natl Acad Sci USA 1975; 72:1007-1011.
38. Belgareh N, Rabut G, Bai S W et al. An evolutionarily conserved NPC subcomplex, which redistributes in part to kinetochores in mammalian cells. J Cell Biol 2001; 154:1147-1160.
39. Griffis ER, Altan N, Lippincott-Schwartz J et al. Nup98 Is a mobile nucleoporin with transcription-dependent dynamics. Mol Biol Cell 2002; 13:1282-1297.
40. Belgareh N, Doye V. Dynamics of nuclear pore distribution in nucleoporin mutant yeast cells. J Cell Biol 1997; 136:747-759.
41. Bucci M, Wente SR. In vivo dynamics of nuclear pore complexes in yeast. J Cell Biol 1997; 136:1185-1199.
42. Haraguchi T, Koujin T, Segura-Totten M et al. BAF is required for emerin assembly into the reforming nuclear envelope. J Cell Sci 2001; 114:4575-4585.
43. Zink D, Cremer T, Saffrich R et al. Structure and dynamics of human interphase chromosome territories in vivo. Hum Genet 1998; 102:241-251.
44. Bornfleth H, Edelmann P, Zink D et al. Quantitative motion analysis of subchromosomal foci in living cells using four-dimensional microscopy. Biophys J 1999; 77:2871-2886.
45. Sadoni N, Langer S, Fauth C et al. Nuclear organization of mammalian genomes. Polar chromosome territories build up functionally distinct higher order compartments. J Cell Biol 1999; 146:1211-1226.
46. Belmont AS. Visualizing chromosome dynamics with GFP. Trends Cell Biol 2001; 11:250-257.
47. Abney JR, Cutler B, Fillbach ML et al. Chromatin dynamics in interphase nuclei and its implications for nuclear structure. J Cell Biol 1997; 137:1459-1468.
48. Tumbar T, Sudlow G, Belmont AS. Large-scale chromatin unfolding and remodeling induced by VP16 acidic activation domain. J Cell Biol 1999; 145:1341-1354.
49. Tumbar T, Belmont AS. Interphase movements of a DNA chromosome region modulated by VP16 transcriptional activator. Nat Cell Biol 2001; 3:134-139.
50. Li G, Sudlow G, Belmont AS. Interphase cell cycle dynamics of a late-replicating, heterochromatic homogeneously staining region: Precise choreography of condensation/decondensation and nuclear positioning. J Cell Biol 1998; 140:975-989.
51. Nigg EA. Assembly-disassembly of the nuclear lamina. Curr Opin Cell Biol 1992; 4:105-109.
52. Gerace L, Blobel G. The nuclear envelope lamina is reversibly depolymerized during mitosis. Cell 1980; 19:277-287.
53. Stick R, Angres B, Lehner CF et al. The fates of chicken nuclear lamin proteins during mitosis: evidence for a reversible redistribution of lamin B2 between inner nuclear membrane and elements of the endoplasmic reticulum. J Cell Biol 1988; 107:397-406.
54. Macaulay C, Meier E, Forbes DJ. Differential mitotic phosphorylation of proteins of the nuclear pore complex. J Biol Chem 1995; 270:254-262.
55. Favreau C, Worman HJ, Wozniak R W et al. Cell cycle-dependent phosphorylation of nucleoporins and nuclear pore membrane protein Gp210. Biochemistry 1996; 35:8035-8044.
56. Alberts B. Molecular Biology of the Cell, xliii, 1294, <1267>. New York: Garland Pub, 1994.
57. Lohka MJ, Masui Y. Formation in vitro of sperm pronuclei and mitotic chromosomes induced by amphibian ooplasmic components. Science 1983; 220:719-721.
58. Lohka MJ, Maller JL. Induction of nuclear envelope breakdown, chromosome condensation, and spindle formation in cell-free extracts. J Cell Biol 1985; 101:518-523.
59. Zeligs JD, Wollman SH. Mitosis in rat thyroid epithelial cells in vivo. I. Ultrastructural changes in cytoplasmic organelles during the mitotic cycle. J Ultrastruct Res 1979; 66:53-77.
60. Bajer A, Molé-Bajer J. Formation of spindle fibers, kinetochore orientation, and behavior of the nuclear envelope during mitosis in endosperm. Chromosoma 1969; 27:448-484.
61. Porter KR, Machado RD. Studies on the endoplasmic reticulum. IV. Its form and distribution during mitosis in cells of onion root tip. J Biophys Biochem Cytol 1960; 7:167-180.
62. Robbins E, Gonatas NK. The ultrastructure of a mammalian cell during the mitotic cell cycle. J Cell Biol 1964; 21:429-463.
63. Yang L, Guan T, Gerace L. Integral membrane proteins of the nuclear envelope are dispersed throughout the endoplasmic reticulum during mitosis. J Cell Biol 1997; 137:1199-1210.
64. Dreier L, Rapoport TA. In vitro formation of the endoplasmic reticulum occurs independently of microtubules by a controlled fusion reaction. J Cell Biol 2000; 148:883-898.
65. Hetzer M, Bilbao-Cortes D, Walther TC et al. GTP hydrolysis by Ran is required for nuclear envelope assembly. Mol Cell 2000; 5:1013-1024.

66. Hetzer M, Meyer HH, Walther TC et al. Distinct AAA-ATPase p97 complexes function in discrete steps of nuclear assembly. Nat Cell Biol 2001; 3:1086-1091.
67. Haraguchi T, Koujin T, Hayakawa T et al. Live fluorescence imaging reveals early recruitment of emerin, LBR, RanBP2, and Nup153 to reforming functional nuclear envelopes. J Cell Sci 2000; 113:779-794.
68. Burke B. The nuclear envelope: filling in gaps. Nat Cell Biol 2001; 3:E273-274.
69. Robinson JT, Wojcik EJ, Sanders MA et al. Cytoplasmic dynein is required for the nuclear attachment and migration of centrosomes during mitosis in *Drosophila*. J Cell Biol 1999; 146:597-608.
70. Gönczy P, Pichler S, Kirkham M et al. Cytoplasmic dynein is required for distinct aspects of MTOC positioning, including centrosome separation, in the one cell stage *Caenorhabditis elegans* embryo. J Cell Biol 1999; 147:135-150.
71. Yoder JH, Han M. Cytoplasmic dynein light intermediate chain is required for discrete aspects of mitosis in Caenorhabditis elegans. Mol Biol Cell 2001; 12:2921-2933.
72. Busson S, Dujardin D, Moreau A et al. Dynein and dynactin are localized to astral microtubules and at cortical sites in mitotic epithelial cells. Curr Biol 1998; 8:541-544.
73. Salina D, Bodoor K, Eckley DM et al. Cytoplasmic dynein as a facilitator of nuclear envelope breakdown. Cell 2002; 108:97-107.
74. Reinsch S, Gönczy P. Mechanisms of nuclear positioning. J Cell Sci 1998; 111:2283-2295.
75. Chaudhary N, Courvalin J-C. Stepwise reassembly of the nuclear envelope at the end of mitosis. J Cell Biol 1993; 122:295-306.
76. Steen RL, Collas P. Mistargeting of B-type lamins at the end of mitosis: Implications on cell survival and regulation of lamins A/C expression. J Cell Biol 2001; 153:621-626.
77. Kiseleva E, Rutherford S, Cotter LM et al. Steps of nuclear pore complex disassembly and reassembly during mitosis in early Drosophila embryos. J Cell Sci 2001; 114:3607-3618.
78. Terasaki M, Campagnola P, Rolls MM et al. A new model for nuclear envelope breakdown. Mol Biol Cell 2001; 12:503-510.
79. Siniossoglou S, Lutzmann M, Santos-Rosa H et al. Structure and assembly of the Nup84p complex. J Cell Biol 2000; 149:41-54.
80. Joseph J, Tan SH, Karpova TS et al. SUMO-1 targets RanGAP1 to kinetochores and mitotic spindles. J Cell Biol 2002; 156:595-602.
81. Bodoor K, Shaikh S, Salina D et al. Sequential recruitment of NPC proteins to the nuclear periphery at the end of mitosis. J Cell Sci 1999; 112:2253-2264.
82. Macaulay C, Forbes DJ. Assembly of the nuclear pore: biochemically distinct steps revealed with NEM, GTP gamma S, and BAPTA. J Cell Biol 1996; 132:5-20.
83. Goldberg MW, Wiese C, Allen TD et al. Dimples, pores, star-rings, and thin rings on growing nuclear envelopes: Evidence for structural intermediates in nuclear pore complex assembly. J Cell Sci 1997; 110:409-420.
84. Wiese C, Goldberg MW, Allen TD et al. Nuclear envelope assembly in *Xenopus* extracts visualized by scanning EM reveals a transport-dependent ‚envelope smoothing‘ event. J Cell Sci 1997; 110:1489-1502.
85. Furukawa K. LAP2 binding protein 1 (L2BP1/BAF) is a candidate mediator of LAP2-chromatin interaction. J Cell Sci 1999; 112:2485-2492.
86. Minc E, Allory Y, Worman HJ et al. Localization and phosphorylation of HP1 proteins during the cell cycle in mammalian cells. Chromosoma 1999; 108:220-234.
87. Kourmouli N, Theodoropoulos PA, Dialynas G et al. Dynamic associations of heterochromatin protein 1 with the nuclear envelope. Embo J 2000; 19:6558-6568.
88. Dechat T, Gotzmann J, Stockinger A et al. Detergent-salt resistance of LAP2alpha in interphase nuclei and phosphorylation-dependent association with chromosomes early in nuclear assembly implies functions in nuclear structure dynamics. Embo J 1998; 17:4887-4902.
89. Dabauvalle MC, Muller E, Ewald A et al. Distribution of emerin during the cell cycle. Eur J Cell Biol 1999; 78:749-756.
90. Nehls S, Snapp EL, Cole NB et al. Dynamics and retention of misfolded proteins in native ER membranes. Nat Cell Biol 2000; 2:288-295.

CHAPTER 3

Targeting and Retention of Proteins in the Inner and Pore Membranes of the Nuclear Envelope

Cecilia Östlund, Wei Wu and Howard J. Worman

Abstract

The targeting of integral proteins to the inner and pore membranes of the nuclear envelope occurs through different mechanisms than the targeting of soluble proteins to the nucleus. Most nuclear integral membrane proteins reach their sites through a diffusion-retention mechanism, where the proteins are inserted into the endoplasmic reticulum membrane during translation, and then laterally diffuse along the endoplasmic reticulum membrane to the pore and inner nuclear membranes. The proteins are then retained at these sites by interactions with other proteins or chromatin. Peripheral proteins of the inner nuclear membrane are imported through the nuclear pore complexes by mechanisms similar to those of other nonmembrane, nuclear proteins. They are then retained at the nuclear envelope through interactions with other proteins or by associations of lipid anchors with membranes.

The nuclear envelope surrounds the cell nucleus and is composed of the nuclear lamina, nuclear pore complexes (NPCs) and nuclear membranes (for reviews see refs. 1 and 2). The nuclear membranes consist of three distinct but interconnected parts, the outer nuclear membrane, the pore membrane and the inner nuclear membrane (INM). The outer membrane is directly continuous with and similar in composition to the endoplasmic reticulum (ER). The pore membranes connect the inner and outer nuclear membranes at the sites of the NPCs, through which proteins and RNA are transported in and out of the nucleus. The INM is associated with chromatin and the nuclear lamina, an intermediate filament network consisting of A-type and B-type lamin proteins. While no proteins have been identified as specific to the outer nuclear membrane, the INM and the pore membrane have their own sets of proteins (Fig. 1). The topic of this review is how proteins are targeted to these nuclear membrane domains.

Targeting of Integral Membrane Proteins to the Inner Nuclear Membrane

Several integral membrane proteins are specifically localized to the INM in interphase cells. The first to be identified was the lamin B receptor (LBR), which has a nucleoplasmic amino-terminal domain followed by a hydrophobic segment with eight putative transmembrane spanning regions.[3,4] Other proteins localized to the INM are the lamina associated polypeptides (LAP) 1 and 2, each having several isoforms, emerin and MAN1.[5-10] Most LAP isoforms (for a review see ref. 11) and emerin[7] have a nucleoplasmic amino-terminal domain, followed

Nuclear Envelope Dynamics in Embryos and Somatic Cells
edited by Philippe Collas. ©2002 Eurekah.com and Kluwer Academic/Plenum Publishers.

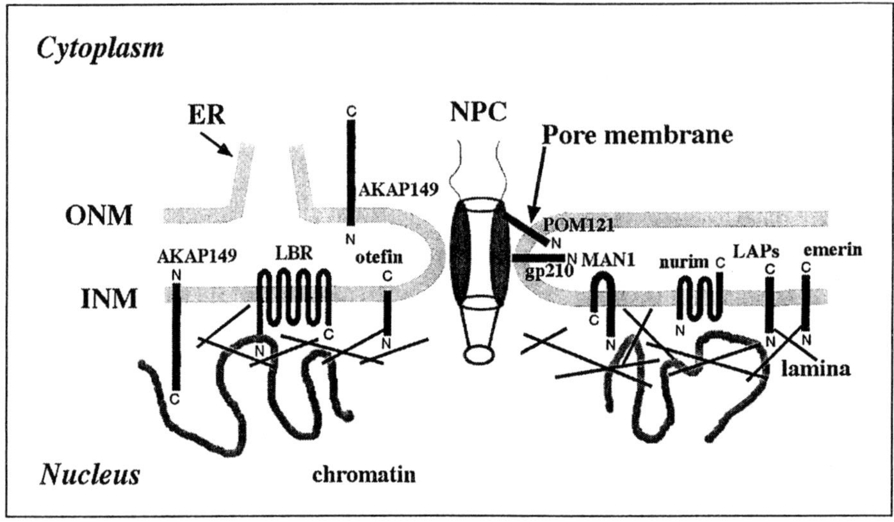

Figure 1. Schematic diagram of the nuclear envelope. The nuclear membranes consist of the outer nuclear membrane (ONM), continuos with the endoplasmic reticulum (ER), the pore membrane and the inner nuclear membrane (INM). The nuclear pore complex (NPC) is associated with the pore membrane. The lamina, chromatin and the best characterized proteins of the pore and inner membrane are also shown.

by one transmembrane domain and a short, luminal tail. MAN1 has two transmembrane segments, and both its amino-terminus and carboxyl-terminus face the nucleoplasm.[10] Interestingly, LAP2, emerin, MAN1 and the peripheral INM protein otefin share a region of sequence similarity of approximately 50 amino acids called the LEM-domain.[10,12]

Nurim, another protein of the INM, was identified using a visual screen of a green fluorescent protein (GFP)-cDNA expression library.[13] Nurim is a membrane protein with five putative transmembrane segments and no large, hydrophilic domains. It is unrelated to the other identified INM proteins. Other proteins recently suggested to localize to the INM include UNC-84, UNCL, the RING-finger binding protein (RFBP) and LUMA.[14-17] The A-kinase anchoring protein AKAP149 has also been shown to partly localize to the INM.[18]

Many studies during the past several years have addressed the question of how integral membrane proteins reach the INM. It is increasingly clear that their targeting is fundamentally different than targeting of soluble proteins to the nucleus. The latter occurs through the central channel of the NPCs, is essential for proteins larger than ~70 kDa which cannot enter the nucleus by diffusion,[19] and is dependent on well-defined nuclear localization sequences (NLSs) most commonly composed of one or two short stretches of basic amino acids (reviewed in ref. 20). The targeting of integral membrane proteins, however, often requires several regions of the protein, and these regions vary between different proteins (Fig. 2). The results of most studies of INM protein targeting are consistent with a diffusion-retention model.[21] In this model, proteins are synthesized on and cotranslationally inserted into the ER membrane. They then diffuse laterally via the pore membrane to the INM, where they are retained and immobilized by binding to other proteins or structures such as the nuclear lamina and chromatin.

LBR was the first integral INM protein for which targeting was studied. The nucleoplasmic, amino-terminal domain of the protein consists of approximately 200 amino acids and has been shown to bind B-type lamins and chromatin.[22-24] The hydrophobic region, with eight putative membrane-spanning segments, shows strong sequence similarity to sterol reductases that are localized to the ER.[25,26] Early studies showed that at least two different targeting/

retention signals were independently sufficient to localize the protein to the INM.[27,28] When the nucleoplasmic amino-terminal region of LBR was fused to the transmembrane domain and luminal carboxyl-terminus of chicken hepatic lectin (CHL), an integral membrane protein of the ER, endosomes and plasma membrane, the chimeric protein localized to the INM.[28] This showed that the amino-terminal region of LBR is sufficient for targeting of a non-nuclear integral protein to the nuclear rim. The first of the eight putative transmembrane spanning regions was also sufficient for targeting to the INM, demonstrating two nonoverlapping targeting signals in LBR.[27]

The nucleoplasmic amino-terminal domains of LAP2, emerin and MAN1 can also target CHL to the INM. Their transmembrane domains, however, lack nuclear targeting information.[29-32] When CHL was fused to the soluble nuclear protein histone H1 or to the NLSs from the SV-40 T antigen or nucleoplasmin, two soluble proteins whose NLSs previously had been shown to target chimeric, soluble proteins to the nucleus, these fusion proteins did not target to the INM, but localized to the ER.[21] These experiments showed that the signals that target integral proteins to the INM are different from those targeting soluble proteins to the nucleus.

As the ER, outer nuclear membrane, pore membrane and INM all are continuous with each other, proteins can potentially move between all these domains by lateral diffusion. To reach the INM, they would have to pass through the pore membrane, where the NPCs are situated. Although the detailed functions of different parts of the NPCs are not well understood, ultrastructural studies have shown the NPCs to contain eight lateral channels adjacent to the membrane.[33,34] The diameter of these channels is approximately 10 nm, and proteins with nucleocytoplasmic domains smaller than 60 kDa can presumably diffuse through these channels.[33] Consistent with this hypothesis, almost all integral proteins found in the INM have nucleoplasmic domains smaller than this size. Only AKAP149, which is only partially localized to the INM, has a larger nucleoplasmic domain. This protein may reach the INM during nuclear envelope reassembly at mitosis.[18] Experiments with chimeric proteins further support the hypothesis that in interphase, INM proteins must pass through the channels of the NPC.[21] When truncated chicken muscle pyruvate kinase (CMPK) is inserted between the LBR amino-terminus and CHL, this integral membrane protein, with a nucleoplasmic domain of approximately 72.5 kDa, remains in the ER, as does a protein where three consecutive amino-terminal domains of LBR (67.5 kDa) is fused to CHL. On the contrary, a protein with only two consecutive amino-terminal LBR domains (45 kDa) fused to CHL is targeted to the INM. Similar results are seen with MAN1. A truncated form of MAN1 containing the nucleoplasmic amino-terminal domain and the first transmembrane segment is targeted to the INM. When CMPK is fused to the amino-terminus of this protein, yielding an integral protein with a nucleoplasmic domain of approximately 100 kDa, this chimeric protein is found in the ER.[32]

If integral proteins reach the INM by lateral diffusion, they must then somehow be retained there or they would just as easily diffuse back to the ER. Experiments using fluorescent recovery after photobleaching (FRAP) and fluorescence loss in photobleaching (FLIP) show that LBR is virtually immobilized when it enters the INM, while it diffuses freely in the ER.[35] When a plasmid encoding a chimeric protein with the amino-terminal and first transmembrane domains of LBR fused to GFP is microinjected into COS-7 cells, LBR-GFP appears initially in the ER, but accumulates in the nuclear envelope over the course of 3 to 10 hours. After 8 hours, its localization is indistinguishable from that of endogenous LBR. At very high levels of expression, LBR-GFP is also present throughout the ER.

In FRAP experiments, a region of the nuclear envelope was bleached, and the recovery of fluorescence, dependent on the ability of LBR-GFP to move laterally within the INM, was followed. A recovery of fluorescence in the bleached region could not be detected until after 20 minutes. When similar experiments were performed in cells with a high expression level of LBR-GFP, with some of the protein present in the ER, LBR-GFP was relatively mobile in this compartment with a diffusion constant (D) of $0.41 \pm 0.1\ \mu m^2/s$. The D of LBR-GFP in the INM could not be calculated due to the high immobile fraction of the protein. A variant of

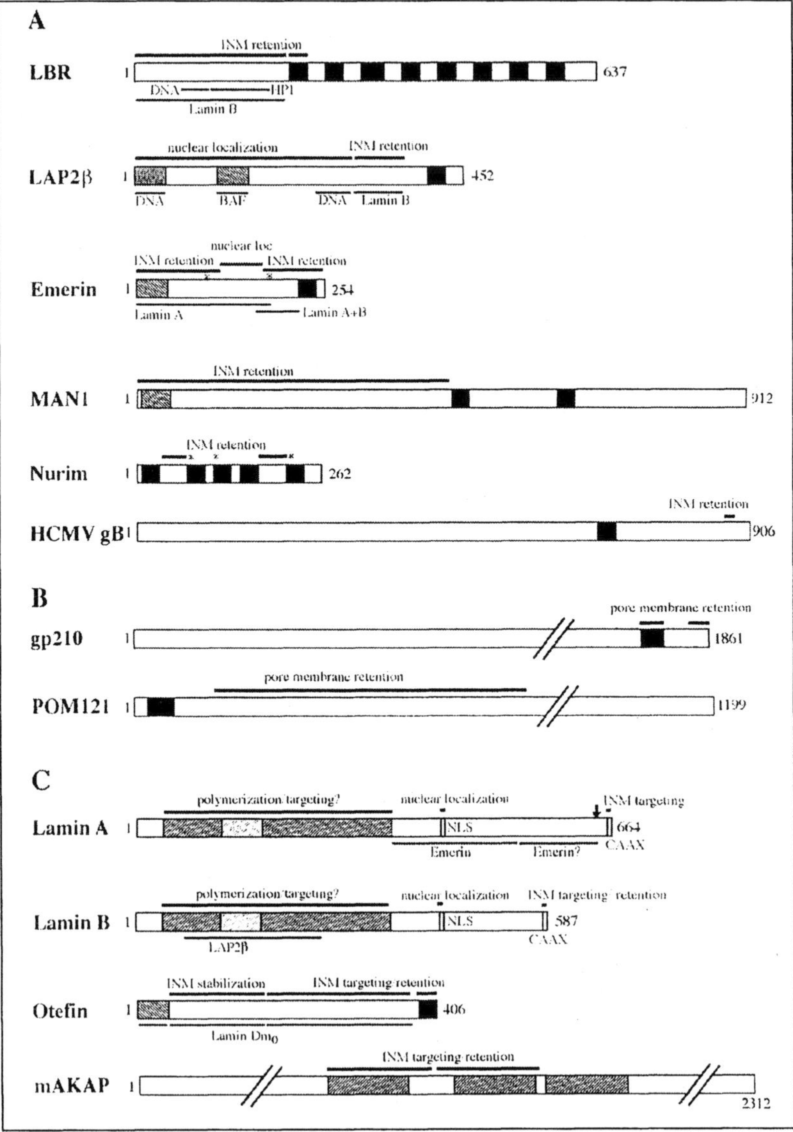

Figure 2. Schematic diagram of nuclear envelope proteins. The amino-termini of the proteins are at the left, with numbers indicating the first and last amino acid of each protein. Bars on top of protein diagrams indicate regions suggested to be involved in nuclear targeting, bars under the diagrams indicate putative regions of protein-protein or protein-DNA interaction. Asterisks indicate point mutations affecting protein localization. (A) Integral INM proteins. Domains of the proteins are represented by: black, putative membrane-spanning regions; stripes, LEM-domains; grey, LEM-like domain. (B) Integral pore membrane proteins. Membrane-spanning regions are indicated in black. (C) Peripheral proteins of the INM. Striped regions of the lamin diagrams indicate rod domains, with the 42 amino acids not found in cytoplasmic intermediate filaments shown in lighter color. NLSs and CAAX-motifs are indicated. Arrow indicates the endoprotease cleavage site for processing prelamin A to lamin A. The striped region in the otefin diagram indicates the LEM-domain and black indicates the hydrophobic region. Striped regions in the mAKAP diagram indicate the spectrin-repeat sequences.

FRAP termed FLIP, where a small area of the cell is repeatedly photobleached and the loss of fluorescence from the whole cell is monitored, was used to further probe the continuity between the INM and the ER. When a region of the ER in an overexpressing cell was bleached, there was a rapid loss of fluorescence from the whole ER, while the fluorescence in the INM remained. This showed that while LBR-GFP rapidly could move between different regions in the ER, it became immobilized upon entry in the INM, and could not reenter the ER. The large immobile fraction of LBR-GFP in the INM indicated binding to a fixed structural component, rather than retention by assembly into multimeric complexes, as the latter would be expected to have some lateral mobility.[35] This structural component could be the lamina and/or chromatin.

FRAP studies of GFP-fusion proteins containing emerin or the amino-terminal domain and first transmembrane segment of MAN1 also support the diffusion-retention model for INM targeting. These proteins also are immobilized in the INM, albeit to a lesser degree than LBR. In the nuclear envelope, fluorescence recovery is nearly complete eight minutes after photobleaching of cells expressing emerin-GFP, giving a D-value of 0.10 ± 0.01 $\mu m^2/s$.[31] The D-value for MAN1-GFP in the nuclear envelope is very similar, 0.12 ± 0.02 $\mu m^2/s$.[32] The diffusion constants of emerin-GFP and MAN1-GFP in the ER are relatively similar to that for LBR-GFP (0.32 $\mu m^2/s \pm 0.01$ for emerin-GFP, 0.28 $\mu m^2/s \pm 0.04$ for MAN1-GFP, 0.41 ± 0.1 $\mu m^2/s$ for LBR-GFP). Retention of emerin in the INM is further supported by the results of FLIP studies, where regions of the cytoplasm repeatedly are bleached. Even after 60 rounds of photobleaching over 80 minutes, when emerin-GFP cannot be detected in the ER, the INM shows substantial fluorescence, indicating that emerin-GFP cannot flow back into the ER from the INM.[31]

B-type lamins and chromatin both are binding partners of the amino-terminal domain of LBR.[22,23,24] LBR mutants with partial deletions of the amino-terminal domain are not capable of binding lamin B, suggesting that structural features of this entire region is important for binding.[22] LBR has also been shown to bind DNA in a "southwestern" blotting assay. Amino acids 71-100 are important for this interaction.[22] In vitro studies have shown that the LBR amino-terminal domain preferentially interacts with the nucleosomal linker region, and that the binding is enhanced by DNA curvature and supercoiling.[36] The region between amino acids 97-174 of LBR also binds to human orthologues of *Drosophila* heterochromatin protein HP1.[24] The importance of chromatin-binding for the INM localization of LBR is further indicated by studies in *Xenopus* oocytes, which lack peripheral chromatin associated with the nuclear membrane.[37] LBR in *Xenopus* oocytes is mainly localized to cytoplasmic membranes, while it is localized to the INM in early embryos, which have peripheral chromatin. Since both cell-types have lamins, these results argue against an importance for interaction between lamins and LBR as a mechanism for LBR targeting. Although a role for lamin-binding in the retention of LBR in the INM cannot be ruled out, accumulating evidence suggests that chromatin may play a major role in this retention. This could explain why the amino-terminal domain of LBR, when expressed alone, shows a diffuse nuclear staining, co-localizing with chromatin, rather than a rim staining, as would be expected if the protein were only retained in the nucleus by interactions with the lamina. It could also explain the observation that LBR accumulates at the edge of decondensing chromosomes during anaphase before lamins are targeted.[38]

Many LAP isoforms have been shown to bind to lamins and chromatin.[6] All LAP2 splice forms share a common amino-terminal domain, which contains two structurally similar regions, the LEM domain (also present in emerin and MAN1) and the LEM-like domain.[12] Analyses of these domains have shown that they are structurally similar to protein-protein interaction domains in bacterial multienzyme complexes.[12,39] This is in agreement with findings that the LEM-like domain (amino acids 1-50) binds to DNA[30,40] and the LEM-domain (amino acids 111-152) interacts with the chromatin associated protein BAF.[40-42] The binding between LAP2β and chromatin is, however, complex, and a region between amino acids 244 and 296 has also been implicated in DNA-binding.[43] A LAP2 mutant containing all these regions (amino acids 1-296) has a diffuse nuclear localization when expressed alone, but is localized to the nuclear rim when fused to the transmembrane spanning region of LAP2β. Neither protein is, contrary to wild-type LAP2β, resistant to extraction with Triton X-100.[6,29] LAP2β binds to B-type lamins via a region in its amino-terminal domain (amino acid 298-370).[30] A chimeric protein containing this region fused to truncated CHL is targeted to the INM and resistant to Triton X-100 extraction.[30] This suggests that although the chromatin-binding domain can target an integral membrane protein to the nucleus, the lamin-binding domain is necessary for stable binding of LAP2β to the INM. A deletion mutant (amino acid 371-452) lacking the amino-terminal region but containing the hydrophobic transmembrane domain and luminal tail localizes to the ER.

Binding to lamins has also been suggested as a retention mechanism for emerin. In a knock-out mouse lacking A-type lamins, emerin is partly mislocalized to the ER while LAP2 localization is unaffected, indicating a role for A-type lamins in anchoring emerin to the INM.[44] The first 188 amino acids of the amino-terminal region of emerin has been shown to bind to lamin A in vitro,[45,46] while amino acids 174-220 are important for the interaction between emerin and A-type and B-type lamins in co-immunoprecipitation experiments.[47] A functionally important interaction between emerin and lamin A is also suggested by the finding that an autosomal dominant form of Emery-Dreifuss muscular dystrophy (EDMD) is caused by mutations in the *LMNA* gene, which encodes the A-type lamins, while a phenotypically identical X-linked form of the disease is caused by mutations in emerin.[7,48] Another hypothesis for emerin immobilization in the INM is its association with chromatin. Both emerin and MAN1 have LEM-domains, which as discussed above are implicated in binding to the chromatin-binding protein BAF. Emerin also accumulates at chromosomes in anaphase independent of lamins, suggesting a direct interaction with chromatin.[49]

Several regions of the amino-terminal domain of emerin (amino acids 1-219) appear to be involved in the INM targeting and retention of this protein. The region between amino acids 117 and 170 is sufficient for a nuclear localization, but is not sufficient for targeting of CHL to the nuclear envelope.[31,50] A portion of emerin containing amino acids 3 to 169 is sufficient for targeting of a transmembrane-spanning protein to the INM.[31,51] These experiments indicate a role for amino acids 3 to 116 in targeting of integral proteins to the INM.[31] A chimeric protein with GFP fused to amino acids 107-254, containing parts of the amino-terminal domain and the transmembrane and luminal regions does, however, target to the nuclear rim.[50] These data show that several regions of emerin are necessary to efficiently target an integral protein to the INM. Partial deletion mutants of the amino-terminal domain of MAN1 also fail to efficiently localize CHL to the INM, suggesting that most of the nucleoplasmic amino-terminal domain of MAN1 is necessary for INM retention.[32]

There are some emerin mutations found in patients with X-linked EDMD that only affect one or a few amino acids. In these instances, the protein is present in normal or somewhat reduced levels in cells. Emerin mutants with a deletion of amino acids 95-99 or a P183H/T missense mutation are partly membrane-bound.[52,53] However, in contrast to wild-type emerin, these mutant proteins are partly soluble in nonionic detergents, such as Triton X-100. Components of the nuclear lamina generally are resistant to such detergents, suggesting a role for amino acids 95–99 and 183 in interactions with the nuclear lamina. Subcellular fractionation shows that these mutant proteins are not confined to the nuclear fraction,[52,53] and when they are expressed in C2C12 cells, immunofluorescence microscopy studies also show them to be partly mislocalized.[47] Interestingly, these two mutations are situated in the regions flanking the nuclear targeting signal between amino acids 117-170, which seems to be important for the targeting of integral protein to the INM and may bind to lamins.

Contrary to the other characterized integral membrane proteins of the INM, nurim lacks a large, hydrophilic nucleoplasmic domain, but has several short stretches of amino acids between transmembrane segments that could extend into the nucleus and possibly interact with nuclear proteins.[13] Deletion experiments have shown that both of the two longest loops following transmembrane segments one and four are important for INM localization, as mutants lacking these regions are localized to the ER. These loops are predicted to face different sides of the membrane. Introduction of three point mutations changing charged residues of the transmembrane domains to leucines also affects targeting and extractability of nurim. A mutation in the second transmembrane domain (D66L) eliminates targeting to the INM and detergent inextractability, while two other mutants, R98L in the third and R217L in the fifth transmembrane domain, have intermediate phenotypes, with the protein partly mislocalizing to the ER and extractable with detergent.[13] Truncated versions of nurim containing two, three or four transmembrane domains are localized to the ER or unstable. These results suggest that multiple regions of nurim, including the transmembrane regions, are important for protein targeting. FRAP studies show that nurim, like LBR, emerin and MAN1, diffuses more slowly in the nuclear envelope than in the ER, with only limited recovery of fluorescence after nine minutes. However, the D66L mutant is freely diffusible in the nuclear envelope.[13] This supports a diffusion-retention mechanism for the targeting of nurim to the INM.

In most cases, the INM targeting domains of integral proteins are multiple. An exception is the integral membrane protein human cytomegalovirus glycoprotein B (HCMV gB), where a short sequence element of the cytoplasmic tail protein is sufficient for targeting to the INM.[54] This virus, like some other herpes viruses, undergoes a maturation phase which requires the translocation of glycoprotein B from the ER to the INM. Studies of chimeric proteins consisting of regions of HCMV gB fused to the transmembrane protein CD8 show that the carboxyl-terminal 42 amino acids, which face the nucleoplasm, are sufficient for targeting of CD8 to the INM.[54] CD8 is normally targeted to the plasma membrane. When these 42 viral amino acids are fused to the soluble, cytoplasmic protein β-galactosidase, the chimeric protein is localized to the cytoplasm, again showing the differences between the targeting of integral and soluble proteins. A region in the middle of the carboxyl-terminal region of glycoprotein B from the related herpes simplex virus type 1 (HSV-1 gB), rather than its extreme end, is sufficient for INM targeting.[54] The INM targeting domains from the two viral proteins exhibit limited homology, containing a conserved amino acid hexamer with the sequence DRLRHR. When the DRLRHR sequence alone is fused to the non-nuclear integral membrane protein CD8, the fusion protein is targeted to the INM. The simultaneous substitution of the four conserved residues **DRLRHR** to alanines and glutamines prevents transport to the INM of HCMV gB, as well as a chimeric protein with the carboxyl-terminus of HCMV gB fused to CD8. Single amino acid substitutions demonstrated that the arginine residues at positions 4 and 6 are essential for function as a targeting signal.

Recently, the A-kinase anchoring protein AKAP149 was shown to partially localize to the INM in addition to the ER and mitochondria.[18] This integral membrane protein of 149 kDa has a short, luminal amino-terminal domain, followed by a transmembrane segment and a large (147 kDa) nucleoplasmic, carboxyl-terminal domain. The size of the carboxyl-terminal domain most likely prevents the protein from reaching the INM during interphase. A portion of cellular AKAP149 may therefore become restricted to the INM upon targeting to chromatin when the nuclear membranes reform after mitosis, while the remainder remains in the ER.[18] Interestingly, AKAP149 is necessary for targeting the type 1 protein phosphatase (PP1) to the INM.[18] PP1 is implicated in the dephosphorylation of B-type lamins necessary for lamin reassembly after mitosis. In nuclear extracts lacking AKAP149, neither PP1 nor B-type lamins localize to the INM.[18] AKAP149 is the first example of a cellular integral protein with a steady-state localization in both the INM and the ER during interphase.

Targeting and Retention of Integral Membrane Proteins to the Pore Membrane

Only two integral proteins, gp210 and POM121, have been identified as specific to the nuclear pore membrane in mammalian cells.[55,56] Similar to integral proteins of the INM, these proteins are believed to be targeted to their sites through a diffusion-retention mechanism.

Both proteins have luminal amino-terminal regions followed by a transmembrane segment and an endoplasmically exposed carboxyl-terminal tail. However, while the luminal region of gp210 contains 95% of the total mass of the protein, the large majority of POM121 faces the endoplasm. The regions involved in retention are also different between the two proteins.

The large, luminal region of gp210 is devoid of any pore membrane sorting determinant, and localizes to the ER when expressed alone.[57] The gp210 transmembrane segment is sufficient for targeting to the pore membrane, as a chimeric protein containing this region fused to the lumenal region of the plasma membrane protein CD8 is targeted to the pore membrane. When the luminal and transmembrane domains of CD8 are fused to the endoplasmic carboxyl-terminus of gp210, the chimeric protein is also in part targeted to the pore membrane, with this targeting being dependent on the last 20 amino acids of gp210. A fraction of the protein is, however, also found in the plasma membrane and Golgi, suggesting that the pore membrane sorting determinant in the carboxyl-terminus of gp210 may be weaker than that in the transmembrane region.[57]

Studies of POM121 fused to GFP showed that amino acids 129-618 of the endoplasmic, carboxyl-terminal region of the protein were sufficient for targeting to the nuclear pores. Amino acids 1-128, comprising the luminal region and transmembrane spanning segment, localized to the ER. The region between amino acids 803 and 1199, which contains XFXFG repeats found in several proteins of the NPCs, did not localize to the NPCs, but were found throughout the cell.[58] FRAP studies on POM121 showed this protein to be stably associated with the NPCs during interphase. Complete recovery of fluorescence in stably transfected cells expressing POM121-yellow fluorescent protein did not occur until 35 hours after photobleaching of the nuclear membrane.[59] Tracking of NPCs also showed that they undergo little independent movement, but rather move as large arrays. The movements of NPCs were correlated with movements of fluorescently labeled B-type lamins, suggesting that lamins and NPCs are part of the same network.[59]

Like the integral proteins of the INM, integral proteins of the pore membrane most likely reach their sites by lateral diffusion after cotranslational insertion into the ER membrane. They are then immobilized and retained in the pore membrane through interactions with other proteins, presumably other proteins of the NPC. As shown by the large endoplasmic region of POM121, these proteins, which do not have to pass through the pores, do not have the size constraints on endoplasmic regions that limit access to the INM.

Targeting of Peripheral Membrane Proteins to the Inner Nuclear Membrane

Several peripheral proteins localized to the nucleoplasmic side of the nuclear envelope have been identified, the best studied of which are the nuclear lamins (for reviews see refs. 1, 60 and 61). Other peripheral proteins of the inner nuclear membrane are otefin, young arrest (YA) and, in some cells types, mAKAP.[62-64] There are two major classes of lamins, A-type and B-type. The B-type lamins appear to be expressed in all somatic cells, while the A-type lamins (lamins A and C, different splice forms both encoded by the *LMNA* gene) are absent from some undifferentiated, hematological and cancer cell types. The lamins belong to the intermediate filament protein family, and have highly conserved alpha-helical rod domains flanked by less conserved head and tail domains. The rod domain is responsible for lamin dimerization, and also required for higher order interactions into polymers. The lamins differ from cytoplasmic intermediate filaments as their carboxyl-terminal domains contain a NLS and, except in lamin C, a CAAX-motif, which gets post-translationally modified by prenylation. The rod domains of the lamins are 42 amino acids longer than those of cytoplasmic intermediate filaments. Lamins, like other nonmembrane nuclear proteins, enter the nucleus through the nuclear pores in a NLS-dependent manner. This was shown in studies of lamins lacking the NLS, which were severely mistargeted and mainly cytoplasmic.[65]

The role of prenylation in lamin targeting is not completely understood, but this lipid modification could help to anchor lamins to the nuclear envelope. However, lamin A prenylation is intriguing since prelamin A, a precursor to lamin A, is prenylated, but this modification is later cleaved off by an endoprotease activity to yield mature lamin A.[66] B-type lamins remain prenylated throughout their life-cycles, while lamin C never gets prenylated. When mutant lamin A and B-type lamins without the CAAX motif are expressed in cells, these proteins do not target rapidly to the nuclear rim, but accumulate in intranuclear aggregates and only later localize at the nuclear rim.[67] Another indication of the importance of CAAX prenylation for membrane association is that B-type lamins, as opposed to lamins A and C, has been suggested to remain membrane bound during mitosis.[68] In recent studies using live cell imaging, B-type lamins seemed, however, not to be associated with membranes during mitosis.[59] The reason for the discrepancy between these results are not yet clear. The mutant B-type lamin lacking the CAAX motif was, similar to wild-type lamins A and C, soluble during mitosis, independent of treatment with detergents.[67] It has been suggested that the prenyl groups direct the lamins to the nuclear membrane where, in the case of lamin A, an endoprotease then cleaves off the region of the protein containing the prenylation (for a review see ref. 69). However, as lamin C is localized to the nuclear membrane and mature lamin A from the previous cell cycle can rebind to the INM after mitosis, there must also be alternative ways of targeting.

Formations of heteropolymers containing different lamins or interactions with other INM proteins may help target nonprenylated lamins. When the tail-region of lamin B1 is expressed alone, it fails to localize to the lamina throughout the cell cycle, suggesting the importance of polymer formation mediated by the rod domain for correct lamin targeting.[70] A region of the rod-domain has also been implicated in the binding of lamin B to LAP2β.[71] A-type lamins with point mutations causing autosomal dominant EDMD located in the rod domain also mislocalize into intranuclear foci when overexpressed in cells.[72,73] The region of lamin A binding to emerin is, however, suggested to lay within the tail-domain.[46]

The *Drosophila* nuclear envelope protein otefin contains 406 amino acids with a calculated molecular mass of 45 kDa. It is a mainly hydrophilic protein, but the last 17 amino acids are hydrophobic, and this stretch resembles the membrane spanning domains of integral membrane proteins.[62] However, extraction experiments using 8 M urea or a buffer of pH 13, show that otefin is a peripheral membrane protein.[74] Otefin is, however, more stably attached to membranes than lamins and is not extracted with high salt or with buffers of pH 11.[74] Transfection experiments in COS-7 cells show that a majority of wild-type otefin is localized to the nuclear rim and resistant to extraction with 1% Triton X-100. The amino-terminal hydrophilic region alone (Δ388-406) has a nucleoplasmic localization and is completely extractable by 1% Triton X-100, showing that the hydrophobic region is important for nuclear envelope targeting. When the hydrophobic region of otefin is fused to a NLS and β-galactosidase, the chimeric protein localizes to the nuclear rim in 60% of transfected cells, but all transfected cells also contain cytoplasmic aggregates of protein. After extraction with Triton X-100, most of the cytoplasmic protein disappears, while 20% of transfected cells still contain some chimeric protein. These data suggest that the carboxyl-terminal hydrophobic region of otefin is crucial for its association with the nuclear envelope, but has only a limited capability of directing the NLS-β-galactosidase protein to this location. The region between amino acid 35 and 172 is important for stabilization of otefin at the INM. When this region is deleted, the protein localizes to the nuclear rim, but is less resistant to Triton X-100 extraction than wild-type. The first 33 amino acids of otefin, a region with homology to the LEM-domain, does not appear to be involved in nuclear envelope targeting. Studies using the two-hybrid system also identified two regions of otefin that bind to the rod domain of *Drosophila* lamin Dm$_0$.[75] Amino acids 35-172 interacted with lamin Dm$_0$, as did a construct consisting of the hydrophilic region of otefin with these amino acids deleted. A suggested mechanism for the targeting of otefin is that it enters the nucleus, either by diffusion or by aid of a putative NLS. It is then sorted to the nuclear envelope by its hydrophobic sequence and retained there by interactions with other nuclear envelope proteins.

The integral membrane protein AKAP149 has been described above. Another AKAP, mAKAP or AKAP100, is a peripheral membrane protein of 255 kDa that is expressed in differentiated striated muscle cells, where it is localized to the nuclear membrane.[64] mAKAP binds to cAMP-dependent protein kinase (PKA) and is believed to be involved in its compartmentalization. Two spectrin repeat-like sequences are necessary and sufficient for the nuclear targeting of mAKAP independent of each other.[64] Spectrin-repeat sequences are found in cytoskeletal proteins such as spectrin and dystrophin and participate in protein-protein interactions with cytoskeletal components such as actin. The identification of interaction partners of mAKAP will be important to understand the targeting mechanism for this protein, and may help answer why it is apparently localized to the INM in some cell-types and to the ER in others.[64]

Conclusion

The large majority of integral proteins of the INM and nuclear pore membranes are believed to reach their final locations by lateral diffusion from the ER membrane. Through binding to chromatin or proteins such as lamins, they are then retained at this location. This hypothesis has been confirmed by several studies where proteins were shown to have a decreased mobility when they reached the INM and pore membrane. The peripheral membrane proteins of the INM are targeted to the nucleus via the same mechanisms as other non-membrane, nuclear proteins. They are then retained at the INM through interactions with other proteins, and in some cases, through hydrophobic interactions with the INM. A better understanding of interactions between different components of the nuclear envelope, and of the structures of nuclear envelope proteins, will further our understanding of these targeting mechanisms and possibly shed light on the role of the nuclear envelope in diseases such as muscular dystrophy[7,48] and lipodystrophy.[76,77]

References

1. Worman HJ, Courvalin J-C. The inner nuclear membrane. J Membr Biol 2000; 177(1):1-11.
2. Wilson KL. The nuclear envelope, muscular dystrophy and gene expression. Trends Cell Biol 2000; 10(4):125-129.
3. Worman HJ, Yuan J, Blobel G et al. A lamin B receptor in the nuclear envelope. Proc Natl Acad Sci USA 1988; 85(22):8531-8534.
4. Worman HJ, Evans CD, Blobel G. The lamin B receptor of the nuclear envelope inner membrane: a polytopic protein with eight potential transmembrane domains. J Cell Biol 1990; 111(4):1535-1542.
5. Senior A, Gerace L. Integral membrane proteins specific to the inner nuclear membrane and associated with the nuclear lamina. J Cell Biol 1988; 107(6):2029-2036.
6. Foisner R, Gerace L. Integral membrane proteins of the nuclear envelope interact with lamins and chromosomes, and binding is modulated by mitotic phosphorylation. Cell 1993; 73(7):1267-1279.
7. Bione S, Maestrini E, Rivella S et al. Identification of a novel X-linked gene responsible for Emery-Dreifuss muscular dystrophy. Nat Genet 1994; 8(4):323-327.
8. Manilal S, Nguyen thi Man, Sewry CA et al. The Emery-Dreifuss muscular dystrophy protein, emerin, is a nuclear membrane protein. Hum Mol Genet 1996; 5(6):801-808.
9. Nagano A, Koga R, Ogawa M et al. Emerin deficiency at the nuclear membrane in patients with Emery-Dreifuss muscular dystrophy. Nat Genet 1996; 12(3):254-259.
10. Lin F, Blake DL, Callebaut I et al. MAN1, an inner nuclear membrane protein that shares the LEM domain with lamina-associated polypeptide 2 and emerin. J Biol Chem 2000; 275(7):4840-4847.
11. Dechat T, Vlcek S, Foisner R. Lamina-associated polypeptide 2 isoforms and related proteins in cell cycle-dependent nuclear structure dynamics. J Struct Biol 2000; 129(2-3):335-345.
12. Laguri C, Gilquin B, Wolff N et al. Structural characterization of the LEM motif common to three human inner nuclear membrane proteins. Structure (Camb) 2001; 9(6):503-511.
13. Rolls MM, Stein PA, Taylor SS et al. A visual screen of a GFP-fusion library identifies a new type of nuclear envelope membrane protein. J Cell Biol 1999; 146(1):29-43.
14. Malone CJ, Fixsen WD, Horwitz HR et al. UNC-84 localizes to the nuclear envelope and is required for nuclear migration and anchoring during *C. elegans* development. Development 1999; 126(14):3171-3181.

15. Fitzgerald J, Kennedy D, Viseshakul N et al. UNCL, the mammalian homologue of UNC-50, is an inner nuclear membrane RNA-binding protein. Brain Res 2000; 877(1):110-123.
16. Mansharamani M, Hewetson A, Chilton BS. Cloning and characterization of an atypical type IV P-type ATPase that binds to the RING motif of RUSH transcription factors. J Biol Chem 2001; 276(5):3641-3649.
17. Dreger M, Bengtsson L, Schöneberg T et al. Nuclear envelope proteomics: novel integral membrane proteins of the inner nuclear membrane. Proc Natl Acad Sci USA 2001; 98(21):11943-11948.
18. Steen RL, Martins SB, Taskén K et al. Recruitment of protein phosphatase 1 to the nuclear envelope by A-kinase anchoring protein AKAP149 is a prerequisite for nuclear lamina assembly. J Cell Biol 2000; 150(6):1251-1262.
19. Paine PL, Moore LC, Horowitz SB. Nuclear envelope permeability. Nature 1975; 254(5496):109-114.
20. Nakielny S, Dreyfuss G. Transport of proteins and RNAs in and out of the nucleus. Cell 1999; 99(7):677-690.
21. Soullam B, Worman HJ. Signals and structural features involved in integral membrane protein targeting to the inner nuclear membrane. J Cell Biol 1995; 130(1):15-27.
22. Ye Q, Worman HJ. Primary structure analysis and lamin B and DNA binding of human LBR, an integral protein of the nuclear envelope inner membrane. J Biol Chem 1994; 269(15):11306-11311.
23. Ye Q, Worman HJ. Interaction between an integral protein of the nuclear envelope inner membrane and human chromodomain proteins homologous to *Drosophila* HP1. J Biol Chem 1996; 271(25):14653-14656.
24. Ye Q, Callebaut I, Pezhman A et al. Domain-specific interactions of human HP1-type chromodomain proteins and inner nuclear membrane protein LBR. J Biol Chem 1997; 272(23):14983-14989.
25. Schuler E, Lin F, Worman HJ. Characterization of the human gene encoding LBR, an integral protein of the nuclear envelope inner membrane. J Biol Chem 1994; 269(15):11312-11317.
26. Holmer L, Pezhman A, Worman HJ. The human LBR/sterol reductase multigene family. Genomics 1998; 54(3):469-476.
27. Smith S, Blobel G. The first membrane spanning region of the lamin B receptor is sufficient for sorting to the inner nuclear membrane. J Cell Biol 1993; 120(3):631-637.
28. Soullam B, Worman HJ. The amino-terminal domain of the lamin B receptor is a nuclear envelope targeting signal. J Cell Biol 1993; 120(5):1093-1100.
29. Furukawa K, Panté N, Aebi U et al. Cloning of a cDNA for lamina-associated polypeptide 2 (LAP2) and identification of regions that specify targeting to the nuclear envelope. EMBO J 1995; 14(8):1626-1636.
30. Furukawa K, Fritze CE, Gerace L. The major nuclear envelope targeting domain of LAP2 coincides with its lamin binding region but is distinct from its chromatin interaction domain. J Biol Chem 1998; 273(7):4213-4219.
31. Östlund C, Ellenberg J, Hallberg E et al. Intracellular trafficking of emerin, the Emery-Dreifuss muscular dystrophy protein. J Cell Sci 1999; 112(11):1709-1719.
32. Wu W, Lin F, Worman HJ. Intracellular trafficking of MAN1, an integral protein of the nuclear envelope inner membrane. J Cell Sci 2002; 115(7):1361-1371.
33. Hinshaw JE, Carragher BO, Milligan RA. Architecture and design of the nuclear pore complex. Cell 1992; 69(7):1133-1141.
34. Akey CW, Radermacher M. Architecture of the *Xenopus* nuclear pore complex revealed by three-dimensional cryo-electron microscopy. J Cell Biol 1993; 122(1):1-19.
35. Ellenberg J, Siggia ED, Moreira JE et al. Nuclear membrane dynamics and reassembly in living cells: targeting of an inner nuclear membrane protein in interphase and mitosis. J Cell Biol 1997; 138(6):1193-1206.
36. Duband-Goulet I, Courvalin J-C. Inner nuclear membrane protein LBR preferentially interacts with DNA secondary structures and nucleosomal linker. Biochemistry 2000; 39(21):6483-6488.
37. Gajewski A, Krohne G. Subcellular distribution of the *Xenopus* p58/lamin B receptor in oocytes and eggs. J Cell Sci 1999; 112(15):2583-2596.
38. Chaudhary N, Courvalin J-C. Stepwise reassembly of the nuclear envelope at the end of mitosis. J Cell Biol 1993; 122(2):295-306.
39. Wolff N, Gilquin B, Courchay K et al. Structural analysis of emerin, an inner nuclear membrane protein mutated in X-linked Emery-Dreifuss muscular dystrophy. FEBS Lett 2001; 501(2-3):171-176.
40. Cai M, Huang Y, Ghirlando R et al. Solution structure of the constant region of nuclear envelope protein LAP2 reveals two LEM-domain structures: one binds BAF and the other binds DNA. EMBO J 2001; 20(16):4399-4407.
41. Furukawa K. LAP2 binding protein 1 (L2BP1/BAF) is a candidate mediator of LAP2-chromatin interaction. J Cell Sci 1999; 112(15):2485-2492.

42. Shumaker DK, Lee KK, Tanhehco YC et al. LAP2 binds to BAF-DNA complexes: requirement for the LEM domain and modulation by variable regions. EMBO J 2001; 20(7):1754-1764.
43. Furukawa K, Glass C, Kondo T. Characterization of the chromatin binding activity of lamina-associated polypeptide (LAP) 2. Biochem Biophys Res Commun 1997; 238(1):240-246.
44. Sullivan T, Escalante-Alcalde D, Bhatt H et al. Loss of A-type lamin expression compromises nuclear envelope integrity leading to muscular dystrophy. J Cell Biol 1999; 147(5):913-919.
45. Clements L, Manilal S, Love DR et al. Direct interaction between emerin and lamin A. Biochem Biophys Res Commun 2000; 267(3):709-714.
46. Sakaki M, Koike H, Takahashi N et al. Interaction between emerin and nuclear lamins. J Biochem (Tokyo) 2001; 129(2):321-327.
47. Fairley EAL, Kendrick-Jones J, Ellis JA. The Emery-Dreifuss muscular dystrophy phenotype arises from aberrant targeting and binding of emerin at the inner nuclear membrane. J Cell Sci 1999; 112(15):2571-2582.
48. Bonne G, Di Barletta MR, Varnous S et al. Mutations in the gene encoding lamin A/C cause autosomal dominant Emery-Dreifuss muscular dystrophy. Nat Genet 1999; 21(3):285-288.
49. Haraguchi T, Koujin T, Hayakawa T et al. Live fluorescence imaging reveals early recruitment of emerin, LBR, RanBP2, and Nup153 to reforming functional nuclear envelopes. J Cell Sci 2000; 113(5):779-794.
50. Tsuchiya Y, Hase A, Ogawa M et al. Distinct regions specify the nuclear membrane targeting of emerin, the responsible protein for Emery-Dreifuss muscular dystrophy. Eur J Biochem 1999; 259(3):859-865.
51. Cartegni L, di Barletta MR, Barresi R et al. Heart-specific localization of emerin: new insights into Emery-Dreifuss muscular dystrophy. Hum Mol Genet 1997; 6(13):2257-2264.
52. Ellis JA, Craxton M, Yates JRW et al. Aberrant intracellular targeting and cell cycle-dependent phosphorylation of emerin contribute to the Emery-Dreifuss muscular dystrophy phenotype. J Cell Sci 1998; 111(6):781-792.
53. Ellis JA, Yates JRW, Kendrick-Jones J et al. Changes at P183 of emerin weaken its protein-protein interactions resulting in X-linked Emery-Dreifuss muscular dystrophy. Hum Genet 1999; 104(3):262-268.
54. Meyer GA, Radsak KD. Identification of a novel signal sequence that targets transmembrane proteins to the nuclear envelope inner membrane. J Biol Chem 2000; 275(6):3857-3866.
55. Gerace L, Ottaviano Y, Kondor-Koch C. Identification of a major polypeptide of the nuclear pore complex. J Cell Biol 1982; 95(3):826-837.
56. Hallberg E, Wozniak RW, Blobel G. An integral membrane protein of the pore membrane domain of the nuclear envelope contains a nucleoporin-like region. J Cell Biol 1993; 122(3):513-521.
57. Wozniak RW, Blobel G. The single transmembrane segment of gp210 is sufficient for sorting to the pore membrane domain of the nuclear envelope. J Cell Biol 1992; 119(6):1441-1449.
58. Söderqvist H, Imreh G, Kihlmark M et al. Intracellular distribution of an integral nuclear pore membrane protein fused to green fluorescent protein—Localization of a targeting domain. Eur J Biochem 1997; 250(3):808-813.
59. Daigle N, Beaudouin J, Hartnell L et al. Nuclear pore complexes form immobile networks and have a very low turnover in live mammalian cells. J Cell Biol 2001; 154(1):71-84.
60. Stuurman N, Heins S, Aebi U. Nuclear lamins: their structure, assembly, and interactions. J Struct Biol 1998; 122(1-2):42-66.
61. Gruenbaum Y, Wilson KL, Harel A et al. Nuclear lamins—Structural proteins with fundamental functions. J Struct Biol 2000; 129(2-3):313-323.
62. Padan R, Nainudel-Epszteyn S, Goitein R et al. Isolation and characterization of the *Drosophila* nuclear envelope otefin cDNA. J Biol Chem 1990; 265(14):7808-7813.
63. Lin H, Wolfner MF. The *Drosophila* maternal-effect gene fs(1)Ya encodes a cell cycle-dependent nuclear envelope component required for embryonic mitosis. Cell 1991; 64(1):49-62.
64. Kapiloff MS, Schillace RV, Westphal AM et al. mAKAP: An A-kinase anchoring protein targeted to the nuclear membrane of differentiated myocytes. J Cell Sci 1999; 112(16):2725-2736.
65. Loewinger L, McKeon F. Mutations in the nuclear lamin proteins resulting in their aberrant assembly in the cytoplasm. EMBO J 1988; 7(8):2301-2309.
66. Sinensky M, Fantle K, Trujillo M et al. The processing pathway of prelamin A. J Cell Sci 1994; 107(1):61-67.
67. Mical TI, Monteiro MJ. The role of sequences unique to nuclear intermediate filaments in the targeting and assembly of human lamin B: evidence for lack of interaction of lamin B with its putative receptor. J Cell Sci 1998; 111(23):3471-3485.
68. Gerace L, Blobel G. The nuclear envelope lamina is reversibly depolymerized during mitosis. Cell 1980; 19(1):277-287.

69. Sinensky M. Functional aspects of polyisoprenoid protein substituents: roles in protein-protein interaction and trafficking. Biochim Biophys Acta 2000; 1529(1-3):203-209.

70. Moir RD, Yoon M, Khuon S et al. Nuclear lamins A and B1: different pathways of assembly during nuclear envelope formation in living cells. J Cell Biol 2000; 151(6):1155-1168.

71. Furukawa K, Kondo T. Identification of the lamina-associated-polypeptide-2-binding domain of B-type lamin. Eur J Biochem 1998; 251(3):729-733.

72. Östlund C, Bonne G, Schwartz K et al. Properties of lamin A mutants found in Emery-Dreifuss muscular dystrophy, cardiomyopathy and Dunnigan-type partial lipodystrophy. J Cell Sci 2001; 114(24):4435-4445.

73. Raharjo WH, Enarson P, Stewart C et al. Nuclear envelope defects associated with *LMNA* mutations causing dilated cardiomyopathy and Emery-Dreifuss muscular dystrophy. J Cell Sci 2001; 114(24):4447-4457.

74. Ashery-Padan R, Weiss AM, Feinstein N et al. Distinct regions specify the targeting of otefin to the nucleoplasmic side of the nuclear envelope. J Biol Chem 1997; 272(4):2493-2499.

75. Goldberg M, Lu H, Stuurman N et al. Interactions among *Drosophila* nuclear envelope proteins lamin, otefin and YA. Mol Cell Biol 1998; 18(7):4315-4323.

76. Cao H, Hegele RA. Nuclear lamin A/C R482Q mutation in Canadian kindreds with Dunnigan-type familial partial lipodystrophy. Hum Mol Genet 2000; 9(1):109-112.

77. Shackleton S, Lloyd DJ, Jackson SNJ et al. *LMNA*, encoding lamin A/C, is mutated in partial lipodystrophy. Nat Genet 2000; 24(2):153-156.

CHAPTER 4

Dynamic Connections of Nuclear Envelope Proteins to Chromatin and the Nuclear Matrix

Roland Foisner

Abstract

The nuclear lamina is a filamentous scaffold structure underneath the inner nuclear membrane and consists of A- and B-type lamins and a number of integral inner nuclear membrane proteins, such as lamin B receptor (LBR), emerin, and various isoforms of lamina-associated polypeptides 1 (LAP1) and LAP2. Lamins, LAP2, emerin and LBR interact with DNA and/or chromosomal proteins, including core histones, BAF, HP1 and HA95, and provide a complex dynamic link between the peripheral lamina and nucleoskeletal structures and chromatin fibers. In addition, components of nuclear pore complexes, such as Nup153 and Tpr, link the nuclear envelope to the nuclear interior. Furthermore, intranuclear complexes of A-type lamins and LAP2α are likely involved in higher order chromatin organization throughout the nucleus. These interactions are tightly regulated in a temporal and spatial manner during the cell cycle and are responsible for the multiple functions of the lamina in dynamic nuclear and chromatin structure organization, in DNA replication, gene transcription, cell cycle progression, and apoptosis.

Introduction

The eukaryotic nucleus contains the chromosomes and is a complex organelle where major cellular processes, such as DNA replication, RNA transcription and processing, and ribosome assembly take place. The function of the nucleus highly depends on its structural organization and the dynamic structural rearrangements occurring in cell differentiation and cell cycle progression [1]. Cellular structures and proteins involved in nuclear architecture are not very well understood except for a few major elements. The nuclear envelope (NE) enwraps the DNA and forms the border between the nucleus and cytoplasm. It is composed of inner and outer nuclear membranes that are separated by the perinuclear space and contain nuclear pore complexes mediating nucleo-cytoplasmic transport. The outer nuclear membrane is continuous with the endoplasmic reticulum and is also directly linked to the inner membrane at sites of nuclear pore complexes (Fig. 1). Despite the continuity of membrane structures, protein and lipid composition and functions of inner and outer membranes are clearly different, most likely due to specific interactions of membrane components with nuclear and cytoplasmic components, respectively. Underneath the inner membrane is a meshwork of nuclear-specific intermediate filaments, termed the nuclear lamina, which includes lamins plus a growing number of lamin-associated proteins, which regulate lamin assembly and function. [1,2] Most of these lamin-binding proteins have been identified as integral membrane proteins of the inner membrane or are tightly associated with the lipid bilayer. [2-4] Biochemically, the nuclear lamina is

Nuclear Envelope Dynamics in Embryos and Somatic Cells
edited by Philippe Collas. ©2002 Eurekah.com and Kluwer Academic/Plenum Publishers.

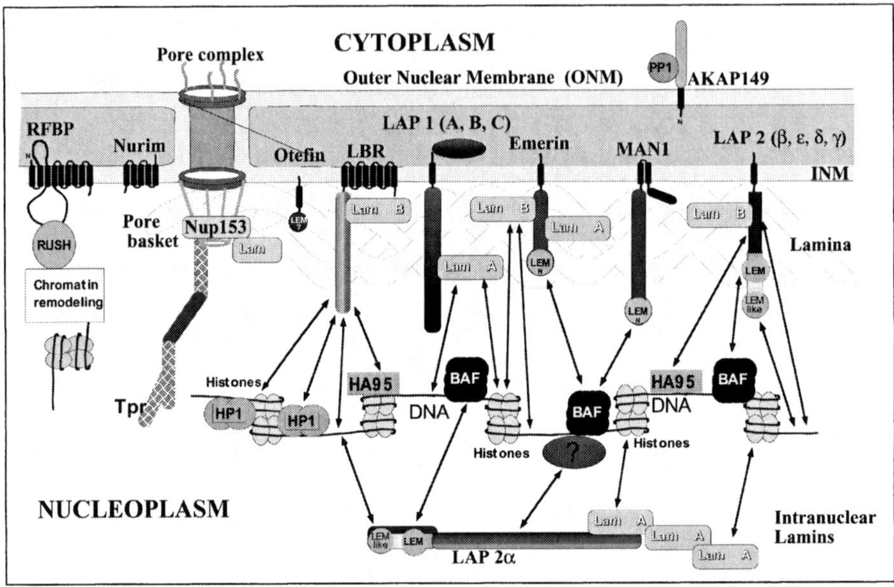

Figure 1. Schematic drawing of the molecular links at the interface between the nuclear envelope and the internal nucleoskeleton/chromatin scaffold. Arrows denote specific interactions of components shown in vitro and/or in vivo. For details see text.

defined as the peripheral nuclear structure that remains insoluble after extraction of nuclei with non-ionic detergents, salt and nucleases.[5,6] However, the lamina is only a subfraction of the detergent-salt-resistant structural framework, which runs throughout the nuclear interior and organizes nuclear space and is often referred to as nucleoskeleton or nuclear matrix.[7] As the visualization of this putative nuclear scaffold has always been hampered by the bulky chromatin mass, it is still under debate, whether there exists a chromatin independent proteinaceous nuclear scaffold or whether intranuclear structures are organized by a complex network of protein-protein, protein-DNA and protein-RNA interactions.[8] In any case, the nuclear scaffold is supposed to provide mechanical stability for nuclear structure, to form a platform for most metabolic nuclear processes, and to organize chromatin in a three-dimensional nuclear space and thus regulate gene expression at the chromatin structure level. Except for the peripheral lamins, the components and molecular organization of the nucleoskeleton are not very well understood.

In this review, I summarize the major components and interactions of the lamina and focus on the interface between the peripheral nuclear envelope and the intranuclear scaffold/chromatin. I will also describe cell cycle-dependent dynamics and potential functions of these interactions, particularly focusing on members of the Lamina-Associated Polypeptide 2 (LAP2) family, whose cell cycle-dependent dynamics have been fairly well characterized in the past years.

Major Components of the Peripheral Nuclear Lamina

The core structure of the nuclear lamina is formed by type V intermediate filament proteins, the lamins.[9] They assemble to a meshwork of tetragonally organized 10 nm filaments underneath the inner nuclear membrane. The number and complexity of lamins has increased during metazoan evolution. Vertebrates have three lamin genes (*LMNA, LMNB1, LMNB2*) encoding at least seven distinct isoforms.[2] B-type lamins are constitutively expressed in cells throughout development and every cell expresses at least one B-type lamin. A-type lamins, comprising lamin A and its smaller splice variant lamin C are only expressed in later stages of development and in differentiated cells.[2]

The assembly and attachment of lamins at the membrane involve several mechanisms and are different between A- and B-type lamins. B-type lamins contain a stable C-terminal farnesyl modification, which is important but not sufficient for targeting and anchoring the protein to the membrane.[10-13] In contrast, lamin A is only transiently farnesylated due to cleavage of the C-terminal residues containing the farnesyl group during protein maturation, and lamin C is never farnesylated.[10] Therefore, B-type lamins are more tightly associated with membrane structures than A-type lamins in mitosis and interphase and are less stably incorporated into the lamina.[14,15]

Homotypic interactions of lamin subunits,[9] hetero-oligomeric interactions of B-type and A-type lamin dimers or oligomers,[11] as well as interactions of lamins with integral and peripheral membrane proteins are essential for the proper assembly of the lamina underneath the nuclear membrane. Most of the lamin-binding proteins are tightly bound to lamins and cofractionate with lamins even after detergent-salt extraction of nuclei or of isolated NE fractions.[5,6,16-18] Therefore, these proteins are considered as genuine components of the nuclear lamina. Among those are (see Fig. 1):

The lamin B receptor (LBR, p58) contains eight transmembrane domains[19] and was found to interact with B-type lamins in vivo and in vitro.[20-22] Since ectopically expressed lamin B1 mutants lacking farnesylation segregate independently of LBR,[12] it was suggested that LBR might also bind to the farnesyl residues of B-type lamins. The hydrophobic domain of LBR shares extensive homology with sterol reductases and exhibits C14 reductase activity, suggesting that the protein might have additional functions in sterol metabolism.[23,24]

Lamina-associated-polypeptide 1 (LAP-1) is a type II integral nuclear membrane protein, containing a nucleoplasmic N-terminus, a single transmembrane spanning region, and a C-terminus located in the luminal space between inner and outer nuclear membrane.[25] LAP1 specifically interacts with A-type lamins in vitro[6] and its nuclear envelope localization depends on the presence of lamin A in vivo.[26] LAP1 exists as three alternatively spliced isoforms, the smallest one, LAP1C, being expressed constitutively, while the larger isoforms, LAP1 A and B, which contain additional domains in the nucleoplasm, are expressed only in differentiated cells like A-type lamins.[25]

Lamina-associated polypeptide 2 (LAP2) is another family of alternatively spliced lamin binding proteins, comprising up to six mammalian isoforms, LAP2α, β, γ, δ, ε, and ζ[27,28] (formerly also called thymopoietins) and three *Xenopus* LAP2 proteins.[29,30] Except for LAP2α and LAP2ζ, all mammalian LAP2 isoforms contain a closely related N-terminal nucleoplasmic domain of variable length and share a single transmembrane spanning region passing the inner nuclear membrane, and a short C-terminus located in the luminal space between inner and outer nuclear membrane.[31] LAP2β possesses the longest nucleoplasmic N-terminus (223aa), LAP2 ε, δ, γ miss regions of 40, 72, and 109 amino acids respectively due to alternative splicing, but are otherwise identical, and LAP2ζ represents a truncated form of LAP2β, missing 190 amino acids of the nucleoplasmic domain as well as the transmembrane and luminal regions. LAP2α is structurally and functionally different from the other isoforms. It shares only the N-terminal 187 amino acids with all the other LAP2 isoforms, but contains a unique C-terminus (506 aa) lacking a transmembrane domain. Unlike LAP2α, LAP2β was found to interact with lamin B in vitro.[6] Its lamin B-interaction domain was located within a 72 amino acid long stretch in the nucleoplasmic region (aa298-370),[32] which is also present in the smaller isoforms LAP2ε and δ, and parts of it in LAP2γ. An interaction of the smaller isoforms with lamin B, however, has not been demonstrated experimentally. LAP2α is unique, as it is located in the nuclear interior[33] and binds specifically intranuclear A-type lamins[34] (see below). Very little is known about the expression patterns of the various LAP2 isoforms. Northern blot analysis and S1-nuclease protection assays revealed that LAP2 mRNAs are ubiquitously found in many tissues and cells of human, mouse and rat origin.[27,28,35] At the protein level, LAP2α and β appear to be the predominantly expressed LAP2 isoforms in mammalian cells,[27,33,36] but a recent proteomics analysis aimed at characterizing novel NE proteins clearly revealed also expression of smaller LAP2 isoforms.[37] While available data on the mammalian LAP2 isoforms indicate an ubiquitous expression, some of the LAP2 homologues, identified in *Xenopus*,[29,30] showed differential expression during development. One of them was found to be expressed only in somatic cells, but was not detected in oocytes, eggs and in early embryos up to the gastrula stage, while a slightly larger putative LAP2 isoform – which has not been cloned yet— was predominantly expressed in *Xenopus* eggs and embryos and was downregulated during embryogenesis.[29]

Emerin and MAN1 are proteins related to LAP2 isoforms. These proteins share a ~40 amino acid long highly homologous structural motif (LEM domain) in their N-termini[17], which consists of a helical turn and two large parallel α-helices connected by a 11 to 12 residue loop.[38,39] Emerin is a ubiquitously expressed type II integral membrane protein of the inner nuclear membrane[18,40] and has been identified as the gene product that is missing or mutated in patients suffering from X-linked Emery-Dreifuss muscular dystrophy (EDMD).[41,42] It binds to both A- and B-type lamins in vitro [43-45] and its nuclear envelope localization is dependent on the expression of lamin A.[46,47] MAN1 is a lamina-associated protein detected by the MAN autoimmune serum,[48] and by sequence analysis is predicted to contain two transmembrane domains.[17] Its interaction with lamins has not been analyzed yet.

A visual screen of a GFP-fusion library[49] and a proteomics approach[37] revealed novel integral membrane proteins termed nurim, with 5 predicted transmembrane domains and only few hydrophilic residues, Unc-84, a protein similar to *Drosophila* Unc-84[50] and a novel protein LUMA with three to four predicted transmembrane domains. Binding of these proteins to lamins has not been analyzed yet.

In addition, a peripheral membrane protein, otefin,[51,52] has been identified as a lamina protein in *Drosophila*.

Lamina Proteins in the Nuclear Interior

Lamins were traditionally regarded as proteins of the nuclear periphery, but with the availability of novel tools and microscopic techniques the concept of intranuclear lamins has recently developed. Since the nuclear membrane forms extensive tubular invaginations projecting deep into the nuclear interior,[53] it is often hard to distinguish whether observed internal lamin structures are still associated with the invaginated nuclear membrane or whether they are truly intranuclear, apart from the nuclear membrane. Nevertheless, specific antibodies,[34,54] several microscopical preparations techniques,[55,56] and the use of expressed green fluorescent protein- (GFP) fusions of lamins,[57,58] have revealed intranuclear lamin structures in foci or along filamentous structures or diffusely distributed throughout the nuclear interior. Although B-type lamins may also localize to intranuclear replication sites[59] and a minor fraction of GFP-lamin B has been detected in stable intranuclear structures by fluorescence recovery after photobleaching (FRAP) analysis,[58] the majority of studies have revealed particularly A-type lamins in the nuclear interior. This observation is consistent with the lack of C-terminal farnesyl modification of mature A-type lamins and the less stable association with the peripheral nuclear membrane and the nuclear lamina as compared with B-type lamins (see above).

Intranuclear A-type lamins may exist in a complex with LAP2α, the only LAP2 isoform not integrated into the membrane. LAP2α is a nucleoskeletal protein, based on its resistance to extraction by detergents and high salt,[33] and was found to directly interact with the C-terminal tail region of mature lamins A and C in vitro.[34] Furthermore, selective disruption of endogenous lamin A structures upon ectopic expression of dominant-negative lamin mutants in Hela cells caused a relocalization of LAP2α to intranuclear lamin A/C aggregates, but had no effect on lamin B, LAP2β, or NuMa.[34] It is still unclear, however, whether lamin A and LAP2α form filaments or other higher order structures of the nuclear scaffold, or whether they exist as smaller complexes involved in the regulation of nuclear processes (see below). It is also not known, whether there is continuity between peripheral and internal nuclear lamin A structures or whether lamin subunits steadily exchange between these two subnuclear compartments.

Several laboratories have reported a transient localization of A-type lamins in the nuclear interior before their assembly into the nuclear lamina. FRAP analyses in GFP-lamin A expressing cells showed that the assembly of lamin A into peripheral nuclear structures is a late event in post-mitotic nuclear reformation,[58] leading to accumulation of the majority of lamins A and C in the nuclear interior in G1 phase.[34,60] Furthermore, microinjected lamin A and/or lamin C were found to first accumulate in nucleoplasmic foci, before the majority was incorporated into the nuclear lamina.[61,62] As non-processed lamin A (missing the farnesyl modification) accumulated in similar intranuclear foci,[63,64] transient intranuclear localization of lamin A might be directly linked to its post-translational processing, but the molecular mechanisms remain unclear. Recently, a novel nuclear protein of unknown cellular function, Narf, has been identified by yeast two-hybrid-screens as a direct and specific interaction partner of unprocessed lamin A.[65]

Farnesylation and C-terminal proteolytic cleavage of A-type lamins during maturation can, however, not be the only reason for their transient accumulation in the nucleus, as intranuclear lamin A found in late stages of post-mitotic nuclear reformation is fully processed, and lamin C, which was also found to accumulate in intranuclear structures,[62] is not processed at all. Thus, other modifications such as (de-) phosphorylation,[66] or specific interactions with still unknown nuclear proteins might also be required for correct targeting of A-type lamins to peripheral as well as intranuclear structures.

Interactions at the Interface Between the Lamina and the Nuclear Scaffold/Chromatin

The transcriptionally silenced or less active and late replicating domains in higher eukaryotic genomes, referred to as heterochromatin, are dynamically associated with the NE.[67-69] This association involves a complex network of specific protein-protein and protein-DNA interactions at the interface of the lamina and the nuclear matrix (Fig. 1). In vitro, A and B-type lamins have been shown to bind directly to matrix/scaffold attachment regions[70,71] and to telomeric and heterochromatic DNA sequences,[72,73] but the physiological relevance of these associations is not clear. However, photo-crosslinking experiments in *Drosophila* cells revealed specific association of interphase lamins with DNA in vivo.[74] Lamins can also interact with and assemble around chromatin, and this is mediated by the lamin rod domain[75] and/or the C-terminal tail domain that binds to core histones.[76,77]

In addition to lamins, many lamin-binding proteins were shown to interact with DNA and/or chromosomal protein (Fig. 1). LBR interacts directly with DNA[22,78] and with human HP1-type chromodomain proteins,[79,80] which function as chromatin modifiers and regulators of gene expression and have been implicated in position effect variegation and heterochromatin organization.[81,82] In line with these findings, microinjected HP1 has been shown to localize transiently at the nuclear periphery in a deacetylation-dependent manner, before it translocates to intranuclear sites.[83] The association of HP1 with the nuclear envelope may be mediated by direct binding to LBR, but recent studies revealed a complex of LBR, HP1 and histones H3/H4, in which histones bind to both LBR and HP1 and mediate the LBR-HP1 interaction.[84] As the interaction of histones with HP1 was found to be affected by methylation of histone at lysine residue 9,[85,86] this could also provide a regulatory mechanisms for chromatin docking at the cellular periphery during cell proliferation and cell differentiation. In cross-linking studies LBR was also found to associate with chromatin-associated HA95, a nuclear protein with high homology to the nuclear A-kinase anchoring protein AKAP95.[87]

LAP2 proteins contain several chromatin and/or DNA binding domains, which are either common to all or unique to some isoforms (Fig. 1). The constant N-terminal region, common to all LAP2 proteins, contains the LEM domain (amino acids 111-152), which was found by yeast two hybrid assay[88] and by biochemical studies[89] to interact with the chromosomal protein

Barrier-to-Autointegration Factor (BAF), a protein that was first identified for its role in retroviral DNA integration.[90,91] Further studies revealed that BAF is a 89-residue protein that is highly conserved in multicellular eukaryotes[92] and binds double stranded DNA non-specifically forming nucleoprotein complexes (dodecamers) between DNA molecules.[93] Since the LEM domain is also present in emerin and MAN1,[17] it can be expected that all these proteins interact with chromatin-associated BAF,[94] but this has not been experimentally tested yet.

Moreover the N-terminal 50 residues of the LAP2 constant region were found by structural studies to contain a LEM-like motif that may bind DNA.[39] In accordance with this findings an N-terminal 85-residue LAP2 fragment was found to associate with chromosomes in vitro.[32]

In addition to the N-terminal chromatin binding domains, common to all LAP2 proteins, in vitro studies revealed a DNA binding region in the LAP2β-specific C-terminus[95], and a region in LAP2α's unique C-terminus was found to mediate chromosome association of LAP2α (the chromosomal binding partner is still unknown). Several studies have indicated that the interaction of the LAP2 isoforms with chromatin, mediated by the different domains, is regulated in a very complex and interdependent manner. For example, LAP2α's unique C-terminal chromatin binding domain was found both essential and sufficient for interaction of the protein with chromosomes during post-mitotic nuclear assembly, while the N-terminal LEM-like and LEM domains were not required at this stage.[96] Furthermore, while LAP2 N-terminal fragments containing both the LEM-like and LEM motif did not interact with chromosomes when overexpressed in cells, and monomeric recombinant fragments did not bind to chromosomes in vitro,[96] GST fusion proteins of the same fragments, which form oligomeres, interacted with chromosomes.[96,32] This suggests that protein oligomerization, achieved by GST in the recombinant fragments or by C-terminal domains downstream of the LEM domain in full length proteins, is required for tight interactions between the LEM domain and BAF and/or between the LEM-like motif and DNA. This hypothesis is further supported by recent in vitro binding studies showing that various *Xenopus* LAP2β-like isoforms, which are identical in their N-terminal part and contain the LEM domain, but differ slightly in their C-terminal regions, varied 9-fold in their affinities for BAF. Thus, the C-terminal unique regions in LAP2 isoforms may regulate the activity of the N-terminal LEM domain.[89] Aside from their diverse interactions with DNA and BAF, LAP2β-has also been identified by cross-linking experiments to associate with HA95, similar to LBR.[87]

It is not clear to what extent lamins and lamin binding proteins contribute to heterochromatin anchorage at the periphery. Considering the ~ ten-fold larger abundance of lamins as compared to most lamin binding proteins, it can be assumed that lamins may have a major role in chromatin association. In line with this observation, it has been shown that the expression of a lamin mutant missing major parts of the rod domain caused relocalization of endogenous lamins and lamin-binding proteins to discrete patches at the nuclear envelope, not overlapping with patches of mutant protein. Despite the redistribution of lamin-binding proteins and pore complexes to patches of endogenous lamins, the position of chromatin was unchanged,[97] suggesting that lamins rather than lamin binding proteins anchor chromatin at the periphery. However, since overexpression of lamin mutants caused major changes in nuclear shape and arrested cell growth, these effects may have been caused by the unphysiological conditions.

The large diversity of interactions of different lamin binding proteins with DNA and with different chromosomal proteins argues for an important role of lamin binding proteins in chromatin organization and anchorage at the NE. These interactions might be important for a "more specialized" cell stage-specific regulation of the chromatin-NE link during cell differentiation and/or cell cycle progression.

Two recently described nuclear pore complex (NPC)-associated proteins might also link the peripheral lamina to the internal nucleoskeleton and mediate chromatin anchorage and organization. Tpr (translocated promoter region) is a constitutive component of filaments extending from the nuclear pore basket structure 100-350 nm into the nucleus[98] in extrachromosomal channel networks.[99] Apart from being involved in mRNA transport,[100,101] yeast Tpr homologues Mlp1 and Mlp2 have been shown to be involved in transcriptionally repressing telomeric genes by tethering telomere-binding factor yKu70 to the perinuclear region.[102] The second candidate for a NPC-associated matrix protein is Nup153, which is a constituent of the nucleoplasmic pore basket[103-105] and has been implicated in nuclear import and export.[106-109] Highly sophisticated FRAP analysis and life cell imaging nicely showed that the nuclear lamina and NPC form a stable elastic network responsible for positioning NPCs in the membrane.[110] Since Nup153 interacts directly with lamin LIII in *Xenopus* oocytes,[111] and NPCs assembled in the absence of Nup153 lack anchorage within the NE,[112] Nup153 may be important for linking the NPC to the lamina. Strikingly however, Nup153 fluorescence recovered much faster than those of other NPC proteins,[110] indicating that Nup153 undergoes a rapid exchange between intranuclear and NPC associated pools. In yeast, overexpression of Nup153 caused the formation of intranuclear membrane lamellae structures[113] suggesting that Nup153 might have additional nuclear binding partners. Zinc-finger motifs in Nup153 may mediate DNA interaction[114] and early association of Nup153 with chromosomes after mitosis in a membrane-independent manner[110,115,116] also suggests that the protein may interact with chromatin.

Potential Functions of Lamina Proteins in Interphase

The molecular and cellular functions of lamins and lamin complexes remain unclear, although functional disruption of lamins in *Drosophila*[117] and *C. elegans*[118] revealed that they are essential for viability. In mice targeted disruption of A-type lamins caused muscular dystrophy, loss of adipose tissue, and early death,[47] while mutations in the human lamin A gene were linked to inherited forms of muscular and lipodystrophies.[41,119-121]

Similar to cytoplasmic intermediate filament networks, the nuclear lamina has been suggested to serve as the structural backbone for the nucleus defining nuclear shape.[2,11,119,120] Consistent with this function, nuclei assembled in vitro under lamin-depleted conditions were rather fragile[122,123] and nuclei of lamin A knockout mice showed a irregular shape.[47]

The complex interactions of lamins and lamin-binding proteins with DNA and with chromatin-associated proteins (histones, HP1, HA95, and BAF) at the nuclear periphery (lamina including membrane proteins) and in the nuclear interior (A-type lamins and LAP2α) suggest functions of these proteins in higher order chromatin organization by providing specific chromatin docking sites at the NE and by structurally organizing chromatin fibers in the 3-dimensional nuclear space. Since higher order chromatin organization is ultimately linked to control of gene expression, lamina proteins might also be involved in this process. In line with this hypothesis, highly silenced human chromosome 18 occupies more peripheral territories in the nucleus as compared to highly active chromosome 19.[124] Furthermore, the lamina protein LBR is found in a complex with HP1 (see above), a protein involved in position effect variegation and control of gene expression.[81,82]

In addition, components of the NE may directly influence transcription by interacting with transcription factors and/or chromatin remodeling complexes. A novel integral membrane protein of the inner membrane (Ring Finger Binding protein, RFBP), which resembles a type IV phospholipid pump, has recently been identified[125] and has been shown to directly interact with RUSH proteins, SWI/SNF transcription factors that model chromatin. Thus, association of chromatin with the NE may regulate transcriptional access directly[119]. Furthermore, several findings have indicated a direct interaction of lamina proteins with E2F transcriptional complexes, which regulate G1-S phase progression in cell cycle by activating transcription of S-phase specific genes.[126] The membrane-bound LAP2β was found by yeast two-hybrid analysis to bind mouse germ cell less (mGCL),[127] which in turn interacts with E2F-associated DP and regulates the cell cycle.[128] As overexpressed LAP2β reduced E2F-dependent reporter activity,[127] it is likely that LAP2β might negatively regulate E2F activity by tethering the transcription complex to the nuclear periphery, a mechanism known also for other transcription factors.[2] In addition, lamins A/C associate directly with the hypophosphorylated, active form of retinoblastoma protein (pRb),[129] which binds E2F and represses transcription of S-phase-specific genes.[130,131] LAP2α, which has been identified as a direct binding partner of A-type lamins[34] might be another functional component of such a complex.

Lamina proteins might also be involved in DNA replication. Nuclei assembled in the absence of lamin B3 in *Xenopus* in vitro nuclear assembly reactions were not able to replicate their DNA,[123,132] but addition of lamin B3 partially restored the phenotype.[133] Similarly, lamin mutants causing nuclear lamina disassembly were shown to inhibit DNA replication,[134-136] and lamin mutants causing a dramatic reorganization of the lamina and lobulated nuclei interfered with DNA replication and cell growth.[97] Interestingly, ectopic expression of lamin-binding LAP2β fragments in mammalian cells inhibited progression into S-phase,[137] while LAP2β mutants added to *Xenopus* in vitro nuclear assembly reactions influenced DNA replication positively.[30] Thus LAP2β might also be involved in DNA-replication either directly or indirectly by affecting lamina assembly.

Lamina proteins might also be involved in controlling apoptosis. In *C. elegans*, for instance, CED-4, a cell death activator, is translocated from mitochondria to the nuclear envelope before caspase activation,[138] suggesting that the lamina provides an attachment site for the apoptotic signaling machinery.[2] Lamins, LAP2α and LAP2β are early targets of apoptosis[139-141] and expression of uncleavable lamin mutants was shown to delay apoptosis for several hours.[141] Furthermore, inhibition of lamin B assembly at the nuclear envelope upon preventing its postmitotic dephosphorylation induced apoptosis in human cells,[142] suggesting that mislocalized lamins actively trigger apoptosis.[143]

Interestingly, the *Drosophila* lamin Dm_o was recently found to be required for a cytoplasmic function in polarized cells, the outgrowth of cytoplasmic extensions from terminal cells of the tracheal system.[144] The molecular mechanisms however remain obscure.

These diverse functions of lamina/matrix proteins may explain how mutations in lamina proteins cause different inheritable human diseases affecting heart and skeletal muscle as well as adipose tissue (laminopathies).[41,119,121] It is conceivable that a disturbance of any of the above described functions of lamins and lamin binding proteins can contribute to the disease to different degrees. Mutations in emerin and lamin A do not only affect these two proteins but may have significant impact on other proteins tightly linked to lamin A structures. Thus, elucidating the function of any potential lamin A binding partner, may provide important clues as to the molecular mechanisms of the disease. Therefore, in view of the recently reported interaction of LAP2α with the C-terminal region of A-type lamins[34] containing many lipodystrophy and EDMD-specific mutations,[145] it is intriguing to speculate that these mutations may effect LAP2α-lamin A interactions and interfere with LAP2α/lamin A functions. As such mutations would predominantly affect structures in the nuclear interior, the disease phenotype would be different from those caused by mutations in emerin or in lamin A affecting mostly lamin A-emerin interactions and functions at the nuclear envelope.

Dynamics and Functions of Lamina-Chromatin Interactions During Mitosis

Multicellular eukaryotes reversibly disassemble the nuclear lamina, NPCs and the nucleoskeleton during mitosis, and the nuclear membrane merges into the endoplasmic reticulum.[146] This process is driven by phosphorylation of lamins, LBR, LAP1 and peripheral as well as intranuclear LAP2 proteins mostly involving mitotic cyclin-dependent kinases (cdk), although other kinases may also play a role.[10,31,66] Mitotic A-type lamins are distributed in the cytoplasm probably as dimers or tetramers, while B-type lamins remain associated with membranes due to their C-terminal farnesyl modification, and probably also due to their interaction with LBR.[21] LAP2β[6] and LAP2α[33] dissociate from lamins and chromosomes in a mitotic phosphorylation dependent manner. There are reported discrepancies depending on the cell systems used as to whether nuclear membrane proteins (LAP2β, LBR and emerin) and ER proteins segregate into distinct membrane structures during mitosis or whether both disperse throughout the endoplasmic reticulum.[147]

Nuclear reassembly requires phosphatase activity and, at least for B-type lamins, has been shown to involve phosphatase PP1.[148] PP1 is targeted to the NE by the membrane integrated A-kinase anchoring protein AKAP149 and this process was shown to correlate with the assembly of B-type lamins.[149] Inhibition of PP1 association with AKAP149 by a peptide containing the PP1 binding domain of AKAP149 resulted in lack of assembly of B-type lamins and apoptosis.[142] Interestingly, however, assembly of A-type lamin was not effected by the peptide in HeLa cells, supporting other studies which show different pathways of assembly of A- and B-type lamins after mitosis.[34,58]

Numerous studies have shown that the assembly of the NE and nuclear structure after cell division requires targeting and assembly of lamins and lamin binding proteins to chromosomes in a temporally and spatially regulated manner (Fig. 2). LAP2α appears to be the first protein among the lamina components to associate with chromosomes in anaphase, clearly before LAP2β-containing membranes accumulate and enclose the decondensing chromosomes.[96] Other membrane proteins, including LBR and emerin accumulate at the chromosomal surface at the same time as LAP2β.[115, 116, 150, 151] Interestingly, initial association of LBR with chromosomes was shown to occur primarily at the peripheral chromosomal regions[115], while LAP2β, emerin and lamins were enriched at more central regions of chromosomes closest to the spindle poles and the midbody.[6,58,115,151,152] This suggested that association of LBR and LAP2β with chromosomes involves different mechanisms. Although a subfraction of B-type lamins might be targeted to the chromosomal surfaces by binding to LAP2β and or LBR and or histones, the bulk of lamin B assembly occurred after accumulation of the membrane proteins at the chromosomal surface.[6,58] There are clear differences in the assembly of B-type and A-type lamins. While lamin B1 assembled around chromosomes at anaphase telophase transition following accumulation of membranes, A-type lamins were targeted to intranuclear sites much later when an intact continuous nuclear envelope had already formed.[58] Lamin A translocation to the nuclear interior might involve interaction with LAP2α[34] (see below).

More recently, the dynamics of the NPC-associated proteins Nup153 and Tpr during mitosis have been tested. Interestingly, Nup153, but not Tpr, was recruited to chromosomes at the same time as LAP2β and LBR,[110,115,116] before accumulation of lamin B.[58] As this interaction is membrane-independent, it seems to be the first step of NPC assembly.

The mechanisms regulating the sequential association of lamina proteins with chromatin are not very well understood. It is particularly intriguing that LAP2α assembles prior to LAP2β, although both proteins contain the LEM and LEM-like motifs in their N-terminal constant regions, known to bind BAF and/or DNA (see above). Several recent observations might help explain this phenomenon. It was shown that the association of LAP2α with chromosomes at early stages of assembly, requires the α-specific C-terminal domain, which is absent in LAP2β.[96] The N-terminal BAF-binding domain, present in both LAP2α and β did not interact with BAF at this stage of nuclear assembly, probably due to post-translational modification of LAP2 and/or BAF or due to an inhibitory effect of the C-terminal regions of LAP2.[89] Therefore, we favor the following model for the initial stages of assembly (Fig. 2). In phase 1 LAP2α associates with chromosomes via its C-terminus, not involving the N-terminal LEM and LEM-like domains. This association triggers conformational changes on the chromosomal surface and/or induces post-translational modification of BAF or other chromosomal proteins and allows binding of the LEM domain to BAF in phase 2. This mediates cross-linking of chromosomal territories by chromosome bound LAP2α, and targeting of LAP2β, and emerin, and thus membrane structures to the chromosomal surface (phase 3).

Despite the fairly detailed analysis of kinetics of chromosome association of various lamina proteins, their specific roles in nuclear assembly is still not clear. The reported contributions of lamins to NE assembly has been controversial.[153] While immunodepletion of lamins from in vitro *Xenopus* nuclear assembly extracts have indicated the formation of a NE in the absence of lamins,[122,123] other studies using *Drosophila*, mammalian and *Xenopus* extracts showed that immunoadsorbtion of lamins inhibited NE assembly.[154-156] As these different results were most

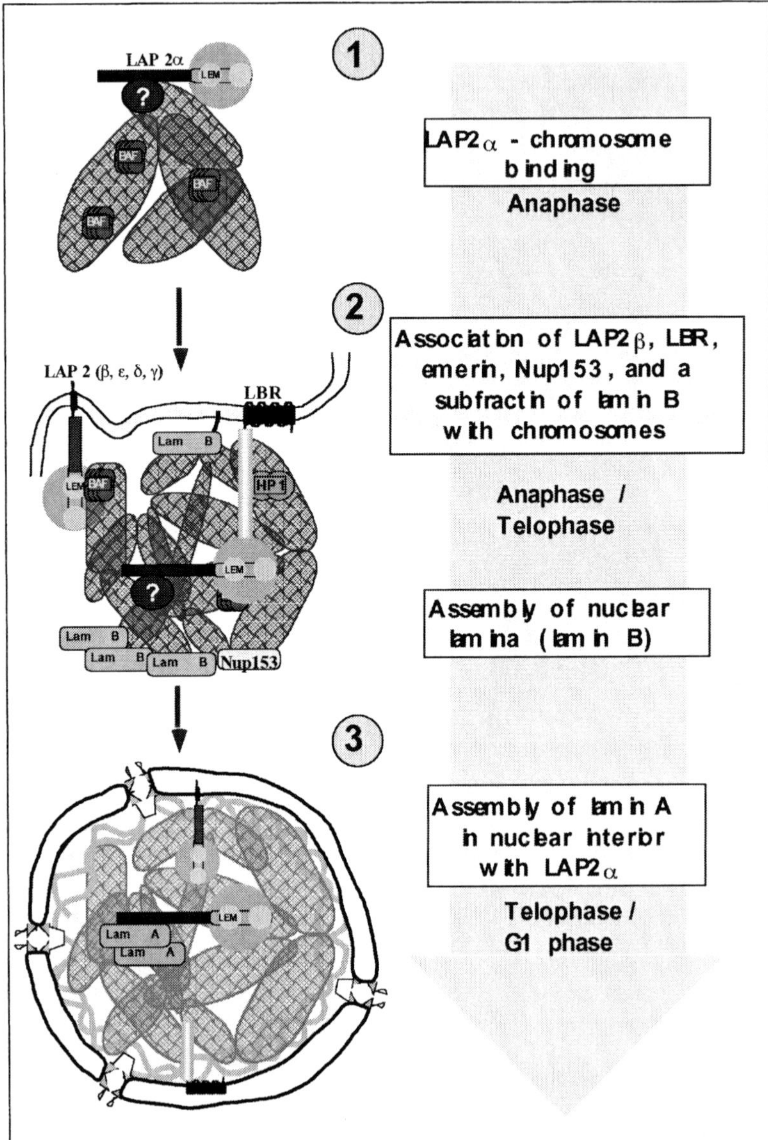

Figure 2. Sequence of protein targeting to chromosomal surfaces during nuclear reassembly following chromatid separation, and responsible interactions. For details see text.

likely caused by the different efficiencies in depleting and/or de-activating lamins by antibodies, Lopez-Soler et al have recently used a peptide containing the C-terminal domain of *Xenopus* lamin B3 in nuclear assembly reactions and found that the peptide inhibits nuclear lamin polymerization and also nuclear membrane assembly around chromatin.[157] Thus *Xenopus* lamin is likely to have important functions in NE assembly.

LAP2β has originally been implicated in targeting membranes to chromosomes[6], but this has not been demonstrated directly. Microinjection of LAP2β's nucleoplasmic domain into mitotic cells did not inhibit nuclear membrane targeting and assembly, but affected nuclear growth during G1-S phase progression.[137] The same fragment had similar effects, when added to *Xenopus* in vitro nuclear assembly reactions.[30] However, all these studies were performed in a background of endogenous LAP2 proteins. Interestingly, an N-terminal LAP2β fragment containing the BAF-binding region was able to inhibit nuclear membrane binding in the *Xenopus* assembly system,[30] but overexpression of the same fragment in mammalian cells had no detectable effect on cell cycle progression.[96] This suggests that different molecular mechanisms of nuclear assembly may exist in early embryonic versus somatic cell divisions. Apart from LAP2β, LBR has also been suggested to be involved in membrane-chromosome interactions in in vitro binding studies using LBR-immunodepleted membrane fractions.[158,159]

The early accumulation of LAP2α around decondensing chromosomes suggested a function of the protein in providing a scaffold for chromatin organization. Considering the existence of at least two chromatin interaction domains in LAP2α, the protein is ideally suited to cross-link chromosomal regions and help to structurally organize chromatin in post-mitotic nuclei. Observations showing that overexpression of a C-terminal chromatin-binding fragment of LAP2α was toxic for the cells and that addition of the same fragment to mammalian in vitro nuclear assembly extracts inhibited NE formation[96] (Vlcek, Korbei and Foisner, submitted), supported this hypothesis, but the mechanisms remain obscure.

Conclusions and Future Prospects

The recent discovery of novel interaction partners for lamins and lamina-associated proteins have changed our view of how these proteins function in the cell nucleus. While former studies revealed mainly structural function, novel findings point towards important functions also in controlling DNA replication, gene transcription and cell cycle progression in a more direct fashion. Nevertheless, I believe that this is just the tip of the iceberg, and more interactions of lamin and lamin-binding proteins, which may be regulated in a development-, differentiation- or cell cycle-specific manner, will be identified in the future. This will give us a much clearer picture of the specific involvement of lamina proteins in these processes and will also allow to completely unravel the molecular mechanisms behind the lamin-related human diseases (laminopathies).

Acknowledgements

Work in the author's laboratory was supported by grants from the Austrian Science Research Fund (FWF) No. P13374 to RF.

References

1. Gotzmann J, Foisner R. Lamins and lamin-binding proteins in functional chromatin organization. Crit Rev Eukaryot Gene Expr 1999; 9:257-265.
2. Cohen M, Lee KK, Wilson KL, et al. Transcriptional repression, apoptosis, human disease and the functional evolution of the nuclear lamina. Trends Biochem Sci 2001; 26:41-47.
3. Foisner R. Inner nuclear membrane proteins and the nuclear lamina. J Cell Sci 2001; 114:3731-3732.
4. Gerace L, Foisner R. Integral membrane proteins and dynamic organization of the nuclear envelope. Trends Cell Biol 1994; 4:127-131.
5. Senior A, Gerace L. Integral membrane proteins specific to the inner nuclear membrane and associated with the nuclear lamina. J Cell Biol 1988; 107:2029-2036.
6. Foisner R, Gerace L. Integral membrane proteins of the nuclear envelope interact with lamins and chromosomes, and binding is modulated by mitotic phosphorylation. Cell 1993; 73:1267-1279.
7. Nickerson J. Experimental observations of a nuclear matrix. J Cell Sci 2001; 114:463-474.
8. Pederson T. Half a century of the "nuclear matrix". Mol Biol Cell 2000; 11:799-805
9. Stuurman N, Heins S, Aebi U. Nuclear lamins: Their structure, assembly, and interactions. J Struct Biol 1998; 122:42-66
10. Moir RD, Spann TP, Goldman RD. The dynamic properties and possible functions of nuclear lamins. Int Rev Cytol 1995; 141:182.
11. Hutchison CJ, Alvarez-Reyes M, Vaughan OA. Lamins in disease: why do ubiquitously expressed nuclear envelope proteins give rise to tissue-specific disease phenotypes? J Cell Sci 2001; 114:9-19.

12. Mical TI, Monteiro MJ. The role of sequences unique to nuclear intermediate filaments in the targeting and assembly of human lamin B: Evidence for lack of interaction of lamin B with its putative receptor. J Cell Sci 1998; 111:3471-3485.

13. Hennekes H, Nigg EA. The role of isoprenylation in membrane attachment of nuclear lamins. A single point mutation prevents proteolytic cleavage of the lamin A precursor and confers membrane binding properties. J Cell Sci 1994; 107:1019-1029.

14. Izumi M, Vaughan OA, Hutchison CJ, et al. Head and/or CaaX domain deletions of lamin proteins disrupt preformed lamin A and C but not lamin B structure in mammalian cells. Mol Biol Cell 2000; 11:4323-4337.

15. Gerace L, Burke B. Functional organization of the nuclear envelope. Annu Rev Cell Biol 1988; 4:335-374.

16. Worman H J, Yuan J, Blobel G, et al. A lamin B receptor in the nuclear envelope. Proc Natl Acad Sci U S A 1988; 85:8531-8534.

17. Lin F, Blake DL, Callebaut I, et al. MAN1, an inner nuclear membrane protein that shares the LEM domain with lamina-associated polypeptide 2 and emerin. J Biol Chem 2000; 275:4840-4847.

18. Manilal S, thi Man N, Sewry CA, et al. The Emery-Dreifuss muscular dystrophy protein, emerin, is a nuclear membrane protein. Hum Mol Genet 1996; 5:801-808.

19. Worman HJ, Evans CD, Blobel G. The lamin B receptor of the nuclear envelope inner membrane: A polytopic protein with eight potential transmembrane domains. J Cell Biol 1990; 111:1535-1542.

20. Simos G, Georgatos SD. The inner nuclear membrane protein p58 associates in vivo with a p58 kinase and the nuclear lamins. EMBO J 1992; 11:4027-4036.

21. Meier J, Georgatos SD. Type B lamins remain associated with the integral nuclear envelope protein p58 during mitosis: implications for nuclear reassembly. EMBO J 1994; 13:1888-1898.

22. Ye Q, Worman HJ. Primary structure analysis and lamin B and DNA binding of human LBR, an integral protein of the nuclear envelope inner membrane. J Biol Chem 1994; 269:11306-11311.

23. Holmer L, Pezhman A, Worman HJ. The human lamin B receptor/sterol reductase multigene family. Genomics 1998; 54:469-476.

24. Silve S, Dupuy PH, Ferrara P, et al. Human lamin B receptor exhibits sterol C14-reductase activity in Saccharomyces cerevisiae. Biochim Biophys Acta 1998; 1392:233-244.

25. Martin L, Crimaudo C, Gerace L. cDNA cloning and characterization of lamina-associated polypeptide 1C (LAP1C), an integral protein of the inner nuclear membrane. J Biol Chem 1995; 270:8822-8828.

26. Powell L, Burke B. Internuclear exchange of an inner nuclear membrane protein (p55) in heterokaryons: In vivo evidence for the interaction of p55 with the nuclear lamina. J Cell Biol 1990; 111:2225-2234.

27. Harris CA, Andryuk PJ, Cline S, et al. Three distinct human thymopoietins are derived from alternatively spliced mRNAs. Proc Natl Acad Sci USA 1994; 91:6283-6287.

28. Berger R, Theodor L, Shoham J, et al. The characterization and localization of the mouse thymopoietin/lamina- associated polypeptide 2 gene and its alternatively spliced products. Genome Res 1996; 6:361-370.

29. Lang C, Paulin-Levasseur M, Gajewski A, et al. Molecular characterization and developmentally regulated expression of *Xenopus* lamina-associated polypeptide 2 (XLAP2). J Cell Sci 1999; 112:749-759.

30. Gant TM, Harris CA, Wilson KL. Roles of LAP2 Proteins in nuclear assembly and DNA replication: Truncated LAP2beta proteins alter lamina assembly, envelope formation, nuclear size, and DNA replication efficiency in *Xenopus laevis* extracts. J Cell Biol 1999; 144:1083-1096.

31. Dechat T, Vlcek S, Foisner R. Review: Lamina-associated polypeptide 2 isoforms and related proteins in cell cycle-dependent nuclear structure dynamics. J Struct Biol 2000; 129:335-345.

32. Furukawa K, Fritze CE, Gerace L. The major nuclear envelope targeting domain of LAP2 coincides with its lamin binding region but is distinct from its chromatin interaction domain. J Biol Chem 1998; 273:4213-4219.

33. Dechat T, Gotzmann J, Stockinger A, et al. Detergent-salt resistance of LAP2alpha in interphase nuclei and phosphorylation-dependent association with chromosomes early in nuclear assembly implies functions in nuclear structure dynamics. EMBO J 1998; 17:4887-4902.

34. Dechat T, Korbei B, Vaughan OA, et al. Lamina-associated polypeptide 2alpha binds intranuclear A-type lamins. J Cell Sci 2000; 113 Pt 19:3473-3484.

35. Theodor L, Shoham J, Berger R, et al. Ubiquitous expression of a cloned murine thymopoietin cDNA. Acta Haematol 1997; 97:153-163.

36. Ishijima Y, Toda T, Matsushita H, et al. Expression of thymopoietin beta/lamina-associated polypeptide 2 (TP beta/LAP2) and its family proteins as revealed by specific antibody induced against recombinant human thymopoietin. Biochem Biophys Res Commun 1996; 226:431-438.

37. Dreger M, Bengtsson L, Schoneberg T, et al. Nuclear envelope proteomics: Novel integral membrane proteins of the inner nuclear membrane. Proc Natl Acad Sci USA 2001; 98:11943-11948.
38. Laguri C, Gilquin B, Wolff N, et al. Structural characterization of the lem motif common to three human inner nuclear membrane proteins. Structure (Camb) 2001; 9:503-511.
39. Cai M, Huang Y, Ghirlando R, et al. Solution structure of the constant region of nuclear envelope protein LAP2 reveals two LEM-domain structures: One binds BAF and the other binds DNA. EMBO J 2001; 20:4399-4407.
40. Ostlund C, Ellenberg J, Hallberg E, et al. Intracellular trafficking of emerin, the Emery-Dreifuss muscular dystrophy protein. J Cell Sci 1999; 112:1709-1719.
41. Morris GE, Manilal S. Heart to heart: From nuclear proteins to Emery-Dreifuss muscular dystrophy. Hum Mol Genet 1999; 8:1847-1851.
42. Nagano A, Koga R, Ogawa M, et al. Emerin deficiency at the nuclear membrane in patients with Emery-Dreyfuss muscular dystrophy. Nature Genet 1996; 12:254-259.
43. Sakaki M, Koike H, Takahashi N, et al. Interaction between emerin and nuclear lamins. J Biochem (Tokyo) 2001; 129:321-327.
44. Fairley EA, Kendrick-Jones J, Ellis JA. The Emery-Dreifuss muscular dystrophy phenotype arises from aberrant targeting and binding of emerin at the inner nuclear membrane. J Cell Sci 1999; 112:2571-2582.
45. Clements L, Manilal S, Love DR, et al. Direct interaction between emerin and lamin A. Biochem Biophys Res Commun 2000; 267:709-714.
46. Vaughan OA, MAlvarez-Reyes M, Bridger JM, et al. Both emerin and lamin C depend on lamin A for localization at the nuclear envelope. J Cell Sci 2001; 114:2577-2590.
47. Sullivan T, Escalante-Alcalde D, Bhatt H, Anver M, Bhat N, Nagashima K et al. Loss of A-type lamin expression compromises nuclear envelope integrity leading to muscular dystrophy. J Cell Biol 1999; 147:913-920.
48. Paulin-Levasseur M, Blake DL, Julien M, et al. The MAN antigens are non-lamin constituents of the nuclear lamina in vertebrate cells. Chromosoma 1996; 104:367-379.
49. Rolls MM, Stein PA, Taylor SS, et al. A visual screen of a GFP-fusion library identifies a new type of nuclear envelope membrane protein. J Cell Biol 1999; 146:29-44.
50. Malone CJ, Fixsen WD, Horvitz HR, et al. UNC-84 localizes to the nuclear envelope and is required for nuclear migration and anchoring during *C. elegans* development. Development 1999; 126:3171-3181.
51. Padan R, Nainudel-Epszteyn S, Goitein R, et al. Isolation and characterization of the Drosophila nuclear envelope otefin cDNA. J Biol Chem 1990; 265:7808-7813.
52. Goldberg M, Lu H, Stuurman N, et al. Interactions among *Drosophila* nuclear envelope proteins lamin, otefin, and YA. Mol Cell Biol 1998; 18:4315-4323.
53. Fricker M, Hollinshead M, White N, et al. Interphase nuclei of many mammalian cell types contain deep, dynamic, tubular membrane-bound invaginations of the nuclear envelope. J Cell Biol 1997; 136:531-544.
54. Jagatheesan G, Thanumalayan S, Muralikrishna B, et al. Colocalization of intranuclear lamin foci with RNA splicing factors. J Cell Sci 1999; 112:4651-4661.
55. Hozak P, Sasseville AM, Raymond Y, et al. Lamin proteins form an internal nucleoskeleton as well as a peripheral lamina in human cells. J Cell Sci 1995; 108:635-644.
56. Neri LM, Raymond Y, Giordano A, et al. Lamin A is part of the internal nucleoskeleton of human erythroleukemia cells. J Cell Physiol 1999; 178:284-295.
57. Broers JL, Machiels BM, van Eys GJ, et al. Dynamics of the nuclear lamina as monitored by GFP-tagged A-type lamins. J Cell Sci 1999; 112:3463-3475.
58. Moir RD, Yoon M, Khuon S, et al. Nuclear lamins A and B1: different pathways of assembly during nuclear envelope formation in living cells. J Cell Biol 2000; 151:1155-1168.
59. Moir RD, Montag-Lowy M, Goldman RD. Dynamic properties of nuclear lamins: Lamin B is associated with sites of DNA replication. J Cell Biol 1994; 125:1201-1212.
60. Bridger JM, Kill IR, O'Farrell M, et al. Internal lamin structures within G1 nuclei of human dermal fibroblasts. J Cell Sci 1993; 104:297-306.
61. Goldman AE, Moir RD, Montag-Lowy M, et al. Pathway of incorporation of microinjected lamin A into the nuclear envelope. J Cell Biol 1992; 119:725-735.
62. Pugh GE, Coates PJ, Lane EB, et al. Distinct nuclear assembly pathways for lamins A and C lead to their increase during quiescence in Swiss 3T3 cells. J Cell Sci 1997; 110:2483-2493.
63. Lutz RJ, Trujillo MA, Denham KS, et al. Nucleoplasmic localization of prelamin A: implications for prenylation-dependent lamin A assembly into the nuclear lamina [published erratum appears in Proc Natl Acad Sci USA 1992 Jun 15;89(12):5699]. Proc Natl Acad Sci USA 1992; 89:3000-3004.

64. Sasseville AM, Raymond Y. Lamin A precursor is localized to intranuclear foci. J Cell Sci 1995; 108:273-285.

65. Barton RM, Worman HJ. Prenylated prelamin A interacts with Narf, a novel nuclear protein. J Biol Chem 1999; 274:30008-30018.

66. Foisner R. Dynamic organisation of intermediate filaments and associated proteins during the cell cycle. BioEssays 1997; 19:297-305.

67. Hari KL, Cook KR, Karpen GH. The *Drosophila* Su(var)2-10 locus regulates chromosome structure and function and encodes a member of the PIAS protein family. Genes Dev 2001; 15:1334-1348.

68. Li G, Sudlow G, Belmont AS. Interphase cell cycle dynamics of a late-replicating, heterochromatic homogeneously staining region: Precise choreography of condensation/decondensation and nuclear positioning. J Cell Biol 1998; 140:975-989.

69. Belmont AS, Zhai Y, Thilenius A. Lamin B distribution and association with peripheral chromatin revealed by optical sectioning and electron microscopy tomography. J Cell Biol 1993; 123:1671-1685.

70. Luderus ME, de Graaf A, Mattia E, et al. Binding of matrix attachment regions to lamin B1. Cell 1992; 70:949-959.

71. Zhao K, Harel A, Stuurman N, et al. Binding of matrix attachment regions to nuclear lamin is mediated by the rod domain and depends on the lamin polymerization state. FEBS Lett 1996; 380:161-164.

72. Shoeman RL, Traub P. The in vitro DNA-binding properties of purified nuclear lamin proteins and vimentin. J Biol Chem 1990; 265:9055-9061.

73. Baricheva EA, Berrios M, Bogachev SS, et al. DNA from *Drosophila melanogaster* beta-heterochromatin binds specifically to nuclear lamins in vitro and the nuclear envelope in situ. Gene 1996; 171:171-176.

74. Rzepecki R, Bogachev SS, Kokoza E, et al. In vivo association of lamins with nucleic acids in *Drosophila melanogaster*. J Cell Sci 1998; 111:121-129.

75. Glass CA, Glass JR, Taniura H, et al. The alpha-helical rod domain of human lamins A and C contains a chromatin binding site. EMBO J 1993; 12:4413-4424.

76. Taniura H, Glass C, Gerace L. A chromatin binding site in the tail domain of nuclear lamins that interacts with core histones. J Cell Biol 1995; 131:33-44.

77. Goldberg M, Harel A, Brandeis M, et al. The tail domain of lamin Dm0 binds histones H2A and H2B. Proc Natl Acad Sci USA 1999; 96:2852-2857.

78. Duband-Goulet I, Courvalin J C. Inner nuclear membrane protein LBR preferentially interacts with DNA secondary structures and nucleosomal linker. Biochemistry 2000; 39:6483-6488.

79. Ye Q, Callebaut I, Pezhman A, et al. Domain-specific interactions of human HP1-type chromodomain proteins and inner nuclear membrane protein LBR. J Biol Chem 1997; 272:14983-14989.

80. Ye Q, Worman J. Interaction between an integral protein of the nuclear envelope inner membrane and human chromodomain proteins homologous to *Drosophila* HP1. J Biol Chem 1996; 271:14653-14656.

81. Cavalli G, Paro R. Chromo-domain proteins: linking chromatin structure to epigenetic regulation. Curr Opin Cell Biol 1998; 10:354-360.

82. Jones DO, Cowell IG, Singh PB. Mammalian chromodomain proteins: Their role in genome organisation and expression. BioEssays 2000; 22:124-137.

83. Kourmouli N, Dialynas G, Petraki C, et al. Binding of heterochromatin protein 1 to the nuclear envelope is regulated by a soluble form of tubulin. J Biol Chem 2001; 276:13007-13014.

84. Polioudaki H, Kourmouli N, Drosou V, et al. Histones H3/H4 form a tight complex with the inner nuclear membrane protein LBR and heterochromatin protein 1. EMBO Rep 2001; 2:920-925.

85. Bannister AJ, Zegerman P, Partridge JF, et al. Selective recognition of methylated lysine 9 on histone H3 by the HP1 chromo domain. Nature 2001; 410:120-124.

86. Lachner M, O'Carroll D, Rea S, et al. Methylation of histone H3 lysine 9 creates a binding site for HP1 proteins. Nature 2001; 410:116-120.

87. Martins SB, Eide T, Steen RL, et al. HA95 is a protein of the chromatin and nuclear matrix regulating nuclear envelope dynamics. J Cell Sci 2000; 113:3703-3713.

88. Furukawa K. LAP2 binding protein 1 (L2BP1/BAF) is a candidate mediator of LAP2- chromatin interaction. J Cell Sci 1999; 112:2485-2492.

89. Shumaker DK, Lee KK, Tanhehco YC, et al. LAP2 binds to BAF-DNA complexes: Requirement for the LEM domain and modulation by variable regions. EMBO J 2001; 20:1754-1764.

90. Chen H, Engelman A. The barrier-to-autointegration protein is a host factor for HIV type 1 integration. Proc Natl Acad Sci USA 1998; 95:15270-15274.

91. Lee MS, Craigie R. A previously unidentified host protein protects retroviral DNA from autointegration. Proc Natl Acad Sci USA 1998; 95:1528-1533.

92. Cai M, Huang Y, Zheng R, et al. Solution structure of the cellular factor BAF responsible for protecting retroviral DNA from autointegration. Nat Struct Biol 1998; 5:903-909.
93. Zheng R, Ghirlando R, Lee MS, et al.. Barrier-to-autointegration factor (BAF) bridges DNA in a discrete, higher-order nucleoprotein complex [In Process Citation]. Proc Natl Acad Sci USA 2000; 97:8997-9002.
94. Wolff N, Gilquin B, Courchay K, et al. Structural analysis of emerin, an inner nuclear membrane protein mutated in X-linked Emery-Dreifuss muscular dystrophy. FEBS Lett 2001; 501:171-176.
95. Furukawa K, Glass C, Kondo T. Characterization of the chromatin binding activity of lamina-associated polypeptide (LAP) 2. Biochem Biophys Res Commun 1997; 238:240-246.
96. Vlcek S, Just H, Dechat T, et al. Functional diversity of LAP2alpha and LAP2beta in postmitotic chromosome association is caused by an alpha-specific nuclear targeting domain. EMBO J 1999; 18:6370-6384.
97. Schirmer EC, Guan T, Gerace L. Involvement of the lamin rod domain in heterotypic lamin interactions important for nuclear organization. J Cell Biol 2001; 153:479-489.
98. Cordes VC, Reidenbach S, Rackwitz HR, et al. Identification of protein p270/Tpr as a constitutive component of the nuclear pore complex-attached intranuclear filaments. J Cell Biol 1997; 136:515-529.
99. Zimowska G, Aris JP, Paddy MR. A *Drosophila* Tpr protein homolog is localized both in the extrachromosomal channel network and to nuclear pore complexes. J Cell Sci 1997; 110:927-944.
100. Bangs P, Burke B, Powers C, et al. Functional analysis of Tpr: identification of nuclear pore complex association and nuclear localization domains and a role in mRNA export. J Cell Biol 1998; 143:1801-1812.
101. Strambio-de-Castillia C, Blobel G, Rout MP. Proteins connecting the nuclear pore complex with the nuclear interior. J Cell Biol 1999; 144:839-855.
102. Galy V, Olivo-Marin JC, Scherthan H, et al. Nuclear pore complexes in the organization of silent telomeric chromatin [see comments]. Nature 2000; 403:108-112.
103. Stoffler D, Fahrenkrog B, Aebi U. The nuclear pore complex: from molecular architecture to functional dynamics. Curr Opin Cell Biol 1999; 11:391-401.
104. Allen TD, Cronshaw JM, Bagley S, et al. The nuclear pore complex: Mediator of translocation between nucleus and cytoplasm. J Cell Sci 2000; 113:1651-1659.
105. Cordes VC, Reidenbach S, Kohler A, et al. Intranuclear filaments containing a nuclear pore complex protein. J Cell Biol 1993; 123:1333-1344.
106. Shah S, Forbes DJ. Separate nuclear import pathways converge on the nucleoporin Nup153 and can be dissected with dominant-negative inhibitors. Curr Biol 1998; 8:1376-1386.
107. Bastos R, Lin A, Enarson M, et al. Targeting and function in mRNA export of nuclear pore complex protein Nup153. J Cell Biol 1996; 134:1141-1156.
108. Moroianu J, Blobel G, Radu A. RanGTP-mediated nuclear export of karyopherin alpha involves its interaction with the nucleoporin Nup153. Proc Natl Acad Sci USA 1997; 94:9699-9704.
109. Nakielny S, Shaikh S, Burke B, et al. Nup153 is an M9-containing mobile nucleoporin with a novel Ran-binding domain. EMBO J 1999; 18:1982-1995.
110. Daigle N, Beaudouin J, Hartnell L, et al. Nuclear pore complexes form immobile networks and have a very low turnover in live mammalian cells. J Cell Biol 2001; 154:71-84.
111. Smythe C, Jenkins HE, Hutchison CJ. Incorporation of the nuclear pore basket protein Nup153 into nuclear pore structures is dependent upon lamina assembly: evidence from cell-free extracts of *Xenopus* eggs. EMBO J 2000; 19:3918-3931.
112. Walther TC, Fornerod M, Pickersgill H, et al. The nucleoporin Nup153 is required for nuclear pore basket formation, nuclear pore complex anchoring and import of a subset of nuclear proteins. EMBO J 2001; 20:5703-5714.
113. Marelli M, Lusk CP, Chan H, et al. A link between the synthesis of nucleoporins and the biogenesis of the nuclear envelope. J Cell Biol 2001; 153:709-724.
114. Sukegawa J, Blobel G. A nuclear pore complex protein that contains zinc finger motifs, binds DNA, and faces the nucleoplasm. Cell 1993; 72:29-38.
115. Haraguchi T, Koujin T, Hayakawa T, et al. Live fluorescence imaging reveals early recruitment of emerin, LBR, RanBP2, and Nup153 to reforming functional nuclear envelopes. J Cell Sci 2000; 113:779-794.
116. Bodoor K, Shaikh S, et al. Sequential recruitment of NPC proteins to the nuclear periphery at the end of mitosis. J Cell Sci 1999; 112:2253-2264.
117. Lenz-Bohme B, Wismar J, Fuchs S, et al. Insertional mutation of the Drosophila nuclear lamin Dm0 gene results in defective nuclear envelopes, clustering of nuclear pore complexes, and accumulation of annulate lamellae. J Cell Biol 1997; 137:1001-1016.

118. Liu J, Ben-Shahar TR, Riemer D, et al. Essential roles for Caenorhabditis elegans lamin gene in nuclear organization, cell cycle progression, and spatial organization of nuclear pore complexes. Mol Biol Cell 2000; 11:3937-3947.
119. Wilson KL, Zastrow MS, Lee KK. Lamins and disease: insights into nuclear infrastructure. Cell 2001; 104:647-650.
120. Wilson KL. The nuclear envelope and disease. Trends Cell Biol. 2000; 10:125-129.
121. Nagano A, Arahata K. Nuclear envelope proteins and associated diseases. Curr Opin Neurol 2000; 13:533-539.
122. Meier J, Campbell KH, Ford CC, et al. The role of lamin LIII in nuclear assembly and DNA replication, in cell- free extracts of *Xenopus* eggs. J Cell Sci 1991; 98:271-279.
123. Newport JW, Wilson KL, Dunphy WG. A lamin-independent pathway for nuclear envelope assembly. J Cell Biol 1990; 111:2247-2259.
124. Croft JA, Bridger JM, Boyle S, et al. Differences in the localization and morphology of chromosomes in the human nucleus. J Cell Biol 1999; 145:1119-1131.
125. Mansharamani M, Hewetson A, Chilton BS. Cloning and characterization of an atypical Type IV P-type ATPase that binds to the RING motif of RUSH transcription factors. J Biol Chem 2001; 276:3641-3649.
126. Lavia P, Jansen-Durr P. E2F target genes and cell-cycle checkpoint control. BioEssays 1999; 21:221-230.
127. Nili E, Cojocaru GS, Kalma Y, et al. Nuclear membrane protein LAP2beta mediates transcriptional repression alone and together with its binding partner GCL (germ-cell-less). J Cell Sci 2001; 114:3297-3307.
128. de la Luna S, Allen KE, Mason SL, La Thangue NB. Integration of a growth-suppressing BTB/POZ domain protein with the DP component of the E2F transcription factor. EMBO J 1999; 18:212-228.
129. Mancini MA, Shan B, Nickerson JA, et al. The retinoblastoma gene product is a cell cycle-dependent, nuclear matrix-associated protein. Proc Natl Acad Sci USA 1994; 91:418-422.
130. Dyson N. The regulation of E2F by pRB-family proteins. Genes Dev 1998; 12:2245-2262.
131. Kaelin WG, Jr. Functions of the retinoblastoma protein. BioEssays 1999; 21:950-958.
132. Jenkins H, Hölman T, Lyon C, et al. Nuclei that lack a lamina accumulate karyophilic proteins and assemble a nuclear matrix. J Cell Sci 1993; 106:275-285.
133. Goldberg M, Jenkins H, Allen T, et al. *Xenopus* lamin B3 has a direct role in the assembly of a replication competent nucleus: evidence from cell-free egg extracts. J Cell Sci 1995; 108:3451-3461.
134. Spann TP, Moir RD, Goldman AE, et al. Disruption of nuclear lamin organization alters the distribution of replication factors and inhibits DNA synthesis. J Cell Biol 1997; 136:1201-1212.
135. Moir RD, Spann TP, Herrmann H, et al. Disruption of nuclear lamin organization blocks the elongation phase of DNA replication. J Cell Biol 2000; 149:1179-1192.
136. Ellis DJ, Jenkins H, Whitfield WG, et al. GST-lamin fusion proteins act as dominant negative mutants in *Xenopus* egg extract and reveal the function of the lamina in DNA replication. J Cell Sci 1997; 110:2507-2518.
137. Yang L, Guan T, Gerace L. Lamin-binding fragment of LAP2 inhibits increase in nuclear volume during the cell cycle and progression into S phase. J Cell Biol 1997; 139:1077-1087.
138. Chen F, Hersh BM, Conradt B, et al. Translocation of *C. elegans* CED-4 to nuclear membranes during programmed cell death. Science 2000; 287:1485-1489.
139. Buendia B, Santa-Maria A, Courvalin JC. Caspase-dependent proteolysis of integral and peripheral proteins of nuclear membranes and nuclear pore complex proteins during apoptosis. J Cell Sci 1999; 112:1743-1753.
140. Gotzmann J, Vlcek S, Foisner R. Caspase-mediated cleavage of the chromosome-binding domain of lamina-associated polypeptide 2 alpha. J Cell Sci 2000; 113:3769-3780.
141. Rao L, Perez D, White E. Lamin proteolysis facilitates nuclear events during apoptosis. J Cell Biol 1996; 135:1441-1455.
142. Steen RL, Collas P. Mistargeting of B-type lamins at the end of mitosis: implications on cell survival and regulation of lamins A/C expression. J Cell Biol 2001; 153:621-626.
143. Burke B. Lamins and apoptosis: a two-way street? J Cell Biol 2001; 153:F5-7.
144. Guillemin K, Williams T, Krasnow MA. A nuclear lamin is required for cytoplasmic organization and egg polarity in Drosophila. Nat Cell Biol 2001; 3:848-851.
145. Bonne G, Mercuri E, Muchir A, et al. Becane HM, Recan D et al. Clinical and molecular genetic spectrum of autosomal dominant Emery-Dreifuss muscular dystrophy due to mutations of the lamin A/C gene [In Process Citation]. Ann Neurol 2000; 48:170-180.
146. Gant TM, Wilson KL. Nuclear assembly. Annu Rev Cell Dev Biol 1997; 13:669-695.

147. Collas I, Courvalin JC. Sorting nuclear membrane proteins at mitosis. Trends Cell Biol 2000; 10:5-8.
148. Thompson LJ, Bollen M, Fields AP. Identification of protein phosphatase 1 as a mitotic lamin phosphatase. J Biol Chem 1997; 272:29693-29697.
149. Steen RL, Martins SB, Tasken K, et al. Recruitment of protein phosphatase 1 to the nuclear envelope by A-kinase anchoring protein AKAP149 is a prerequisite for nuclear lamina assembly. J Cell Biol 2000; 150:1251-1262.
150. Ellenberg J, Siggia ED, Moreira JE, et al. Nuclear membrane dynamics and reassembly in living cells: targeting of an inner nuclear membrane protein in interphase and mitosis. J Cell Biol 1997; 138:1193-1206.
151. Dabauvalle MC, Muller E, Ewald A, et al. Distribution of emerin during the cell cycle. Eur J Cell Biol 1999; 78:749-756.
152. Buendia B, Courvalin JC. Domain-specific disassembly and reassembly of nuclear membranes during mitosis. Exp Cell Res 1997; 230:133-144.
153. Lourim D, Krohne G. Lamin-dependent nuclear envelope reassembly following mitosis. Trends Cell Biol 1994; 4:324-318.
154. Dabauvalle MC, Loos K, Merkert H, et al. Spontaneous assembly of pore complex-containing membranes ("annulate lamellae") in *Xenopus* egg extract in the absence of chromatin. J Cell Biol 1991; 112:1073-1082.
155. Ulitzur N, Harel A, Feinstein N, et al. Lamin activity is essential for nuclear envelope assembly in a Drosophila embryo cell-free extract. J Cell Biol 1992; 119:17-25.
156. Burke B, Gerace L. A cell free system to study reassembly of the nuclear envelope at the end of mitosis. Cell 1986; 44:639-652.
157. Lopez-Soler RI, Moir RD, Spann TP, et al. A role for nuclear lamins in nuclear envelope assembly. J Cell Biol 2001; 154:61-70.
158. Pyrpasopoulou A, Meier J, Maison C, et al. The lamin B receptor (LBR) provides essential chromatin docking sites at the nuclear envelope. EMBO J 1996; 15:7108-7119.
159. Collas P, Courvalin J-C, Poccia D. Targeting of Membranes to sea urchin sperm chromatin is mediated by an LBR-like integral membrane protein. J Cell Biol 1996; 135:1715-1725.

CHAPTER 5

Role of Ran GTPase in Nuclear Envelope Assembly

Chuanmao Zhang and Paul R. Clarke

Abstract

Ran, a small GTPase that is highly conserved in eukaryotes, plays crucial roles in nuclear structure and function throughout the cell division cycle. During interphase, Ran-GTP is generated in the nucleus by the chromatin-bound guanine nucleotide exchange factor, RCC1. Ran-GTP determines the directionality of nucleocytoplasmic transport by controlling the stability of specific complexes formed between cargo proteins carrying specific targeting signals and transport proteins of the importin/exportin family. During mitosis Ran is dispersed in the cytoplasm, but localised generation of Ran-GTP in the vicinity of chromatin releases proteins required for mitotic spindle assembly from inhibitory complexes with importins. Recent advances in the use of cell-free systems made from *Xenopus* eggs have demonstrated an additional role for Ran in the control of nuclear envelope (NE) assembly. Both generation of Ran-GTP by RCC1 and GTP hydrolysis on Ran stimulated by RanGAP1 are required. In *Xenopus* egg extracts and mammalian cell extracts, Sepharose beads coated with Ran induce the formation of functional NEs containing nuclear pore complexes in the absence of chromatin. Ran induces NE assembly through importin-β and a mutation in importin-β that blocks interaction with nucleoporins containing FxFG repeats does not support NE assembly. However, the target of importin-β and the mechanism by which Ran controls NE assembly is unclear. We propose that Ran plays a central role in coordinating changes in NE structure, nuclear transport and mitotic spindle assembly during the cell division cycle.

Background

Ran GTPase and Its Regulators

Ran, a member of the Ras small GTPase superfamily, is highly conserved in apparently all eukaryotic cells from primitive *Giardia lamblia* to yeast, plants and vertebrates.[1] The activity of Ran is determined both by its localisation and its guanine-nucleotide bound state. During interphase, Ran is concentrated mainly in the nucleus, but during the open mitosis of vertebrate cells in which the nuclear envelope (NE) breaks down, Ran is dispersed throughout the cell. At the end of mitosis, when the NE reforms during telophase, Ran is relocalised to chromatin.[2] Like other GTPases, Ran exists in GTP and GDP bound states that interact differently with its regulators and effectors (reviewed in refs. 3, 4). The intrinsic GTPase activity of Ran is very low, but it is stimulated by the interaction of a GTPase-activating protein (RanGAP1) located at the cytoplasmic face of the nuclear pore and in the cytoplasm. Ran GTPase activity is further stimulated by the predominantly cytoplasmic Ran binding protein 1 (RanBP1) and by Ran-binding domains of Ran binding protein 2 (RanBP2/Nup358), a component of the

Nuclear Envelope Dynamics in Embryos and Somatic Cells

cytoplasmic filaments of the nuclear pore complex (NPC). These activities ensure that the relatively low concentration of Ran in the cytoplasm is predominantly GDP-bound. By contrast, a high concentration of Ran-GTP is generated in the nucleus by RCC1 bound to chromatin (reviewed in refs. 1, 4). Generation of Ran-GTP in the nucleus is also promoted by the nucleotide-destabilising factor Mog1, which may act together with Ran-GTP binding proteins to dissociate GDP from Ran and reloading with GTP.[5]

Role of Ran in Nucleocytoplasmic Transport

Compartmentalisation of the factors that regulate the nucleotide bound state of Ran produces a steep gradient of Ran-GTP concentrations across the NE.[6,7,8,9] One function of the gradient is to determine the directionality of nucleocytoplasmic transport by controlling the assembly and disassembly of complexes formed between transported cargoes and proteins of the importin/exportin family that act as receptors for targeting sequences on the cargo (Fig. 1).[6,10] In the nucleus, interaction of Ran-GTP with the exportin Crm1 (the target of leptomycin B) causes the assembly of complexes with proteins containing a leucine-rich nuclear export signal (NES) and their export from the nucleus through nuclear pores. In the cytoplasm, import complexes are formed between importins and karyophilic cargo. Proteins carrying a lysine-rich nuclear localisation signal (NLS) interact with importin-β through an adaptor, importin-α. Importin-β also interacts directly with some cargos, whereas other importin family members play more specialised roles in the transport of specific proteins that have distinct signal sequences. After translocation through the nuclear pore, import complexes are dissociated in the nucleoplasm by Ran-GTP, which binds importin-β and ejects the cargo. During translocation through the pore, transport cargoes may interact with nuclear pore complex proteins (nucleoporins) in transient interactions controlled by Ran.[11]

Role of Ran in Mitotic Spindle Assembly

Disruption of Ran or its interacting proteins suggested a requirement in cell cycle progression in the genetically amenable yeasts, as well as in cultured mammalian cells.[12] However, it was difficult in these systems to distinguish secondary effects that might have been due to defects in nuclear transport. In 1999, the breakthrough came from a number of groups utilising *Xenopus* egg extracts as a model cell-free system suitable for biochemical analysis.[2,13-16] Changing the balance of Ran-GTP and Ran-GDP in mitotic egg extracts disrupted spindle assembly, a process that occurs in the absence of nuclear compartmentalisation and is therefore distinct from nucleocytoplasmic transport. Ran is present in *Xenopus* egg extracts at a high concentration (1-2 μM) and judging by its interaction with specific binding proteins, is predominantly GDP-bound.[9,17] When the concentration of Ran-GTP is increased by adding the exchange factor RCC1, or Ran mutants (G19V, Q69L or L43E) that are deficient in GTPase activity and are thereby stabilised in the GTP-bound state, microtubule assembly is promoted throughout the extract, resulting in ectopic asters containing typical centrosome-associated proteins such as γ-tubulin, NuMA and Xgrip109.[2,13-16] With incubation, these asters may form spindle-like structures in the absence of chromatin or centrioles, albeit smaller than proper spindles formed following addition of sperm heads.[13,14]

The effect of Ran-GTP on spindle assembly is mediated, at least in part, through the same factors involved in nuclear protein import, namely the importins. Nachury et al showed that depletion of importin-β from extracts using RanQ69L causes widespread microtubule polymerisation that is suppressed by exogenous importin-β.[18] One of the factors inhibited by importin-β is NuMA, a microtubule associated protein (MAP) that is essential for spindle assembly. Similarly, Wiese et al showed that the induction of asters by RanL43E or a fragment of NuMA (NuMA tail II) is suppressed by importin-β.[19] Gruss et al showed that importin-α also inhibits aster formation in mitotic extract and this effect can be overcome by exogenous TPX2, a microtubule-associated protein that targets the motor protein XKlp2 to microtubules.[20] Together, these results suggest that importins play roles in controlling spindle assembly in mitotic *Xenopus* egg extracts by binding to and suppressing the activities of spindle assembly

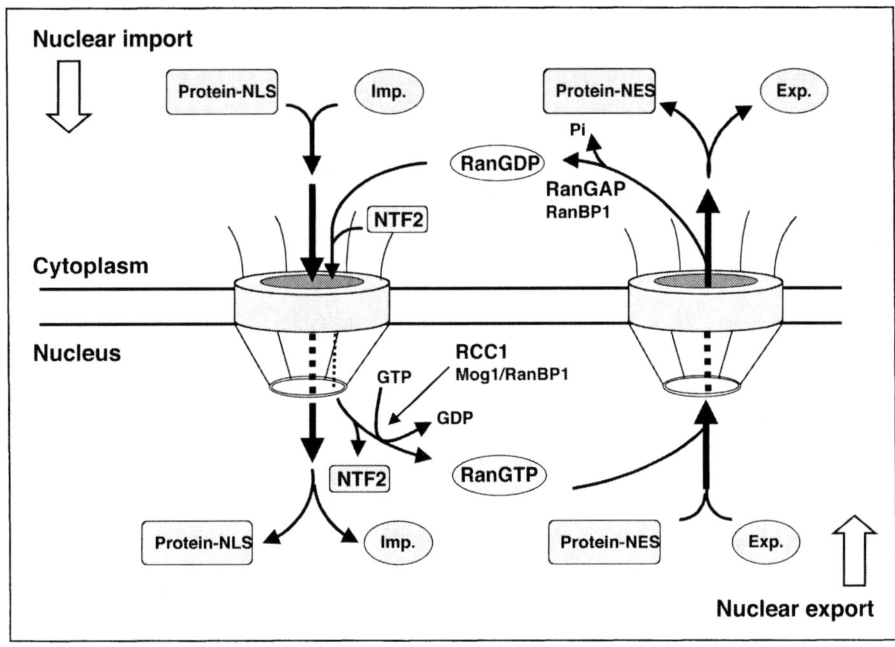

Figure 1. Ran directs nucleocytoplasmic transport. Ran shuttles across the nuclear envelope via the nuclear pores, but is concentrated in the nucleus by active import involving NTF2. In the nucleus, a high concentration of Ran-GTP is generated by guanine nucleotide exchange catalysed by RCC1 bound to chromatin. Mog1, together with RanBP1 or other Ran-GTP binding proteins, promotes nucleotide exchange. Ran-GTP causes the dissociation of imported complexes containing proteins with nuclear localisation signals (NLS) by binding to importin-β (Imp.) and ejecting the cargo. Conversely, binding of Ran-GTP to exportin/Crm1 (Exp.) promotes assembly of export complexes containing proteins with nuclear export signals (NES). In the cytoplasm, Ran-GTP is removed by activation of Ran GTPase activity by RanGAP and RanBP1, and export complexes are dissociated. The importins and exportins are recycled by specific mechanisms back across the pore (not shown). In addition to this basic mechanism, other members of the importin family mediate the transport of specific cargoes.

factors such as NuMA and TPX2 that are present throughout the extracts.[21] Increased Ran-GTP levels dissociate the factors from importins and cause widespread aster formation.

Chromatin has a positional effect on spindle formation by decreasing the catastrophe rate and increasing the rescue frequency of dynamic microtubules, thereby promoting the elongation of spindle microtubules specifically towards the chromatin. Ran-GTP not only stabilises microtubule elongation, but also regulates motor proteins involved in spindle dynamics.[22,23] Generation of Ran-GTP by RCC1 in the locality of chromatin during mitosis could thereby account for the stabilising influence of chromatin on spindles (Fig. 2).[21] The marking of chromatin by local generation of Ran-GTP may be particularly important in an open system such as the *Xenopus* early embryo in which there are no positional constraints on the orientation of the spindle. In this system, a cloud of Ran-GTP is generated around chromatin in mitosis, whereas Ran dispersed in the extract is predominantly GDP bound.[9]

In somatic cells, the role of Ran during mitosis may be more complex. While RCC1 is localised to mitotic chromosomes in mammalian cells (Moore, Zhang and Clarke, unpublished), RanGAP1, modified by SUMO-1, and RanBP2 are present on the mitotic spindle, suggesting that hydrolysis of GTP by Ran may be localised there.[24,25] Consistent with a requirement for GTP hydrolysis by Ran, in the nematode worm *Caenorhabditis elegans*, RanGAP is required for spindle assembly and chromosome positioning.[26,27] Targeting of Ran or

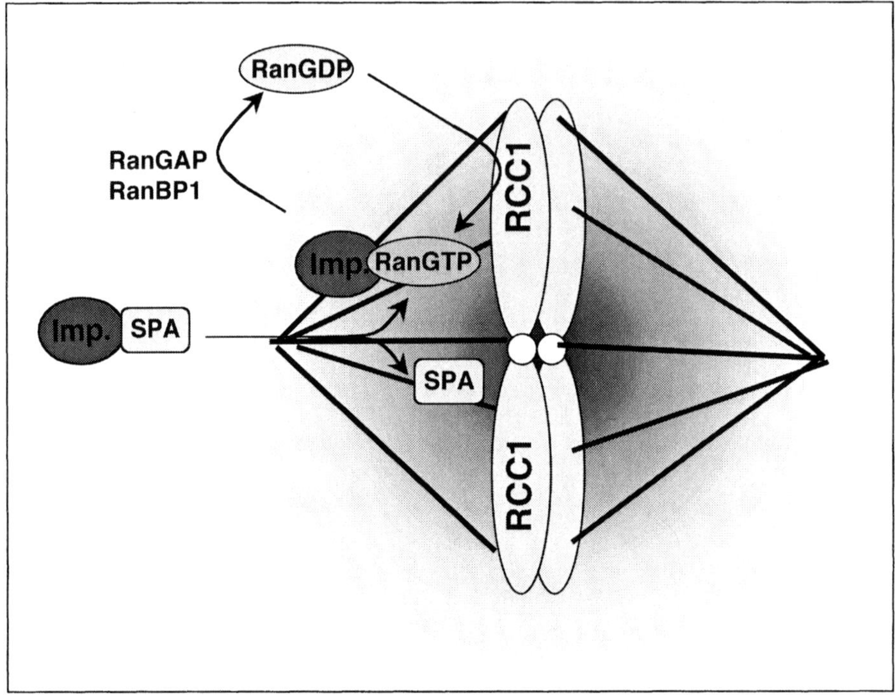

Figure 2. Ran directs mitotic spindle assembly. Ran-GTP releases spindle-promoting activities (SPA) from inhibited complexes with importins to promote mitotic spindle assembly. It is proposed that chromosomal localisation of RCC1 generates Ran-GTP locally, stabilising microtubules in the vicinity of chromatin.

Ran-interacting proteins to the spindle microtubules, centrosomes or kinetochore regions of chromosomes may permit localised activity of Ran.[2,27]

In micro-organisms such as yeast with a closed mitosis, there is no evidence at present of localised generation of Ran-GTP or GTP hydrolysis by Ran. In these cells, spindle assembly and chromosome alignment are contained within the nucleus throughout mitosis. Ran-GTP may not used as a positional marker in yeast, since there are not the same logistical requirements for spindle orientation and chromosome attachment as in a large vertebrate cell. Maintenance of a high Ran-GTP concentration throughout the nucleus may be sufficient.

Control of Nuclear Envelope Assembly by Ran

At the end of mitosis, the mitotic spindle is disassembled and microtubules return to interphase dynamics. In cells with an open mitosis, the NE is re-assembled around chromatin, nuclear pore complexes are reformed, nuclear transport is restarted and the distinct environment of the nucleus is re-established. Recently, it has become apparent that Ran also plays a critical role in the assembly of the NE in cells with an open mitosis, and a function in NE expansion is likely to be conserved in other species such as yeast that have a closed mitosis.

Early Evidence from Xenopus Egg Extracts

Interphase *Xenopus* egg extracts provide a model system to study the assembly of the vertebrate nucleus. Nuclear assembly is usually initiated by the addition of sperm chromatin, which first undergoes decondensation, a process that involves the exchange of basic proteins for histones mediated by nucleoplasmin. Subsequently, membrane vesicles bind to chromatin and fuse to form a double membrane, nuclear pore complexes are assembled and nuclear growth occurs (Fig. 3). The mechanism of NE assembly is not well understood, but vesicle fusion is inhibited by GTPγS, suggesting the involvement of a GTPase, or by N-ethylmaleimide (NEM), a reagent that reacts with thiol groups on proteins.[28] Vesicle fusion may be a prerequisite for nuclear pore complex (NPC) assembly, since GTPγS or NEM prevent NPC formation, while NPC formation can be uncoupled from membrane formation, since the metal cation chelator BAPTA results in the formation of a NE without NPCs.[28,29] The initial recruitment of vesicles to chromatin may involve binding between chromatin or lamins and integral membrane proteins that become constituents of the inner membrane.[30]

In interphase *Xenopus* egg extracts, Ran is required for nuclear assembly from sperm chromatin and the establishment of DNA replication, a process that is disrupted by dominant Ran mutants that are deficient in GTP binding or GTP hydrolysis.[2,17,31,32] Depletion of RCC1, the guanine nucleotide exchange factor for Ran, or addition of excess RanBP1, which opposes RCC1, also prevent proper nuclear assembly.[33-35] RanBP1 addition only disrupts nuclear assembly when added at early in the assembly reaction and does not directly inhibit nuclear protein import.[34] Together, these results suggested that both generation of Ran-GTP from Ran-GDP by RCC1 and GTP hydrolysis by Ran are required during an early stage of nuclear assembly. Using Field Emission In–lens Scanning Electron Microscopy (FEISEM) in collaboration with M.W. Goldberg and T.D. Allen (Paterson Institute, Manchester) we found that excess RanBP1 resulted in defective NEs that were highly convoluted, suggesting that Ran and RCC1 might play a role in NE assembly.[34]

For us, the first clear indication of a role for Ran in NE assembly came from FEISEM studies in which it was apparent that addition of Ran-GDP promoted NE assembly, whereas RanQ69L, locked in the GTP-bound form, inhibited the fusion of membrane vesicles on the surface of chromatin (Zhang, Goldberg, Allen & Clarke, unpublished). By contrast, RanT24N, which fails to bind GTP and inhibits RCC1, caused the formation of small, rounded up nuclei with highly convoluted NEs very similar to those formed when RanBP1 is added (Fig. 3). These results suggested that both generation of Ran-GTP by RCC1 and GTP hydrolysis by Ran play a role in NE assembly. We also found that Ran is concentrated on sperm chromatin,[2] suggesting that concentration of Ran might be involved in the induction of the process. However, manipulation of the Ran system caused significant changes in the morphology of chromatin in these experiments, so it remained possible that the primary effect of Ran was on chromatin structure rather than a direct role in vesicle attachment and fusion.

NE Assembly Around Ran Beads

To test the hypothesis that concentration of Ran might play a direct role in NE formation, we examined whether Ran expressed as a fusion with glutathione-S-transferase (GST) and coupled to glutathione-Sepharose beads could induce NE assembly in the absence of chromatin. Indeed, the beads rapidly accumulated membrane vesicles that fused to form a continuous lipid layer that incorporated nucleoporins.[36] Using electron microscopy, nuclear pore complexes were apparent crossing a double membrane, indicating that a complete NE was assembled. The envelopes were functional, since a fluorescent dextran that is too large to diffuse across the nuclear pores was excluded, whereas a karyophilic protein containing an NLS was concentrated within the beads. In other words, simply concentrating Ran on the surface of beads was sufficient to induce membrane vesicle binding and fusion, as well as the assembly of NPCs and the subsequent initiation of nucleocytoplasmic transport. In contrast to Ran, beads coated with the mutants RanT24N or RanQ69L failed to make intact NEs, indicating that both loading of Ran with GTP and GTP hydrolysis are required for envelope assembly around Ran beads.

The ability of Ran concentrated on the surface of beads to induce the formation of NEs is not restricted to *Xenopus* egg extracts, since extracts prepared from mitotic vertebrate cells also work.[37] Using human cell extracts, we were able to show that if beads coated with Ran-GDP are used, then RCC1 is required to generate Ran-GTP, whereas beads coated with Ran-GTP

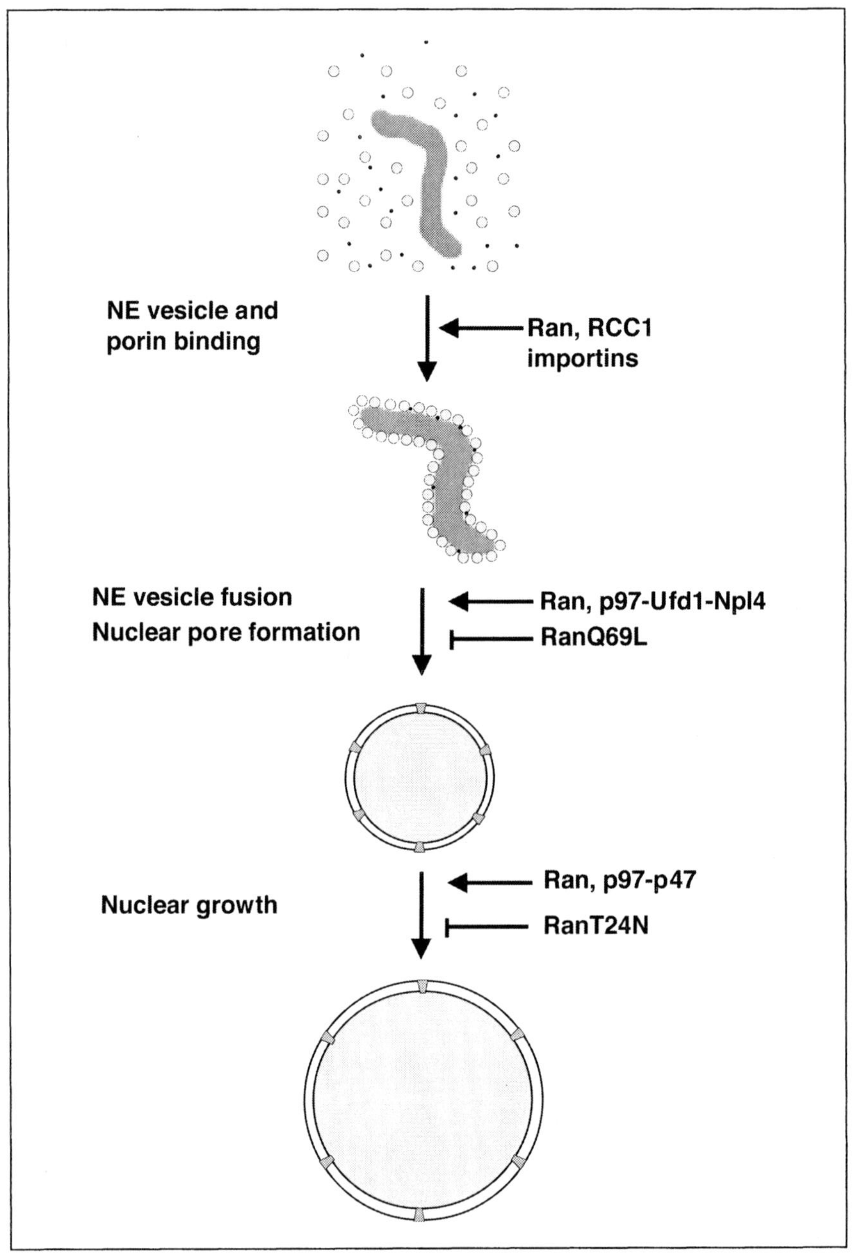

Figure 3. Involvement of Ran at several stages of NE formation around sperm chromatin in *Xenopus* egg extracts. Ran is required for vesicle recruitment to chromatin, fusion to form an intact NE and the assembly of nuclear pore complexes. Ran acts in vesicle recruitment through importin-β which is proposed to bind to a component of the vesicles. Ran may bind to chromatin prior to vesicle recruitment and attracts RCC1, thereby generating Ran-GTP locally on the chromatin and initiating the process. The AAA-ATPase complexes p97-Ufd1-Npl4 and p97-p47 are required at distinct stages of NE formation and expansion. The relationship between p97 complexes and Ran is unknown.

do not require RCC1. Conversely, if RanGAP is inhibited with an antibody, then vesicle binding and fusion are reduced, showing that GTP hydrolysis on Ran is involved.

This work showed that concentration of Ran is sufficient to induce complete NE assembly, and that this process did not require the presence of chromatin. This is consistent with studies in which stacks of NE-like membranes containing pore complexes, called annulate lamellae, can be formed in *Xenopus* egg extracts.[38] Indeed, increased concentrations of Ran induce annulate lamellae formation in this system (Zhang and Clarke, unpublished).

A Novel Assay for NE Assembly Around Chromatin Demonstrates a Role for Ran

Hetzer et al[39] addressed the possible role of Ran in NE assembly using an assay in which chromatin was first decondensed by treatment with a heat stable fraction of extract containing nucleoplasm, then vesicles labelled with two different coloured lipophilic dyes were incubated with the decondensed chromatin with the addition of a soluble extract fraction. Membrane formation was assayed by the mixing of the two colours as fusion between the two vesicle populations occurred. Hetzer et al found that addition of RanQ69L or RanT24N inhibited vesicle fusion in this assay. When extracts were depleted of RCC1 using RanT24N, exogenous RCC1 or Ran-GTP, but not Ran-GDP, were able to overcome the inhibition of vesicle fusion around chromatin assembled on λ DNA, indicating that generation of Ran-GTP by RCC1 is required. However, Ran bound with the nonhydrolysable GTP analogue GTPγS failed to rescue RCC1 depletion, indicating that GTP hydrolysis by Ran is also required. Immunodepletion experiments also showed that Ran itself is essential for membrane fusion and nucleoporin incorporation.

The AAA-ATPase p97 is Required for NE Formation

Hetzer et al[40] have developed their assay to characterise the role of additional factors in membrane fusion. An AAA-ATPase, p97, plays a role in endoplasmic reticulum (ER) membrane fusion in complex with an adaptor protein p47 that interacts with the t-SNARE syntaxin. Hetzer et al found that p97-p47 also plays a role in the expansion of the NE when the nucleus grows after the initial enclosure of the NE. p97, in a different complex with Ufd1 and Npl4, is also required at an earlier stage of NE formation when membrane vesicles form membrane structures reminiscent of ER tubules, then fuse to make a complete NE. This mechanism is likely to be conserved in other eukaryotes, since the homologue of p97 in *Saccharomyces cerevisiae*, Cdc48p, plays a role in ER fusion[41] and Ndl4p is involved in nuclear protein transport and maintenance of nuclear envelope structure[42]. The relationship between the role of the Ran system in NE assembly and the function of p97 complexes will be of interest.

A Role for Importin-β in NE Assembly

One immediate question has been: What are the targets of Ran during NE assembly? To investigate the possible role of Ran-interacting proteins such as importin-β, we used an assay in which NE assembly around Sepharose beads coated with Ran was studied in *Xenopus* egg extracts depleted of Ran-binding proteins (ΔRanBP extracts) using RanQ69L, a mutant defective in GTPase activity and therefore locked in the GTP-bound form.[43] ΔRanBP extracts were deficient in ability to promote membrane vesicle recruitment and fusion to form continuous membranes around Ran beads, but NE assembly was restored by addition of 5 μM importin-β, a concentration similar to that of the endogenous protein in nondepleted extracts. By contrast, other related import factors were ineffective.

A truncated protein (importin$^{\beta1\text{-}409}$) that lacks importin-α binding activity[44] also restored NE assembly around Ran-beads in ΔRanBP extracts, indicating that importin-β does not function in NE assembly by interaction through importin-α with karyophilic proteins carrying Lys-rich NLS motifs. By contrast, importin$^{\beta45\text{-}462}$, which lacks the Ran-binding region was not functional. These results suggest that importin-β acts as an adaptor that recruits target proteins to Ran to initiate NE formation.

In addition to transported cargoes and Ran, importin-β interacts directly with protein components of the nuclear pore (nucleoporins) containing FxFG (Phe-x-Phe-Gly, where x is usually Ser, Gly or Ala) repeats. An importin-β mutant in which Ile178 is changed to Asp (I178D), which decreases the affinity of importin-β for FxFG nucleoporins but not nucleoporins containing GLFG (Gly-Leu-Phe-Gly) repeats or transport cargoes.[45] This mutation inhibited the ability of importin-β to promote NE assembly around Ran beads, suggesting that importin-β acts by recruiting FxFG nucleoporins to Ran. Furthermore, beads coated with importin-β proteins also appear to form continuous envelopes around them, and this activity is again abolished by the I178D mutation. Pretreatment of importin-β with N-ethylmaleimide (NEM), which reacts with cysteine residues on importin-β and inactivates it,[46] abolished the ability to induce NE formation. Importin-β may therefore account, at least in part, for the NEM-sensitivity of NE assembly.

These results indicate that importin-β plays a role during NE assembly induced by Ran through interaction with FxFG domains on nucleoporins. However, the target(s) of importin-β remain to be identified and there may be more than one type of interaction between Ran, importin-β and nucleoporins. The role of the putative importin-β interacting protein will need to be tested in vesicle recruitment to chromatin.

Is the Function of Ran in NE Structure Common to All Eukaryotic Cells?

The high degree of similarity of components of the Ran system amongst eukaryotes suggests that control of NE formation by Ran is likely to have been conserved during evolution. Ran beads assemble functional NEs in human somatic cell extracts[37] as in *Xenopus* egg extracts[36], showing conservation of function between vertebrates. In the nematode *C. elegans*, inhibition of Ran, RanGAP or RanBP1 expression by RNAi treatment of embryos disrupts spindle structure and prevents proper nuclear assembly following mitosis.[26] Bamba et al[27] have further characterised the role of Ran in *C. elegans*, showing that Ran localises to kinetochores during metaphase/anaphase, relocalising to the periphery of the reforming nucleus at telophase. Depletion of Ran by RNAi treatment prevents NE assembly, as well as causing defects in mitotic spindle positioning.[27] In *Drosophila*, importin-β plays a role in NE integrity: injection of a dominant mutant of importin-β blocks NE assembly in cleavage stage embryos.[47]

Ran may also be required for NE integrity even in organisms that have a closed mitosis in which the NE is not disassembled. In the fission yeast *Schizosaccahromyces pombe*, a temperature sensitive mutation of the RCC1 homologue Pim1p produces fragmentation of the NE at exit from mitosis, although the NE normally remains intact during mitosis in this species[48] Generation of Ran-GTP by Pim1p may be required at this point in the cell cycle because additional NE needs to be formed during nuclear division. It can be envisaged that the expansion of the NE during mitosis in yeast and the formation of the NE at the end of an open mitosis are closely related processes that are likely to both involve the fusion of dispersed vesicles and/or the feeding in of new membranes to the NE via the ER. Thus, Ran is likely be play a role in NE formation in all eukaryotes.

A Model for the Role of Ran in NE Assembly

Using evidence derived primarily from studies using *Xenopus* egg extracts, a model for the role of Ran in multiple stages during NE assembly can be proposed (Fig. 4) Ran is concentrated on chromatin prior to NE assembly by an unknown mechanism and recruits RCC1 which generates Ran-GTP locally. Recruitment of RCC1 to chromatin may be a specialized mechanism to initiate NE assembly following fertilization of the egg, whereas in somatic cells, RCC1 may be present on chromatin throughout mitosis. Ran-GTP recruits vesicles to the surface of chromatin, acting through importin-β. Vesicle fusion requires further activity of Ran and involves the ATPase complex, p97-Ufd1-Npl4. NPC assembly and the initiation of nuclear transport are also promoted by Ran. NE expansion involves p97-p47.

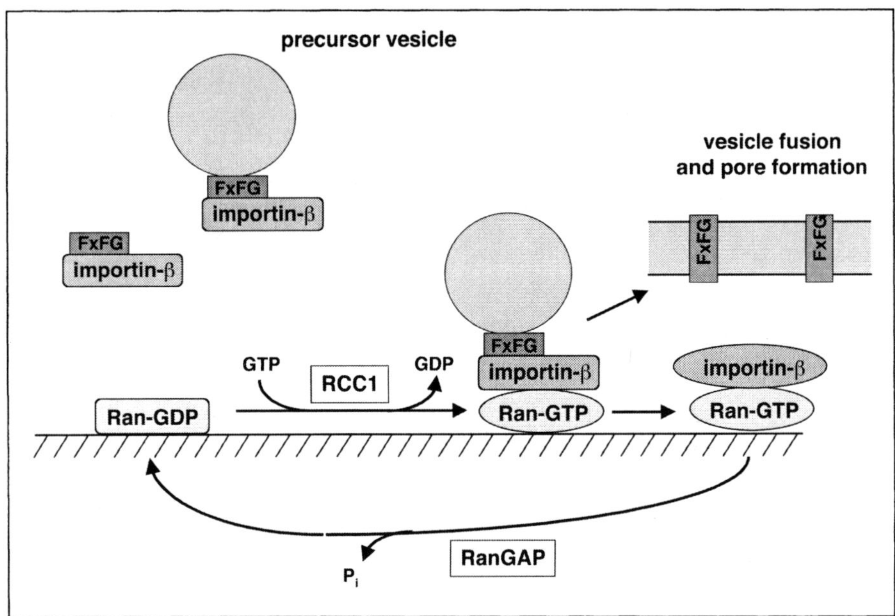

Figure 4. A model for the role of Ran in vesicle recruitment during NE assembly Ran-GTP is generated at the surface of chromatin by RCC1. Ran-GTP recruits importin-β complexes with FxFG nucleoporins present on vesicles or as soluble proteins. Dissociation of importin-β from nucleoporins released the proteins locally, permitting complex assembly and vesicle fusion to form an intact membrane. Ran-GTP/importin-β complexes are recycled by GTP-hydrolysis promoted by RanGAP.

Ran-GTP may recruit membrane vesicles to chromatin through binding to target proteins such as nucleoporins through importin-β. In one model, a transient complex would be formed between Ran-GTP, importin-β and a nucleoporin, then the nucleoporin would be released locally to promote assembly of a precursor complex. GTP hydrolysis by Ran would release importin-β, but may also play a role in vesicle fusion (Fig. 4).

In vertebrate cells undergoing mitosis, the majority of Ran molecules are excluded from the chromosomes and dispersed into the cytoplasm.[2,49] Relocalisation of Ran to chromatin at the end of mitosis may co-ordinate the initiation of NE assembly with disassembly of the mitotic spindle. The function of Ran in this transition is likely to be coupled to changes in the activity of cyclin-dependent protein kinases and other activities that control the progression of the cell cycle. It remains to be seen if regulation of the activity of components of the Ran system plays an active role in changes in microtubule dynamics and NE structure during mitosis.

References

1. Clarke PR, Zhang C. Ran GTPase: a master regulator of nuclear structure and function during the eukaryotic cell division cycle? Trends Cell Biol 2001; 11:366-371.
2. Zhang C, Hughes M, Clarke PR. Ran-GTP stabilises microtubule asters and inhibits nuclear assembly in *Xenopus* egg extracts. J Cell Sci 1999; 112:2453-2461.
3. Avis JA, Clarke PR. Ran, a GTPase involved in nuclear processes: its regulators and effectors. J Cell Sci 1996; 109:2423-2427.
4. Künzler M, Hurt E. Targeting of Ran: variation on a common theme? J Cell Sci 2001; 114:3233-3241.
5. Nicolás F, Moore WJ, Zhang C et al. XMog1, a nuclear Ran-binding protein in *Xenopus*, is a functional homologue of *Schizosaccharomyces pombe* Mog1p that co-operates with RanBP1 to control generation of Ran-GTP. J Cell Science 2001; 114:3013-3023.

6. Görlich D, Kutay U. Transport between the cell nucleus and the cytoplasm. Annu Rev Cell Dev Biol 1999; 15:607-660.
7. Nachury MV, Weis K. The direction of transport through the nuclear pore can be inverted. Proc Natl Acad Sci USA 1999; 96:9622-9627.
8. Smith AE, Slepchenko BM, Schaff JC et al. Systems analysis of Ran transport. Science 2002; 295:488-491.
9. Kalab P, Weis K, Heald R. Visualization of a Ran-GTP gradient in interphase and mitotic *Xenopus* egg extracts. Science 2002; 295:2452-2456.
10. Kuersten S, Ohno M, Mattaj IW. Nucleocytoplasmic transport: Ran, beta and beyond. Trends Cell Biol 2001; 11:497-503.
11. Stewart M, Baker RP, Bayliss R et al. Molecular mechanism of translocation through nuclear pore complexes during nuclear protein import. FEBS Lett 2001; 498:145-149.
12. Sazer S. The search for the primary function of the Ran GTPase continues. Trends Cell Biol 1996; 6:81-85.
13. Carazo-Salas RE, Guarguaglini G, Gruss OJ et al. Generation of GTP-bound Ran by RCC1 is required for chromatin-induced mitotic spindle formation. Nature 1999; 400:178-181.
14. Kalab P, Pu RT, Dasso M. The Ran GTPase regulates mitotic spindle assembly. Curr Biol 1999; 9:481-484.
15. Ohba T, Nakamura M, Nishitani H et al. Self-organisation of microtubule asters induced in *Xenopus* egg extracts by GTP-bound Ran. Science 1999; 284:1356-1359.
16. Wilde A, Zheng Y. Stimulation of microtubule aster formation and spindle assembly by the small GTPase Ran. Science 1999; 284:1362-1365.
17. Hughes M, Zhang C, Avis JM et al. The role of Ran GTPase in nuclear assembly and DNA replication: Characterisation of the effects of Ran mutants. J Cell Sci 1998; 111:3017-3026.
18. Nachury MV, Maresca TJ, Salmon WC et al Importin β is a mitotic target of the small GTPase Ran in spindle assembly. Cell 2001; 104:95-106.
19. Wiese C, Wilde A, Moore MS et al. Role of importin-β in coupling Ran to downstream targets in microtubule assembly. Science 2001; 291:653-656.
20. Gruss OJ et al. Ran induces spindle assembly by reversing the inhibitory effect of importin α on TPX2 activity. Cell 2001; 104:83-93.
21. Dasso M. Running on Ran: Nuclear transport and the mitotic spindle. Cell 2001; 104:321-324.
22. Carazo-Salas RE, Gruss OL, I.W. M, Karsenti E. Ran-GTP coordinates regulation of microtubule nucleation and dynamics during mitotic-spindle assembly. Nature Cell Biol 2001; 3:228-234.
23. Wilde A, Lizarraga SB, Zhang L et al. Ran stimulates spindle assembly by altering microtubule dynamics and the balance of motor activities. Nature Cell Biol 2001; 3:221-227.
24. Matunis MJ, Coutavas E, Blobel G. A novel ubiquitin-like modification modulates the partitioning of the Ran-GTPase-activating protein RanGAP1 between the cytosol and the nuclear pore complex. J Cell Biol 1996; 135:1457-1470.
25. Joseph J, Tan SH, Karpova TS et al. SUMO-1 targets RanGAP1 to kinetochores and mitotic spindles. J Cell Biol 2002; 156:595-602.
26. Gönczy P, Echeverri G., Oegema K et al. Functional genomic analysis of cell division in *C. elegans* using RNAi of genes on chromosome III. Nature 2000; 408:331-336.
27. Bamba C, Bobinnec Y, Fukuda M et al. The GTPase Ran regulates chromosome positioning and nuclear envelope assembly in vivo. Curr Biol 2002; 12:503-507.
28. Macaulay C, Forbes DJ. Assembly of the nuclear pore: biochemically distinct steps revealed with NEM, GTP gamma S, and BAPTA. J Cell Biol 1996; 132:5-20.
29. Goldberg MW, Wiese C, Allen TD et al. Dimples, pores, star-rings, and thin rings on growing nuclear envelopes: evidence for structural intermediates in nuclear pore complex assembly. J Cell Sci 1997; 110:409-420.
30. Vasu SK, Forbes DJ. Nuclear pores and nuclear assembly. Curr Opin Cell Biol 2001; 13:363-375.
31. Dasso M, Seki T, Azuma Y et al. A mutant form of the Ran/TC4 protein disrupts nuclear function in *Xenopus* laevis egg extracts by inhibiting the RCC1 protein, a regulator of chromosome condensation. EMBO J 1994; 13:5732-5744.
32. Kornbluth S, Dasso M, Newport J. Evidence for a dual role for TC4 protein in regulating nuclear structure and cell cycle progression. J Cell Biol 1994; 125:705-719.
33. Dasso M, Nishitani H, Kornbluth S et al. RCC1, a regulator of mitosis, is essential for DNA replication. Mol Cell Biol 1992; 12:3337-3345.
34. Nicolás FJ, Zhang C, Hughes M et al. *Xenopus* Ran-binding protein 1: Molecular interactions and effects on nuclear assembly in *Xenopus* egg extracts. J Cell Sci 1997; 110:3019-3030.
35. Pu RT, Dasso M. The balance of RanBP1 and RCC1 is critical for nuclear assembly and nuclear transport. Mol Biol Cell 1997; 8:1955-1970.

36. Zhang C, Clarke PR. Chromatin-independent nuclear envelope assembly induced by Ran GTPase in *Xenopus* egg extracts. Science 2000; 288:1429-1432.
37. Zhang C, Clarke PR. The roles of Ran-GTP and Ran-GDP in precursor vesicle recruitment and fusion during nuclear envelope assembly in a human cell-free system. Curr. Biol 2001; 11:208-212.
38. Dabauvalle MC, Loos K, Merkert H et al. Spontaneous assembly of pore complex-containing membranes (annulate lamellae) in *Xenopus* egg extract in the absence of chromatin. J Cell Biol 1993; 112:1073-1082.
39. Hetzer M, Bilbao-Cortés D, Walter TC et al. GTP hydrolysis by Ran is required for nuclear envelope assembly. Mol Cell 2000; 5:1013-1024.
40. Hetzer M, Meyer HH, Walther TC et al. Distinct AAA-ATPase p97 complexes function in discrete steps of nuclear assembly. Nat Cell Biol 2001; 3:1086-1091.
41. Latterich M, Frohlich KU, Schekman R. Membrane fusion and the cell cycle: Cdc48p participates in the fusion of ER membranes. Cell 1995; 82:885-893.
42. DeHoratius C, Silver PA. Nuclear transport defects and nuclear envelope alterations are associated with mutation of the *Saccharomyces cerevisiae* NPL4 gene. Mol Biol Cell 1996; 7:1835-1855.
43. Zhang C, Hutchins JRA, Mühlhäusser P et al. Role of importin-β in the control of nuclear envelope assembly by Ran. Curr Biol 2002; 12:498-502.
44. Kutay U, Izaurralde E, Bischoff FR et al. Dominant-negative mutants of importin-beta block multiple pathways of import and export through the nuclear pore complex. EMBO J 1997; 16:1153-1163.
45. Bayliss R, Littlewood T, Stewart M. Structural basis for the interaction between FxFG nucleoporin repeats and importin-beta in nuclear trafficking. Cell 2000; 102(1):99-108.
46. Chi NC, Adam SA. Functional domains in nuclear import factor p97 for binding the nuclear localization sequence receptor and the nuclear pore. Mol Biol Cell 1997; 8:945-956.
47. Timinszky G et al. The importin-beta P446L dominant-negative mutant protein loses RanGTP binding ability and blocks the formation of intact nuclear envelope. J Cell Sci 2002; 115:1675-1687.
48. Demeter J, Morphew M, Sazer S. A mutation in the RCC1-related protein pim1 results in nuclear envelope fragmentation in fission yeast. Proc Natl Acad Sci USA 1995; 92:1436-1440.
49. Ren M, Drivas G, D'Eustachio P et al. Ran/TC4: A small nuclear GTP-binding protein that regulates DNA synthesis. J Cell Biol 1993; 120:313-323.

CHAPTER 6

Mitotic Control of Nuclear Pore Complex Assembly

Khaldon Bodoor and Brian Burke

Introduction

The interface between the nucleus and cytoplasm is defined by the nuclear envelope (NE) (for reviews see refs. 1 and 2), a selective barrier that plays an essential role in the maintenance of the unique biochemical identities of these two compartments. The most prominent features of the NE are the inner and outer nuclear membranes (INM and ONM) which together enclose the perinuclear space (PNS). While the ONM is biochemically and functionally similar to the endoplasmic reticulum (ER), the INM contains a characteristic set of integral membrane proteins and is closely associated with the underlying chromatin. Annular junctions between the INM and ONM create aqueous channels connecting the nucleus and cytoplasm. These channels, or nuclear pores, contain nuclear pore complexes (NPCs), extremely elaborate multi-subunit structures that regulate the trafficking of macromolecules across the NE (reviewed in ref. 3). Mammalian cells, such as fibroblasts and heptaocytes, will typically contain 2000-4000 NPCs.[2] The annular junctions between the ONM and INM form the pore membrane domain (POM), highly curved regions of the nuclear membranes that are enriched in several NPC-specific membrane proteins. Notwithstanding their biochemical and functional differences, the INM, ONM, POM and ER form a single continuous membrane system with the PNS representing an extension of the ER lumen.

The Nuclear Lamina

The final major structural element of the NE is the nuclear lamina.[2,4] This is a thin (~20nm in mammalian somatic cells) yet insoluble protein meshwork that lines the nuclear face of the INM and which maintains extensive interactions with INM proteins, NPCs and chromatin components. The lamina is composed primarily of type V intermediate filament family members known as A- and B-type nuclear lamins (reviewed in ref. 5). While B-type lamins are found in all cell types, A-type lamins are absent from cells of the early embryo.[6,7] In the mouse, A-type lamins do not appear until about midway through gestation. Although evidently not essential proteins, defects in A-type lamins have been linked to several human diseases.[8]

The Inner Nuclear Membrane

So far about a dozen integral proteins have been localized to the inner nuclear membrane. These proteins include lamina-associated polypeptide-1 (LAP1) and LAP2 family members, lamin B receptor (LBR), emerin, nurim, and MAN1.[9-14] With the exception of nurim, all of these proteins are arranged with their amino-terminal domains facing the nucleoplasm and bind to nuclear lamins and/or chromatin (for a review see ref. 15). It is generally agreed that localization of integral proteins to the INM involves a mechanism of selective retention. Membrane proteins that are mobile within the ER membrane are able to access the INM, probably

Nuclear Envelope Dynamics in Embryos and Somatic Cells
edited by Philippe Collas. ©2002 Eurekah.com and Kluwer Academic/Plenum Publishers.

via the POM at the periphery of each NPC. However, only those that can interact with nuclear components such as lamins or chromatin are retained within the INM. The only exception to this rule is nurim, a small polytopic membrane protein the bulk of which resides within the lipid bilayer. It is likely that nurim will be found to bind to the transmembrane segment of another INM protein(s). Binding of INM proteins to chromatin may be mediated by interactions with several chromatin proteins, including HP1, BAF (for barrier to autointegration factor), and HA95.[16,17] Nuclear lamins may also function in this capacity since they are capable of binding both INM proteins and chromatin.

Nuclear Pore Complexes

High resolution EM analyses have provided a consensus view of NPC organization.[18,19] The central framework of the NPC consists of a massive symmetrical structure lying at the level of the nuclear membranes. This framework is composed of eight multi-subunit spokes sandwiched between apparently identical cytoplasmic and nuclear rings.[20-22] A segment of each spoke extends across the POM into the PNS and in this way, functions as a linker between nuclear membranes and the NPC. Consistent with this, the large N-terminal domain of the NPC membrane protein gp210 (see below) has been localized to the luminal regions of these spokes.[23,24]

The central framework or ring-spoke complex embraces a central channel that mediates transport of macromolecules up to 26nm in diameter, provided that they contain an appropriate import/export signal.[20,25,26] The precise features of this central channel, and the mechanisms by which signal bearing molecules and their cognate receptors traverse it have yet to be firmly established. Additional peripheral structures are associated with both nuclear and cytoplasmic faces of the ring-spoke complex.[27] Eight flexible filaments, containing docking sites for import substrates, extend about 100nm into the cytoplasm.[28,29] On the nuclear face of the NPC eight filaments ar attached proximally to the ring-spoke complex. They are joined at their distal ends by a 30-50nm ring forming a basket-like structure that extends about 50-100nm in to the nucleoplasm.[27,30] Like the cytoplasmic filaments, the nuclear basket contains docking sites for transport substrates.[31]

Biophysical measurements indicate the mass of amphibian oocyte NPCs to be approximately 125 MDa, about 30 times that of a eukaryotic ribosome.[32] Based partly on this figure it has been suggested that the vertebrate NPC may be composed of multiple copies of 50 or more distinct protein subunits (nucleoporins or nups).[33] Consistent with this estimate, the yeast NPC, with a mass of about 66MDa, contains approximately 30 different nucleoporins.[34] To date, 21 vertebrate nucleoporins have been characterized in detail.[35,36] Assuming a high copy number (i.e., 16 or 32 copies/NPC), these could account for about 50% of the NPC mass. The properties of these vertebrate nucleoporins are summarized in Table 1.

A feature shared by several nucleoporins is the presence of FG repeats (FXFG, GLFG, and FG; single letter code. "X" is any amino acid with a small or polar side chain).[37-41] These repeats are considered to play an essential role in nucleocytoplasmic transport since they mediate binding to import and export receptors.[42] A subset of the FG nups is asymmetrically localized within the NPC. For example, Nup358 and Nup214 are localized to the cytoplasmic filaments[38-40,43] while Nup153 and Nup98 are components of the nuclear basket.[41,44,45] Another group of FG nups which comprise the Nup62 complex (containing Nup62, Nup58, Nup54 and Nup45) localizes to the central channel region of the NPC[46,47] and is accessible from both the nuclear and cytoplasmic faces. The distribution of FG- containing nups across the length of the NPC, combined with their direct binding to transport receptors, have suggested models in which cargo molecules could be transported across the NPC via a series of association and dissociation reactions (for a model see ref. 48). Other equally compelling models based upon Brownian exclusion[34] or phase partitioning[49] have also been proposed.

Table 1. The properties of vertebrate Nucleoporins

Nup	Location	Motifs	Timing of Reassembly	Comments During Mitosis
Nup62	Central channel	FXFG	Early telophase. After membrane recruitment. Before Nup214.	Complex with Nup58, Nup54, and Nup45. Modified with O-GluNAc (46, 47, 60, 90)
Nup58	Central channel	FXFG	Early telophase. After	Complex with Nup62 membrane recruitment. (46, 47, 60)
Nup45	Central channel	FXFG	Early telophase. After membrane recruitment.	Complex with Nup62. Splice variant of Nup58 (46, 47, 60)
Nup54	Central channel	FXFG	Early telophase. After	Complex with Nup62 membrane recruitment. (46, 47, 60)
Nup50	Nuclear	FXFG	Dependent upon Nup153?	Complex with Nup153 (84, 85)
Nup84 (88)	Cytoplasmic	Coiled-coil C-terminal domain.	Presumably mid-late telophase. Concomitant with Nup214. After p62 complex.	Complex with Nup214 in interphase and mitosis. Phosphorylated in mitosis. (43, 91)
Nup93	Nuclear		Dependent upon Nup153.	Complex with Nup205, Nup188, and Nup62 complex (76)
Nup96	Nuclear	GLFG	Early-mid-anaphase? Concomitant with Nup133	Complex with Nup107, Nup133, Nup160. Derived from p186 (33, 80)
Nup98	Nuclear	GLFG	Mid-late anaphase. Dependent upon Nup153.	Derived from p186, complex with Tpr (33, 45)
Nup107	Nuclear and cytoplasmic	Leucine zipper	Early-mid-anaphase? Concomitant with Nup 133.	Complex with Nup96, Nup133, Nup160, sec13 Binds to kinetochores in mitosis (58, 78, 80)
POM121	Pore membrane domain	FXFG, transmembrane domain	Early-mid anaphase.	(87)
Nup133	Nuclear ?		Early-mid anaphase.	Complex with Nup107, Nup96, Nup160, Sec13 (58, 80)
Nup153	Nuclear basket	FXFG, 4 zinc fingers	Early-mid anaphase.	Complex with Nup50 (41, 44)
Nup155	Nuclear and cytoplasmic		Telophase?	(92)
Nup160	Nuclear		Early-mid-anaphase? Concomitant with Nup 133.	Complex with Nup133, Nup107, Nup96, sec13 (80)

continued on next page

Table 1. Cont.

Nup	Location	Motifs	Timing of Reassembly	Comments During Mitosis
Nup160	Nuclear		Early-mid-anaphase? Concomitant with Nup 133.	Complex with Nup133, Nup107, Nup96, sec13 (80)
Nup188	Nuclear		Dependent upon Nup 153?	Complex with Nup205 and Nup93 (93)
Nup205	Nuclear		Dependent upon Nup153?	Complex with Nup93 (76)
Gp210	Pore membrane domain	Transmembrane N-glycosylation sites	Late telophase/early G1. After Nup214.	N-glycosylated (24, 86)
Nup214	Cytoplasmic filaments	FXFG	Mid-late telophase. After Nup62 complex. Efficient assembly into NPC requires Nup98.	Complex with Nup84 Modified with O-GluNAc (38, 94)
Tpr (265 kDa)	Nuclear basket/ filaments	Large-coiled coil domain	Early G1.	Associated with Nup98 (95, 96)
Nup358	Cytoplsmic fibrils	FXFG, zinc fingers	Early telophase.	(39, 40)

Dynamics of the Nuclear Envelope During Mitosis

Progression through mitosis requires that condensed chromosomes within the nucleus gain access to the microtubules of the mitotic spindle. In higher cells this is accomplished by the breakdown of the NE and is the hallmark of an "open" mitosis. In vertebrates, NE breakdown (NEB) involves all of the major elements of the NE. The lamina depolymerizes to yield soluble A-type lamins and membrane-associated B-type lamins, and NPCs are disassembled (for a review see ref. 1). In mammalian somatic cells, nuclear membrane proteins, including those of the INM and NPC, become mobile and disperse, apparently uniformly, throughout the peripheral ER. All of these disassembled and dispersed components are subsequently used in the reformation of a new NE in each daughter cell.

Nuclear Envelope Breakdown

Events leading to NEB are initiated during prophase and involve the depolymerization and detachment of the nuclear lamina and gradual disassembly of NPCs.[50] Subsequent dynein and microtubule-dependent disruption of the nuclear membranes defines the start of prometaphase.[51,52] By metaphase, when the condensed chromatids are finally aligned along the spindle equator, NE breakdown and dispersal is complete. In vitro studies of NEB in *Xenopus* egg extracts reveals that NPC disassembly is an ordered process.[53] It begins with the disappearance of the cytoplasmic filaments, followed by the cytoplasmic ring, and results in the progressive unmasking other internal NPC structures. This ultimately leads to the transient appearance of small pores or fenstrae within the nuclear membranes. NPC disassembly is completed prior to any visible changes in nuclear membrane organization, suggesting that these are two distinct processes. Ultrastructural analyses of *Drosophila* early embryos reinforce the view that NPC disassembly (and subsequent reassembly) proceeds via a series of bona fide structural intermediates.[54]

NEB is regulated in large part by the cell cycle-dependent phosphorylation of NE components[1] including lamins, INM proteins and nucleoporins. In the case of the latter, Nup358, Nup214, Nup153, Nup98, and Nup84 have all been shown to be hyperphosphorylated during mitosis.[55-57] The pore membrane protein gp210 also undergoes mitosis-specific phosphorylation[56] (will be discussed in detail later). NPC disassembly does not, however, go to completion. Instead, many nucleoporins remain associated with their interphase partners during mitosis and represent units for NPC disassembly and reassembly.[1] For example, Nup214 remains associated with Nup84 (Nup88), the Nup62 complex (containing Nups62/58/54/45) remains intact and Nup107 remains associated with Nup133.[44,55,58-60] (See Table 1 for a list of Nup complexes). The Nup107/Nup133 complex is highly unusual in that instead of distributing uniformly throughout the mitotic cytoplasm, it concentrates at kinetochores. This association with kinetochores is subsequently reversed at the end of anaphase when nuclear reformation commences. As NPCs disassemble they must leave behind fenstrae in the nuclear membranes. It has recently been proposed that this is a precipitating step in nuclear membrane dispersal. This suggestion is based upon the finding that dynein and microtubule dependent deformation of the NE during prophase places the nuclear membranes under tension. This in turn may induce catastrophic expansion of fenstrae leading to nuclear membrane disruption.[51,52]

Nuclear Envelope Reformation

At the end of mitosis, all of the disassembled NE components are recycled to form a nucleus in each daughter cell. NE reformation commences in mid-late anaphase and is finally completed in early G1.[61] In mammalian somatic cells recruitment of nuclear membrane components to the chromatin surfaces is one of the earliest events in nuclear reformation. Observations on both live and fixed cells, reveal that proteins such as LBR and LAP2 are found to concentrate along the polar and lateral margins of the chromatids during late anaphase.[55,62-64] Since these proteins are mobile within a continuous ER network during mitosis, their recruitment presumably reflects the appearance of specific binding sites at the chromatin surfaces.[63,64] In this way INM protein recruitment may drive the enwrapment of the chromatin by the nuclear membranes. Furthermore, binding to chromatin is likely to represent the mechanism by which the INM domain is initially established. It is self evident that nuclear membrane formation requires extensive membrane fusion in order to form a sealed NE. Mattaj and co-workers[65] have shown that this is under the control of two separate membrane fusion systems based upon the p97 AAA ATPase. This had previously been shown to regulate post mitotic reformation of the Golgi apparatus.[66]

Concomitant with nuclear membrane recruitment, B-type lamins begin to associate with the nuclear periphery. Work by Collas and colleagues has demonstrated that B-type lamin assembly is regulated by chromatin associated protein phosphatase 1 (PP1) and a membrane associated PP1 binding protein (AKAP 149).[67,68] The regulation of A-type lamin assembly is less well understood, and while some early (late anaphase) recruitment can be detected, the bulk of the A-type lamins return to the NE very late in mitosis or even in early G1.[69]

NPC Assembly

NPC reassembly within the double nuclear membranes represents an intriguing topological problem. There are two ways in which an NPC might be incorporated into the NE at the end of mitosis. The first, and conceptually simplest method, would be to assemble the central region of the ring-spoke complex on the chromatin surface and then to surround this by flattened membrane cisternae. These would then form the INM and ONM as well as the POM. All that would be required is fusion to form sealed membranes around each NPC. The alternative is to create a largely continuous double membrane employing the appropriate p97 fusion apparatus and then to insert the NPC, or NPC subunits. If this mechanism is employed, then in order to create an aqueous channel between the nucleus and cytoplasm, as well as the POM, it is necessary to induce an additional fusion event between the lumenal faces of the INM and ONM. Studies carried out in several laboratories and employing inhibitors of NPC assembly have demonstrated quite convincingly that NPCs are indeed assembled into continuous membranes and do require fusion between the INM and ONM.[70,71] While the identity of the fusion apparatus is unknown, the large NPC membrane protein gp210 has often been suggested as a candidate. This is based upon its large lumenal domain and the presence of sequences similar to fusogenic peptides seen in certain viral envelope glycoproteins.

Observations by Allen, Goldberg, Wilson and colleagues using high resolution scanning EM techniques, have revealed what appear to be a series of membrane-associated NPC assembly intermediates in the NEs of nuclei assembled in vitro in *Xenopus* egg extracts.[72] The most striking aspect of this work is the initial appearance of small holes in intact nuclear membranes with the impression that NPC assembly commences within the nucleus. It is only as the NPC apparently "matures" that more elaborate structures become evident. Recently Wilson and co-workers have found that gp210 plays an important role in the initial dilation of the membrane channel to accommodate the assembling NPC.[73]

Despite the fact that NPC assembly occurs within intact membranes, the first nucleoporins to be recruited to the nuclear periphery are soluble proteins.[55] This recruitment commences in early- to mid-anaphase, prior to the appearance of nuclear membranes and INM components (Fig. 1). One of these "early" proteins, Nup153, is a component of the basket structure on the nuclear face of the NPC. Nup153 is a member of the FG repeat family of nucleoporins and is extensively modified with O-linked N-acetylglucosamine (O-GlcNAc). Immuno-EM analyses of anaphase cells have confirmed that Nup153 associates with membrane-free regions of chromatin. Indeed, purified chromosomes incubated in mitotic cytosol will bind Nup153. The same is also true of Nup98, a nucleoporin that contains GLFG repeats, and which resides on the nuclear face of the NPC, probably within the proximal regions of the basket structure. Immunofluorescence microscopy reveals that its recruitment either coincides with or follows very closely that of Nup153. As outlined in Figure 1, and described more comprehensively below, stepwise recruitment of several other nucleoporins has also been observed, both in fixed and live cells.[55,74]

The assembly of several other nucleoporins into the NPC is dependent upon the presence of Nup153. This was demonstrated in a series of in vitro nuclear assembly experiments, utilizing *Xenopus* egg extracts immunologically depleted of Nup153. It was found by Mattaj and co-workers[75] that in the absence of Nup153, recruitment of Nup93, Nup98 and Tpr was abolished. Since Nup93 exists in a complex with Nup205 and Nup188, it is reasonable to predict that assembly of these two proteins into the NPC is also dependent upon Nup153. A similar possibility exists for Nup96. This protein is derived from the same high molecular weight precursor as Nup98 and is associated with Nup98 during interphase.[33] However, it has yet to be determined if this interaction is maintained in mitotic cells.

Nup93, Nup98 and Tpr reside on the nucleoplasmic face of the NPC, and it is possible therefore that Nup153 is required for the establishment of assembly sites for each these proteins. While this is likely the case for Nup98 (and probably Nup93) the situation with Tpr is less clear. In contrast to Nup98, Tpr is recruited from the cytoplasm to the nuclear face of the NE very late in mitosis, at the end of telophase/early G1, almost certainly after the NE is sealed.[55,74] It is likely therefore that assembly of Tpr, but not that of Nup98, requires a functional import system. Mattaj and co-workers[75] have shown that nuclei assembled in the absence of Nup153 are incapable of importing proteins bearing conventional nuclear localization sequences. It is possible therefore, that failure to assemble Tpr into the NPC is a reflection of this import block. An additional finding by Mattaj and co-workers is that in the absence of Nup153, NPCs become mobile within the plane of the nuclear membranes. The suggestion is that Nup153, or a protein whose assembly is dependent upon Nup153, interacts with the immobile nuclear lamina. Indeed the N-terminal domain of Nup153 does in fact contain a binding site for A-type lamins (Paul Eanrson and Brian Burke, unpublished results). In related studies, Forbes and co-workers[76] found that adsorption of Nup93 from egg extracts resulted in a decline in the efficiency of NPC assembly in vitro. Whether those NPCs that did assemble, utilized residual Nup93 or whether NPCs can assemble without this protein is not clear.

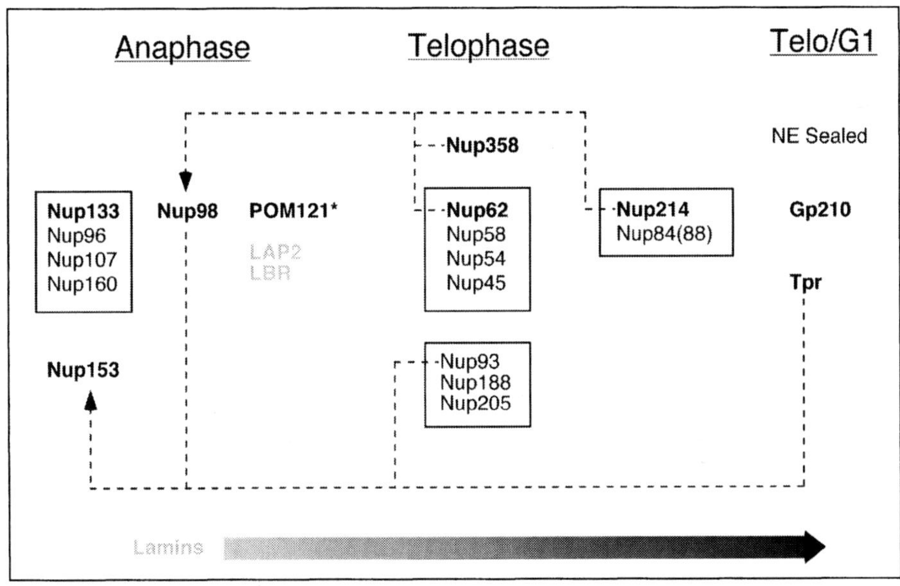

Figure 1. Sequential recruitment of nucleoporins to the NE at the end of mitosis. Boxes indicate nucleoporin sub-complexes. Dashed lines indicate dependence of assembly on either Nup98 or Nup153. Relative recruitment times have been determined experimentally for those proteins in bold. Recruitment times are inferred for those in plain text. Non-NPC proteins are indicated in gray. Nuclear lamin recruitment is an extended process (shaded arrow) that commences in anaphase with B-type lamins and continues in to early G1.

Van Deursen and colleagues[77] have eliminated Nup98 by homologous recombination in ES cells and have derived mice harboring the deleted allele. The strategy employed left Nup96 synthesis unaffected. Nup98 -/- mice are inviable. They exhibit profoundly retarded embryonic development and invariably die in utero. It has, however, been possible to derive viable Nup98 null cells form arrested embryos, indicating that Nup98 is not essential in terms of cell survival. However, these cells have reduced growth rates relative to wild-type and are compromised in terms of nuclear protein import. While NPCs still form in Nup98 null cells, they display a proliferation of annulate lammelae (AL) that are enriched in certain cytoplasmically oriented nucleoporins (i.e., nup358, nup214, nup84 and nup62). The distributions of other NPC components that reside in the nuclear basket (i.e., nup153, nup93, and nup50) are not affected by Nup98 deletion. Taking all of these results together, it is apparent that Nup98 assembly is dependent upon Nup153, but not vice versa.

Another soluble protein whose recruitment to the nuclear periphery precedes that of nuclear membranes is Nup107. Unlike Nup153, Nup107 contains neither FG repeats nor O-GlcNAc and is localized on both faces of the NPC, most likely as a component of the nuclear and cytoplasmic rings.[58,78] Nup107 is found in a NPC subcomplex with Nup96, mammalian sec13, and p37.[33] A similar complex (Nup84, Nup85, Nup120, Nup133, Nup145-C, Seh1 and sec13) has been described in yeast.[79] Recently, Doye and co-workers[58] identified a new member of this complex, Nup133. This complex is very stably integrated within the NPC. This again contrasts with Nup153, which will exchange rapidly between interphase NPCs with a half time of less than one minute. Consistent with its association with Nup107, Nup133 is also localized to both faces of the NPC and remains associated with Nup107 during cell division.[58] However, a recent study by Forbes and co-workers indicate that Nup133 is accessible only from the nuclear side of the pore.[80] The behavior of this complex as cell progresses through mitosis is very intriguing. After NEB, a fraction of Nup107/Nup133 is found at kinetochores. Whether this complex has any specific function at these sites is not at all clear. Later, in early anaphase these two proteins redistribute to the surfaces of newly segregated chromatids, roughly coincident with the recruitment of Nup153. The fact that some nucleoporins associate with chromatin in the absence of membranes is reminiscent of an earlier study by Sheehan and co-workers.[81] They demonstrated, by electron microscopy, the presence of membrane-free "pre-pores" around the surfaces of sperm chromatin following extended incubation in *Xenopus* egg extracts. It is possible that these "pre-pores" might actually represent a subset of chromatin bound Nups, including Nup153, Nup98, Nup107 and Nup133 that might define sites for the formation of the early membrane associated NPC assembly intermediates documented by Goldberg and colleagues.[72]

The essential N-terminal domain of Nup153 (amino acids 1-610) contains two distinct, but overlapping, targeting functions. The first of these, amino acids 3-144, confers localization to the nuclear face of the NE but not specifically to NPCs. The second region, amino acids 39-339, directs incorporation into the NPC.[82] It is obvious that in order to carry out this NPC-targeting function, this 300-residue segment of Nup153 must interact with one or more additional nucleoporins. Forbes and co-workers[80] have used this region of Nup153 in pulldown assays to search for interacting proteins in *Xenopus* egg extracts. Their results identified a pentameric complex containing, Nup96, Nup107, Nup133, sec13, and a novel nucleoporin, Nup160. Nup160 has no FG repeats, localizes to the nuclear side of the NPC, and together with Nup133, plays an essential role in mRNA export.[80] Surprisingly, a similar set of proteins (i.e., Nup96, Nup107, Nup133, Nup160, and Sec13) were identified when they used the carboxy-terminal domain of Nup98 in similar pulldown assays. While Nup153 binds directly to Nup160, Nup98 binding to the Nup160 complex is mediated by Nup96.[83] As described above it has been established that Nup133 and Nup107 are recruited very early to the reform-ing NE. Since the Nup160 complex was originally recovered from meiotic extracts, it is reason-able to conclude that it remains intact in mitotic cells. Consequently the early recruitment of Nup133 most likely signifies early recruitment of the complex as a whole. It has yet to be determined however, how Nup153 and Nup98 might influence the targeting of the Nup160 complex to chromatin and vice versa in late anaphase cells.

Nup153 interacts with another nucleoporin, Nup50[84,85] which has been shown to play a role in the export of macromolecules from the nucleus. Deletion of the Nup50 gene by homologous recombination in ES cells is associated with late embryonic lethality in homozygous null mice. However, Nup50 null cells are evidently viable and exhibit no obvious defects in Nup153 localization.[85] The inference is that Nup50 is not essential for Nup153 assembly. Whether Nup50 localization is dependent upon Nup153 has yet to be determined, although this will most likely prove be the case.

The vertebrate NPC contains only two known integral membrane proteins, gp210[24,86] and POM121.[87] Either one or both of these proteins has to be involved in the formation of the early membrane associated assembly intermediates. In mammalian somatic cells, these two proteins are free to diffuse throughout a continuous ER network during mitosis. Just as is the case with the INM proteins, gp210 and POM121 will only concentrate at the nuclear periph-ery as chromatin associated binding partners become available.[55,63,64] POM121 is recruited to the reforming nuclear envelope as early as late anaphase, in parallel with INM proteins such as LAP2 and LBR.[55] In this way, POM121 recruitment may be directed by chromatin bound nucleoporins such as Nup153, Nup98, Nup107/133 which contribute to the basket structure and to the nuclear face of the NPC. Thus, POM121 is present at the right place and right time to direct the formation of the ring-spoke complex of the NPC. This notion is reinforced by the finding that proteins of the central channel (the Nup62 complex) and cytoplasmic filaments (Nup214 and its associated protein, Nup84) do not accumulate at the nuclear periphery until after POM121 and membranes. Based on these observations we would predict that other pro-teins of the NPC central framework, Nup155 for instance, should not concentrate at the NE until after POM121. A recent study by Elllenberg and co-workers[88] support such a view of POM121 function. Employing photobleaching methods, they were able to show that POM121 becomes effectively immobile within the reforming NE and indicate that there is negligible exchange of POM121 between NPCs once the cells enter G1 (less than once per cell cycle). The conclusion is that POM121 is an extremely stable component of the ring-spoke complex.

Gp210 and POM121 have distinct topologies in the NE. Most of the mass of gp210 is located within the PNS and only a small portion of it is exposed to the cytoplasm.[23,89] In contrast, most of the POM12 mass is exposed to the cytoplasm and is available to interact with soluble nucleoporins. The exposed COOH-terminal domain of gp210 has been suggested to function as an anchor for components of the NPC.[23,89] In support of this, transfection experiments revealed that this domain is capable of targeting a reporter protein to the NPC.[89] Gp210 is also specifically phosphorylated during mitosis at a serine residue located in the COOH-terminal domain, and phosphorylation of this residue could potentially be involved in dissociating NPCs from nuclear membranes.[56] Surprisingly, however, gp210 only begins to concentrate in the reforming NE towards the end of telophase, much later than POM121[55,62] (Fig. 2). At first sight this not consistent with a role for gp210 in the early stages of pore formation. Nevertheless it is clear that gp210 does contribute to pore dilation, a relatively early event in NPC assembly.[73] Heterokaryon analyses suggest that in contrast to POM121, gp210 is a mobile NPC component (i.e., it can exchange between NPCs). This view is supported by the finding that at any one time during interphase, about 30% of gp210 is free in the peripheral ER (Khaldon Bodoor and Brian Burke, unpublished results). These results can best be rationalized by the suggestion that gp210 instead of being a structural component of the NPC, functions as an assembly factor, NPC chaperonin, that can exchange between NPCs during the assembly process.

The sequential recruitment of NPC proteins to the nuclear periphery at the end of mitosis is supported by ultrastructural studies of Goldberg and co-workers.[72] They have used high reolution scanning EM to observe NE and NPC formation in *Xenopus* egg extracts and more recently in *Drosophila* embryos. They were able to identify a number of potential NPC assembly intermediates, and have been able to place these into a logical sequence based partly on time of appearance and partly on complexity. The first sign of NPC assembly is the appearance of a dimple within regions of intact nuclear membranes. The dimple is then converted into a nuclear membrane fenestra featuring sharp edges and an average diameter of about 45nm. This is significantly smaller than a mature nuclear pore which has a diameter of about 70nm. The formation and stabilization of the small pore must involve the activities of both lumenal and cytoplasmically disposed proteins such as gp210 and/or POM121 as well as other soluble nucleoporins. Construction of an octagonal "star ring" then follows, one subunit at a time. On top of the star ring, a thin ring is formed. Additional subunits are added to the "thin ring", which becomes the cytoplasmic ring. Finally, eight cytoplasmic filaments are added to the cytoplasmic ring. The order of recruitment of a subset of nucleoporins correlates, for the most part, with the sequential addition of individual subunits. However, other stages in NPC assembly, in particular, nuclear basket formation, have not been detected and thus, correlation of such structural intermediates with nucleoporin recruitment is not yet possible. We would predict however, that basket protein assembly should precede formation of the membrane dimple and small pore.

When Does the NPC Become Functional?

Some soluble nucleoporins, such as Nup153, Nup107, Nup358, Nup62, and Nup214 appear to accumulate at the nuclear periphery prior to the formation of active NPCs.[55,58,74] Others such as Tpr are recruited late and thus may need the presence of functional NPCs to cross the NE.[55,74] In yeast a remarkably large number of nucleoporins are functionally redundant, and in mammalian systems, Nup50 and Nup98 are clearly not essential proteins (at least at the cellular level). It is still not yet known which NPC proteins constitute the minimal assembly to form a transport-competent NPC. However, by determining when, during mitosis, the nuclear membranes become sealed we can define a point at which NPCs must be active. Through the use of differential permeabilization or in vivo imaging it is evident that the NE becomes impermeable to large macromolecules in late-telophase/early G1, coincident with the recruitment of Tpr and gp210.[55,74]

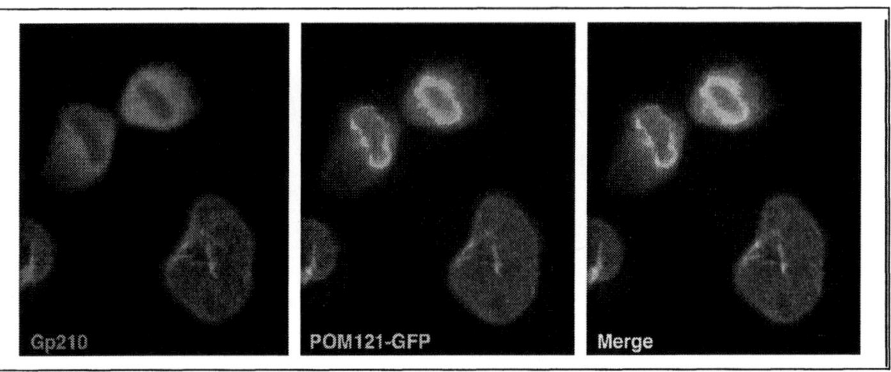

Figure 2. Differences in behavior of two NPC membrane proteins at the end of mitosis. POM121 accumulates at the nuclear periphery prior to Gp210.

Summary

Perhaps 50% of vertebrate nucleoporins have now been identified and characterized. The next challenge is to obtain an improved picture of the functional interactions of these proteins within the 3-D architecture of the NPC. This will be essential if we are to find the "Holy Grail" of nucleocytoplasmic transport, the mechanism of translocation through the NPC. It has become clear that NPCs contain discreet nucleoporin subcomplexes that are frequently preserved in mitotic cells. Further defining the molecular details of NPC disassembly/reassembly during mitosis should provide us with a wealth of information on how the different NPC subunits associate. The sequential recruitment of NPC proteins at the end of mitosis combined with powerful in vitro assembly systems now makes possible investigations into how the presence or absence of a particular nucleoporin might affect the assembly of others. A priority will now be to employ high resolution imaging methods, including both electron microscopy and atomic force microscopy, to define the structural contributions of individual NPC subunits. These types of studies are now getting underway.

References

1. Gant TM, Wilson KL. Nuclear assembly. Annu Rev Cell Dev Biol 1997; 13:669-695.
2. Gerace L, Burke B. Functional organization of the nuclear envelope. Ann Rev Cell Biol 1988; 4:335-374.
3. Mattaj IW, Englmeier L. Nucleocytoplasmic transport: The soluble phase. Annu Rev Biochem 1998; 67:265-306.
4. Aebi U, Cohn JB, Buhle L, et al. The nuclear lamina is a meshwork of intermediate type filaments. Nature 1986; 323:560-564.
5. Stuurman N, Heins S, Aebi U. Nuclear lamins: their structure, assembly, and interactions. J Struct Biol 1998; 122:42-66.
6. Roeber RA, Weber K, Osborn M. Differential timing of lamin A/C expression in the various organs of the mouse embryo and the young animal: A developmental study. Development 1989; 105:365-378.
7. Stewart C, Burke B. Teratocarcinoma stem cells and early mouse embryos contain only a single major lamin polypeptide closely resembling lamin B. Cell 1987; 51:383-392.
8. Mounkes LC, Burke B, Stewart CL. The A-type lamins: Nuclear structural proteins as a focus for muscular dystrophy and cardiovascular diseases. Trends Cardiovasc Med 2001; 11:280-285.
9. Furukawa K, Panté N, Aebi U, et al. Cloning of a cDNA for lamina-associate polypeptide 2 (LAP2) and identification of regions that specify targeting to the nuclear envelope. EMBO J 1995; 14:1626-1636.
10. Lin F, Blake DL, Callebaut I, et al. MAN1, an inner nuclear membrane protein that shares the LEM domain with lamina-associated polypeptide 2 and emerin. J Biol Chem 2000; 275:4840-4847.

11. Martin L, Crimaudo C, Gerace L. cDNA cloning and characterization of lamina-associated polypeptide 1C (LAP1C), an integral protein of the inner nuclear membrane. J Biol Chem 1995; 270:8822-8828.

12. Rolls MM, Stein PA, Taylor SS, et al. A visual screen of a GFP-fusion library identifies a new type of nuclear envelope membrane protein. J Cell Biol 1999; 146:29-44.

13. Senior A, Gerace L. Integral membrane proteins specific to the inner nuclear membrane and associated with the nuclear lamina. J Cell Biol 1988; 107:2029-2036.

14. Worman HJ, Evans CD, Blobel G. The lamin B receptor of the nuclear envelope inner membrane: A polytopic protein with eight potential transmembrane domains. J Cell Biol 1990; 111:1535-1542.

15. Gruenbaum Y, Wilson KL, Harel A, et al. Review: Nuclear lamins—Structural proteins with fundamental functions. J Struct Biol 2000; 129:313-323.

16. Worman HJ, Courvalin JC. The inner nuclear membrane. J Membr Biol 2000; 177:1-11.

17. Georgatos SD. The inner nuclear membrane: simple, or very complex? Embo J 2001; 20:2989-2994.

18. Goldberg MW, Allen TD. Structural and functional organization of the nuclear envelope. Curr Opin Cell Biol 1995; 7:301-309.

19. Panté N, Aebi U. Molecular dissection of the nuclear pore complex. Crit Rev Biochem Mol Biol 1996; 31:153-199.

20. Akey CW, Radermacher M. Architecture of the *Xenopus* nuclear pore complex revealed by three dimensional cryo-electron microscopy. J Cell Biol 1993;122:1-20.

21. Hinshaw JE, Carragher BO, Milligan RA. Architecture and design of the nuclear pore complex. Cell 1992; 69:1133-1141.

22. Yang Q, Rout MP, Akey CW. Three-dimensional architecture of the isolated yeast nuclear pore complex: Functional and evolutionary implications. Mol Cell 1998; 1:223-234.

23. Greber UF, Senior A, Gerace L. A major glycoprotein of the nuclear pore complex is a membrane-spanning polypeptide with a large lumenal domain and a small cytoplasmic tail. EMBO J 1990; 9:1495-1502.

24. Wozniak RW, Bartnik E, Blobel G. primary structure analysis of an integral membrane glycoprotein of the nuclear pore. J Cell Biol 1989; 108:2083-2092.

25. Dworetzky SI, Lanford RE, Feldherr CM. The effects of variations in the number and sequence of of targeting signals on nuclear uptake. J Cell Biol 1988; 107:1279-1287.

26. Feldherr C, Kallenbach E, Schultz N. Movement of a karyophilic protein through the nuclear pores of oocytes. J Cell Biol 1984; 99:2216-2222.

27. Jarnik M, Aebi U. Toward a more complete 3-D structure of the nuclear pore complex. J Struct Biol 1991; 107:291-308.

28. Panté N, Aebi U. Sequential binding of import ligands to distinct nucleopore regions during their nuclear import. Science 1996; 273:1729-1732.

29. Richardson WD, Mills AD, Dilworth SM, et al. Nuclear protein migration involves two steps: Rapid binding at the nuclear envelope followed by slower translocation through the nuclear pores. Cell 1988; 52:655-664.

30. Goldberg MW, Allen TD. High resolution scanning electron microscopy of the nuclear envelope: Demonstration of a new, regular, fibrous lattice attached to the baskets of the nucleoplasmic face of the nuclear pores. J Cell Biol 1992; 119:1429-1440.

31. Kiseleva E, Goldberg MW, Daneholt B, et al. RNP export is mediated by structural reorganiztion of the nuclear pore basket. J Mol Biol 1996; 260:304-311.

32. Reichelt R, Holzenburg A, Buhle EL, et al. Correlation between structure and mass distribution of the nuclear pore complex and of distinct pore components. J Cell Biol 1990; 110:883-894.

33. Fontoura BM, Blobel G, Matunis MJ. A conserved biogenesis pathway for nucleoporins: proteolytic processing of a 186-kilodalton precursor generates Nup98 and the novel nucleoporin, Nup96. J Cell Biol 1999; 144:1097-1112.

34. Rout MP, Aitchison JD, Suprapto A, et al. The yeast nuclear pore complex: composition, architecture, and transport mechanism. J Cell Biol 2000; 148:635-651.

35. Bastos R, Panté N, Burke B. Nuclear pore complex proteins. Int Rev Cytol 1995; 162B:257-302.

36. Stoffler D, Fahrenkrog B, Aebi U. The nuclear pore complex: from molecular architecture to functional dynamics. Curr Opin Cell Biol 1999; 11:391-401.

37. Davis LI, Blobel G. Identification and charactarization of a nuclear pore complex protein. Cell 1986; 45:699-709.

38. Kraemer D, Wozniak RW, Blobel G, et al. The human CAN protein, a putative oncogene product associated with myeloid leukemogenesis, is a nuclear pore complex protein that faces the cytoplasm. Proc Natl Acad Sci USA 1994; 91:1519-1523.

39. Wu J, Matunis MJ, Kraemer D, et al. Nup358, a cytoplasmically exposed nucleoporin with peptide repeats, Ran-GTP binding sites, zinc fingers, a cyclophilin A homologous domain, and a leucine-rich region. J Biol Chem 1995; 270:14209-14213.

40. Yokoyama N, Hayashi N, Seki T, et al. A giant nucleopore protein that binds Ran/TC4. Nature 1995; 376:184-188.

41. Sukegawa J, Blobel G. A nuclear pore complex protein that contains zinc fingers and faces the nucleoplasm. Cell 1993; 72:29-38.

42. Rexach M, Blobel G. Protein import into nuclei: association and dissociation reactions involving transport substrate, transport factors and nucleoporins. Cell 1995; 83:683-692.

43. Bastos R, Ribas de Pouplana L, Enarson M, et al. Nup84, a novel mammalian nucleoporin that is associated with CAN/Nup214 on the cytoplasmic face of the NPC. J Cell Biol 1997; 137:989-1000.

44. Panté N, Bastos R, McMorrow I, et al. Interactions and three-dimensional localization of a group of nuclear pore complex proteins. J Cell Biol 1994; 126:603-617.

45. Radu A, Moore MS, Blobel G. The peptide repeat domain of nucleoporin Nup98 functions as a docking site in transport across the nuclear pore complex. Cell 1995; 81:215-222.

46. Guan T, Müller S, Klier G, et al. Structural analysis of the p62 complex, an assembly of O-linked glycoproteins that localizes near the central gated channel of the nuclear pore complex. Mol Biol Cell 1995; 6:1591-1603.

47. Hu T, Guan T, Gerace L. The molecular and functional characterization of the p62 complex, an assembly of nuclear pore complex glycoproteins. J Cell Biol 1996; 134:589-601.

48. Ben-Efraim I, Gerace L. Gradient of increasing affinity of importin beta for nucleoporins along the pathway of nuclear import. J Cell Biol 2001; 152:411-417.

49. Ribbeck K, Gorlich D. The permeability barrier of nuclear pore complexes appears to operate via hydrophobic exclusion. Embo J 2002; 21:2664-2671.

50. Lee KK, Gruenbaum Y, Spann P, et al. *C. elegans* nuclear envelope proteins emerin, MAN1, lamin, and nucleoporins reveal unique timing of nuclear envelope breakdown during mitosis. Mol Biol Cell 2000; 11:3089-99.

51. Beaudouin J, Gerlich D, Daigle N, et al. Nuclear envelope breakdown proceeds by microtubule-induced tearing of the lamina. Cell 2002; 108:83-96.

52. Salina D, Bodoor K, Eckley DM, et al. Cytoplasmic dynein as a facilitator of nuclear envelope breakdown. Cell 2002; 108:97-107.

53. Cotter LA, Goldberg MW, Allen TD. Nuclear pore complex disassembly and nuclear envelope breakdown during mitosis may occur by both nuclear envelope vesicularisation and dispersion throughout the endoplasmic reticulum. Scanning 1998; 20:250-251.

54. Kiseleva E, Rutherford S, Cotter LM, et al. Steps of nuclear pore complex disassembly and reassembly during mitosis in early *Drosophila* embryos. J Cell Sci 2001; 114:3607-3618.

55. Bodoor K, Shaikh S, Salina D, et al. Sequential recruitment of NPC proteins to the nuclear periphery at the end of mitosis. J Cell Sci 1999; 112:2253-2264.

56. Favreau C, Worman HJ, Wozniak RW, et al. Cell cycle-dependent phosphorylation of nucleoporins and nuclear pore membrane glycoprotein gp210. Biochemistry 1996; 35:8035-8044.

57. Macaulay C, Meier E, Forbes DJ. Differential mitotic phosphorylation of proteins of the nuclear pore complex. J Biol Chem 1995; 270:254-262.

58. Belgareh N, Rabut G, Bai SW, et al. An evolutionarily conserved NPC subcomplex, which redistributes in part to kinetochores in mammalian cells. J Cell Biol 2001; 154:1147-1160.

59. Dabauvalle M-C, Loos K, Scheer U. Identification of a soluble precursor complex essential for nuclear pore assembly in vitro. Chromosoma 1990; 100:56-66.

60. Finlay D, Meier E, Bradley P, et al. A complex of nuclear pore proteins required for pore function. J Cell Biol 1991; 114:169-183.

61. Maul G. Nuclear pore complexes: elimination and reconstruction during mitosis. J Cell Biol 1977; 74:492-500.

62. Chaudhary N, Courvalin JC. Stepwise reassembly of the nuclear envelope at the end of mitosis. J Cell Biol 1993; 122:295-306.

63. Ellenberg J, Siggia ED, Moreira JE, et al. Nuclear membrane dynamics and reassembly in living cells: Targeting of an inner nuclear membrane protein in interphase and mitosis. J Cell Biol 1997; 138:1193-1206.

64. Yang L, Guan T, Gerace L. Integral membrane proteins of the nuclear envelope are dispersed throughout the endoplamic reticulum during mitosis. J Cell Biol 1997; 137:1199-1210.

65. Hetzer M, Meyer HH, Walther TC, et al. Distinct AAA-ATPase p97 complexes function in discrete steps of nuclear assembly. Nat Cell Biol 2001; 3:1086-1091.

66. Kondo H, Rabouille C, Newman R, et al. p47 is a cofactor for p97-mediated membrane fusion. Nature 1997; 388:75-78.

67. Steen RL, Martins SB, Tasken K, et al. Recruitment of protein phosphatase 1 to the nuclear envelope by A-kinase anchoring protein AKAP149 is a prerequisite for nuclear lamina assembly. J Cell Biol 2000; 150:1251-1262.

68. Steen RL, Collas P. Mistargeting of B-type lamins at the end of mitosis. Implications on cell survival and regulation of lamins a/c expression. J Cell Biol 2001; 153:621-626.

69. Moir RD, Yoon M, Khuon S, et al. Nuclear lamins A and B1: Different pathways of assembly during nuclear envelope formation in living cells. J Cell Biol 2000; 151:1155-1168.

70. Macaulay C, Forbes DJ. Assembly of the nuclear pore: biochemically distinct steps revealed with NEM, GTPgS and BAPTA. J Cell Biol 1996;132:5-20.

71. Wiese C, Goldberg MW, Allen TD, et al. Nuclear envelope assembly in *Xenopus* extracts visualized by scanning EM reveals a transport-dependent 'envelope smoothing' event. J Cell Sci 1997; 110:1489-1502.

72. Goldberg MW, Wiese C, Allen TD, et al. Dimples, pores, star-rings and thin rings on growing nuclear envelopes: evidence for structural intermediates in nuclear pore complex assembly. J Cell Sci 1997; 110:409-420.

73. Drummond SP, Wilson KL. Interference with the cytoplasmic tail of gp210 disrupts "close apposition" of nuclear membranes and blocks nuclear pore dilation. J Cell Biol 2002; 158:53-62.

74. Haraguchi T, Koujin T, Hayakawa T, et al. Live fluorescence imaging reveals early recruitment of emerin, LBR, RanBP2, and Nup153 to reforming functional nuclear envelopes. J Cell Sci 2000; 113:779-794.

75. Walther TC, Fornerod M, Pickersgill H, et al. The nucleoporin Nup153 is required for nuclear pore basket formation, nuclear pore complex anchoring and import of a subset of nuclear proteins. Embo J 2001; 20:5703-5714.

76. Grandi P, Dang T, Panté N, et al.. Nup93, a vertebrate homologue of yeast Nic96p, forms a complex with a novel 205-kDa protein and is required for correct nuclear pore assembly. Mol Biol Cell 1997; 8:2017-2038.

77. Wu X, Kasper LH, Mantcheva RT, et al. Disruption of the FG nucleoporin NUP98 causes selective changes in nuclear pore complex stoichiometry and function. Proc Natl Acad Sci USA 2001; 98:3191-3196.

78. Radu A, Blobel G, Wozniak RW. Nup107 is a novel nuclear pore complex protein that contains a leucine zipper. J Biol Chem 1994; 269:17600-17605.

79. Siniossoglou S, Wimmer C, Rieger M, et al. A novel complex of nucleoporins, which includes Sec13p and a Sec13p homolog, is essential for normal nuclear pores. Cell 1996; 84:265-275.

80. Vasu S, Shah S, Orjalo A, et al. Novel vertebrate nucleoporins Nup133 and Nup160 play a role in mRNA export. J Cell Biol 2001; 155:339-354.

81. Sheehan MA, Mills AD, Sleeman AM, et al. Steps in the assembly of replication-competent nuclei in a cell-free system. J Cell Biol 1988; 106:1-12.

82. Enarson P, Enarson M, Bastos R, et al. Amino-terminal sequences that direct nucleoporin Nup153 to the inner surface of the nuclear envelope. Chromosoma 1998; 107:228-236.

83. Fontoura BM, Dales S, Blobel G, et al. The nucleoporin Nup98 associates with the intranuclear filamentous protein network of TPR. Proc Natl Acad Sci USA 2001; 98:3208-3213.

84. Guan T, Kehlenbach RH, Schirmer EC, et al. Nup50, a nucleoplasmically oriented nucleoporin with a role in nuclear protein export. Mol Cell Biol 2000; 20:5619-5630.

85. Smitherman M, Lee K, Swanger J, et al. Characterization and targeted disruption of murine Nup50, a p27(Kip1)- interacting component of the nuclear pore complex. Mol Cell Biol 2000; 20:5631-5642.

86. Gerace L, Ottaviano Y, Kondor-Koch C. Identification of a major polypeptide of the nuclear pore complex. J Cell Biol 1982; 95:826-837.

87. Hallberg E, Wozniak RW, Blobel G. An integral membrane protein of the pore membrane domain of the nuclear envelope contains a nucleoporin-like region. J Cell Biol 1993; 122:513-522.

88. Daigle N, Beaudouin J, Hartnell L, et al. Nuclear pore complexes form immobile networks and have a very low turnover in live mammalian cells. J Cell Biol 2001; 154:71-84.

89. Wozniak RW, Blobel G. The single transmembrane segment of gp210 is sufficient for sorting to the pore membrane domain of the nuclear envelope. J Cell Biol 1992; 119:1441-1449.

90. Starr CM, D'Onofrio M, Park MK, et al. Primary sequence and heterologous expression of nuclear pore glycoprotein p62. J Cell Biol 1990; 110:1861-1871.

91. Fornerod M, van Deursen J, van Baal S, et al. The human homologue of CRM1 is in a dynamic subcomplex with CAN/Nup214 and a novel nuclear pore component Nup88. EMBO J 1997; 16:807-816.

92. Radu A, Blobel G, Wozniak RW. Nup155 is a novel nuclear pore complex protein that contains neither repetitive sequence motifs nor reacts with WGA. J Cell Biol 1993; 121:1-9.

93. Miller BR, Powers M, Park M, et al. Identification of a new vertebrate nucleoporin, Nup188, with the use of a novel organelle trap assay. Mol Biol Cell 2000; 11:3381-3396.
94. von Lindern M, van Baal S, Wiegant J, et al. Grosveld G. *can*, a putative oncogene associated with myeloid leukemogenesis, may be activated by fusion of its 3' half to different genes: Characterization of the *set* gene. Mol Cell Biol 1992; 12:3346-3355.
95. Byrd DA, Sweet DJ, Pante N, et al. Tpr, a large coiled coil protein whose amino terminus is involved in activation of oncogenic kinases, is localized to the cytoplasmic surface of the nuclear pore complex. J Cell Biol 1994; 127:1515-1526.
96. Mitchell PJ, Cooper CS. The human tpr gene encodes a protein of 2094 amino acids that has extensive coiled-coil regions and an acidic C-terminal domain. Oncogene 1992; 7:2329-2333.

CHAPTER 7

Structure, Function and Biogenesis of the Nuclear Envelope in the Yeast *Saccharomyces cerevisiae*

George Simos

Introduction

As in all eukaryotes, the yeast nuclear envelope (NE) serves as the boundary between the nuclear and the cytoplasmic compartments. However, the NE is anything but a passive barrier and participates actively in vital processes such as the transport of macromolecules between the two compartments, nuclear division and chromatin organization. The basic structure of the NE is common in all nucleated cells. It is composed of two distinct lipid bilayers, the inner and outer nuclear membranes (INM and ONM, respectively), which enclose a flattened cisternal space continuous with the lumen of the endoplasmic reticulum (ER). The ONM is also a continuation of the ER membrane and performs rough ER functions. In fact, the ONM in yeast represents the bulk of the rough ER and is studded with ribosomes. At multiple locations the ONM bends sharply and fuses to the INM, which faces the nucleoplasm and is in close contact with the chromatin, creating cylindrically shaped apertures, the nuclear pores. These pores are occupied by the nuclear pore complexes (NPC), massive proteinaceous assemblies that function as channels and mediate all transport between the nucleoplasm and the cytoplasm.

Despite the similarities in the basic architecture, there are several differences between the yeast NE and the NE of higher eukaryotes. In multicellular organisms the INM is lined with a filamentous network called the nuclear lamina. The nuclear lamina confers mechanical stability to the nuclear membrane and contributes to its organization and interaction with the peripheral chromatin. However, the nuclear lamins, the protein components of the nuclear lamina, which are present in all metazoa, are notably absent from yeast. Furthermore, the INM of higher eukaryotes contains a unique set of integral membrane proteins, most of which also interact with the nuclear lamina. These proteins have no apparent homologues in yeast. Another striking difference between yeast and higher eukaryotic organisms is that the yeast NE remains intact throughout the cell cycle. During cell division, yeast cells undergo a "closed" mitosis in contrast to the "open" mitosis encountered in all multicellular species. An "open" mitosis is characterized by NE breakdown, which allows the invasion of the centrosome-emanated microtubules into the nuclear space and the formation of the mitotic spindle, causing at the same time the intermixing of nuclear and cytosolic contents. The "closed" mitosis of single-cell eukaryotes such as yeast proceeds without disassembly of the NE and depends on the formation of an intranuclear spindle apparatus. This is mediated by the spindle pole bodies (SPBs), the microtubule organizing centers of yeast, which unlike the metazoan centrosomes, are embedded into the yeast nuclear envelope.

Nuclear Envelope Dynamics in Embryos and Somatic Cells
edited by Philippe Collas. ©2002 Eurekah.com and Kluwer Academic/Plenum Publishers.

Closed mitosis and the SPB are particular features of the yeast NE and will not be discussed further in this Chapter. On the other hand, the functions and most likely the biogenesis of the nuclear membrane and NPC appear to be conserved in all eukaryotes. Information obtained from the yeast system on these aspects is likely to be applicable to all organisms and, as such, will be presented in the following sections.

Overview of the Yeast NPC and its Function in Transport

A common architecture characterizes NPCs from all eukaryotes.[1-7] The NPC in all organisms studied so far appears to contain three separate structural elements all of which display an apparent eight fold rotational symmetry: the central core, the cytoplasmic fibrils and the nuclear basket. The central core is the part of the NPC embedded in the double nuclear membrane and comprises eight spokes encircling the central plug (or transporter) and two rings, one at each side of the membrane. The center of this symmetrical structure harbors the channel through which all transport in and out of the nucleus occurs. The functional diameter of this channel for diffusion is 9-10 nm. However, for signal-directed transport the functional diameter of the channel can increase up to 26 nm, or even 39 nm,[8] in order to accommodate passage of particles as large as the ribosomal subunits.

At the cytoplasmic side of the NPC, eight 50 nm long fibrils emanate from the spoke-ring complex and stretch out into the cytoplasm. Eight longer and thinner filaments also extend into the nucleoplasm but their distal ends connect to a small ring forming the nuclear pore basket. When compared to their vertebrate counterparts, the yeast NPCs are significantly smaller both in mass and in volume. The estimated molecular masses are 60 MDa for *S. cerevisiae* and 125 MDa for vertebrates. The central cores are approximately 120 nm (diameter) by 80 nm (height) and 96 nm by 38 nm, respectively. The yeast nuclear filaments are also shorter. These differences in size and mass are a reflection of a simpler yeast NPC central core structure. It is likely that the yeast NPC represents the minimal structure that can support active and signal-mediated nucleocytoplasmic transport.

Before discussing in more detail the elements that make up the yeast NPC, I will briefly introduce the current models for the nucleocytoplasmic transport mechanisms, which have been recently and extensively reviewed.[3,9-17] Nuclear import as well as nuclear export of proteins and RNA occurring through the NPCs is usually an active, carrier-mediated process. Proteins which are destined to enter the nucleus are recognized and bound in the cytoplasm by receptors (named importins or karyopherins) that mediate targeting and translocation through the NPC. Inside the nucleus, the receptor releases its import cargo and is then recycled to the cytoplasm. Importins can either associate directly with their transport cargo recognizing a nuclear localization sequence (NLS) or require adaptor molecules as is the case for the "classical" NLS recognized by the heterodimer importin α/importin β. All the importins identified so far, are members of the same protein family (named importin β family after its founding member) and contain a characteristic Ran-GTP binding domain. The principles of nuclear protein import also apply to export of proteins and RNA from the nucleus. Importin β family members have been shown to be involved in nuclear export processes and were subsequently termed exportins. One of these, CRM1 (Xpo1p in yeast), is the export receptor for the leucine-rich nuclear export signal (NES) contained in many different proteins. CRM1 not only mediates nuclear export of proteins but also of RNA-protein complexes. Thus, it is possible that RNAs can be exported from the nucleus by using as adaptors RNA-binding proteins containing a NES. This appears to be true for at least a subset of RNAs, such as viral mRNAs and snRNAs. On the other hand, tRNA can be exported from the nucleus via its direct interaction with the exportin Xpo-t/Los1p, although alternative tRNA export pathways have also been suggested in yeast.[18]

Nucleocytoplasmic transport is thought to be centrally regulated by the small GTPase Ran and its effectors.[19-21] Current models propose that nuclear Ran is at the GTP-bound form, while in the cytoplasm Ran-GDP predominates. This is due to the exclusive nuclear localization of the Ran nucleotide exchange factor RCC1 (Prp20p in yeast) and the cytoplasmic location of the GTPase activating protein RanGAP1 (Rna1p in yeast). Binding of Ran-GTP to an importin can trigger the dissociation of the importin/import substrate complex and its release from the NPC, while, on the other hand, binding to an exportin promotes the association with the corresponding export cargo and the formation of an export competent complex associating with the NPC. Hydrolysis of the Ran-bound GTP, which should occur only in the cytoplasm, can then cause the dissociation of Ran from the exportin and the release of the corresponding export substrate. Therefore, the Ran system appears to control the directionality of the transport processes across the NPC. Intriguingly, vertebrate Ran has also been recently implicated in mitotic spindle assembly and the re-assembly of the nuclear envelope at the end of mitosis (reviewed in refs. 5, 22 and 23; see also other Chapters of this volume).

Although the model described above is both simple and apparently universal, it is not directly applicable to all nucleocytoplasmic transport processes. There have been several reports on proteins that can enter the nucleus without the help of importins.[12,24-26] These proteins may have acquired the capacity to directly associate with components of the NPC in a way that mimics the interaction between importins and NPC, and therefore triggers their translocation across the nuclear membrane. The involvement of the known exportins and Ran in two major transport processes, namely nuclear export of mRNA and ribosomal subunits, is controversial.[9,12,27] Both mRNA and rRNA are leaving the nucleus in association with many proteins as large ribonucleoprotein particles (RNPs) or pre-ribosomal subunits, respectively. Nuclear export of these particles is clearly more complex than transport of single proteins, and this may not simply be due to the size differences between these two types of cargo. The passage of an RNA through the NPC is likely to represent an integral part of its maturation process and as such it may strongly depend on upstream i.e., nuclear or even downstream i.e., cytoplasmic RNA processing events. Indeed, a tight link has been demonstrated between mRNA splicing and mRNA export (reviewed in ref. 17). An integration of pre-ribosomal subunit maturation and nuclear export, which may involve the exportin Xpo1p, has also been recently suggested.[28-32]

Composition and Structure-Function Relationships of the Yeast NPC

While the karyopherins and Ran constitute the mobile elements of the nuclear transport machinery, the NPCs provide the stationary phase. An intensive effort from many laboratories using biochemical, immunological and genetic techniques culminated by the successful purification of intact NPCs[33] has led to the identification of what can be now considered as the complete repertoire of the yeast NPC protein components, or nucleoporins (nups). Approximately 30 distinct polypeptides are thought to compose the whole yeast NPC. This may seem as a very small number for such a huge complex but, taking into account the symmetry displayed by the NPCs, each individual nucleoporin may occur in 16 to 32 copies per NPC. By analogy as well as by direct experimentation,[34] the vertebrate NPC is thought to contain up to 50 different nups, half of which have been so far identified. At least 65% of the yeast nups have direct orthologues in vertebrates.

The 30 yeast nucleoporins fall into three partially overlapping groups based on their sequence and functional properties: FG nucleoporins, non-FG nucleoporins and pore membrane proteins (POMs). The FG-nucleoporin class contains 12 members. All of them contain at least one domain with GLFG, FXFG or FG amino acid repeat motifs. The FG-domains function as interaction sites for soluble nucleocytoplasmic transport factors such as members of the karyopherin family (reviewed in ref. 3), NTF2, the nuclear import factor of Ran, and the specific mRNA export factors of the TAP/Mex67p family.[35-38] The FG-domains are thought to occur peripherically exposed throughout the NPC and to serve as docking sites for the transport factors during their passage through the nuclear pore. The non-FG nucleoporins, that lack any obvious FG repeats, are the components of the structural core of the NPC. They are estimated to provide almost 70% of the total mass of the NPC and are thought to constitute the central framework onto which the FG-nucleoporins are attached. Two of them, Nup170 and Nup188 have been shown to be required for establishing the functional resting diameter of the yeast NPC central transport channel.[39]

Only three of the yeast NPC proteins are integral membrane proteins (POMs) and because of their tight membrane association are assumed to be responsible for anchoring the NPC to the nuclear membrane. Two of them, Pom152p and Pom34p are specific for the NPC and are non-essential while the third one, Ndc1p is essential and also a component of the SPB, required for its duplication.[33,40] None of them has homologues in vertebrates. A small number of other integral membrane proteins functionally or genetically interact but are not physically associated with the NPC. The role of these proteins in the biogenesis of the nuclear membrane and the NPC is discussed at the end of this Chapter.

Many of the yeast nucleoporins are non-essential proteins. However, almost all of the non-essential nucleoporins become necessary for viability if a non-lethal mutation is introduced to another component of the NPC.[41] Actually, the genetic screens based on synthetic lethality proved to be very successful in identifying many of the yeast nucleoporins. It therefore seems that the network of interactions between the various nups is so tight that removal of a single component can often be tolerated without much structural or functional damage. Nevertheless, the simultaneous removal of two NPC components can become often catastrophic, especially when these two proteins belong to the same sub-structure.

Yeast nucleoporins are generally characterized by their organization into biochemically stable NPC subcomplexes. This trend is now becoming apparent also in higher eukaryotes, as more vertebrate nucleoporins are being characterised. It is logical to assume that a particle as big and as symmetric as the NPC is built up not from individual components but rather from smaller soluble pre-formed assemblies. At least four major discrete NPC subcomplexes have been isolated and characterised from yeast (Fig. 1).

The first one is the "Nsp1p" complex which has been localised close to the central gated channel and on both sides of the NPC.[33,42] It is composed of four essential nucleoporins: three of the FG-type, Nsp1p, Nup57p and Nup49, and one without repeats, Nic96p.[43,44] All three of these FG-nups have been found to associate with soluble transport factors, members of the karyopherin family (reviewed in ref. 3) and, indeed, mutations in these nups cause defects in both import and export processes (reviewed in ref. 45). The role of Nic96p in this complex is thought to be the anchoring of the FG components to the central core of the NPC.

Interestingly, Nic96p also physically interacts with other non-FG nups that are part of the structural framework of the NPC such as Pom152, Nup188p[46,47] and Nup192p.[48] Mutations in Nic96p and/or Nup192p cause a decrease in the total number of NPCs per nucleus.[49] Furthermore, severe structural alterations of the NPC and the NE were observed in cells carrying a dominant *nup188* mutant allele or following depletion of Pom152 in cells lacking Nup188p.[46]

The Nsp1p complex is conserved in evolution and its mammalian counterpart is the p62 complex composed of p62, p54, p58 and Nup93 (homologues of Nsp1p, Nup57p Nup49p, and Nic96p, respectively). Closely resembling the situation in yeast, Nup93 forms a separate sub-complex with the vertebrate nucleoporins Nup188 and Nup205 (homologues of yeast Nup188 and Nup192, respectively).[5,41,50,51] Taken together, the Nsp1p complex represents a conserved NPC sub-structure most likely involved both in the translocation of transport complexes across the central transporter and in NPC assembly.

A pool of Nsp1p, distinct from the one found in the described Nsp1p complex, also associates with what is referred to as the "Nup82p-complex".[2,3,5,33,41,42,45,52-57] This complex, which is localized exclusively at the cytoplasmic side of the NPC and may constitute part of the cytoplasmic fibers, consists of nucleoporins Nup82p and Nup159p. Nup82p also interacts physically with Nup116, which in its turn, binds to the mRNA export factor Gle2p. The association of Nup116-Gle2p to the core of the "Nup82p-complex" may be dynamic as the former two proteins are also localized to the nuclear side of the NPC.

Two features described for the "Nsp1p-complex" also apply to the "Nup82p-complex". First, the non-FG component Nup82p serves as a docking site for the repeat-containing constituents of the complex (Nsp1p, Nup159p and Nup116), which can bind to various transport factors and probably facilitate their translocation through the NPC. Second, a

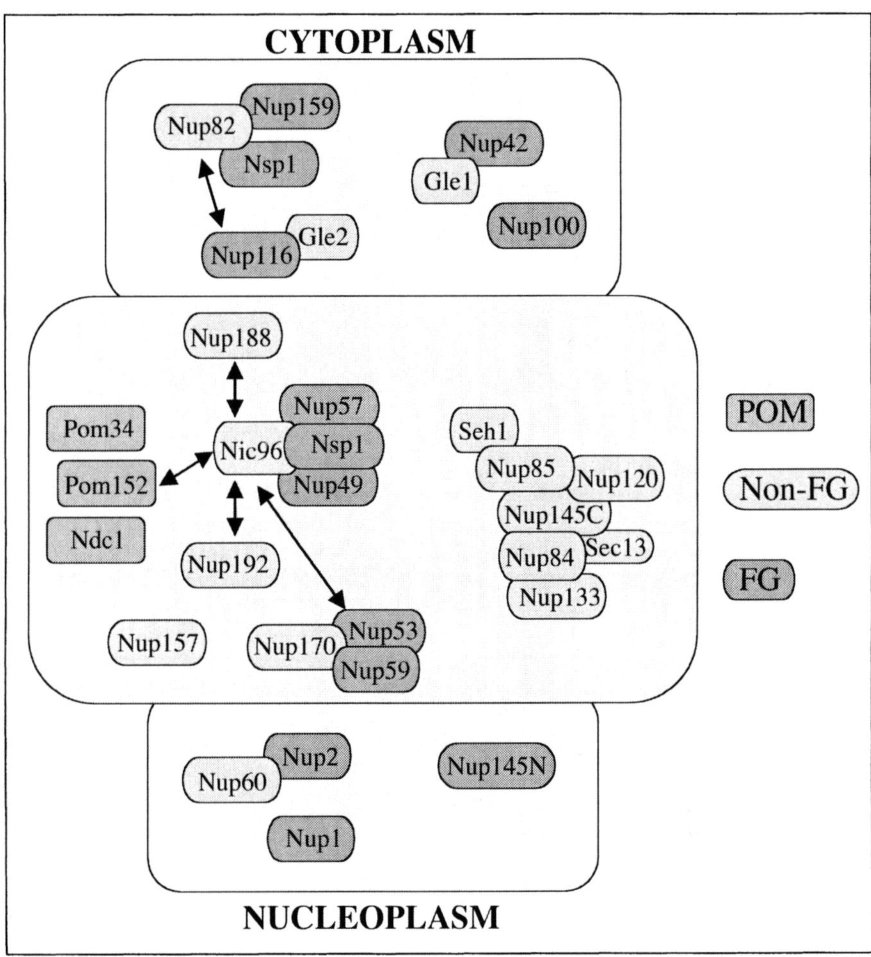

Figure 1. The nucleoporins of the yeast NPC are organized into complexes and are divided into three groups depending on their relative location on the NPC structure. Top: nucleoporins with an exclusive or biased localization at the cytoplasmic side of the NPC. Middle: nups with a symmetrical localization at both sides of the NPC. Bottom: nups with an exclusive or biased localization at the nucleoplasmic side of the NPC. (adapted from ref. 33). Nucleoporins represented by overlapping boxes belong to stable biochemical complexes. Double arrows indicate weaker physical interactions. Different shades of grey are used to represent integral membrane proteins of the NPC (POMs), FG nucleoporins and non-FG nucleoporins, as indicated.

functionally and structurally related complex also exists in vertebrates composed by Nup214/ CAN, Nup84 and, probably, p62. However, unlike the "Nsp1p-complex", Nup82p and its partners appear to be more important for mRNA export rather than protein import. Furthermore, mutations in any of these proteins also cause structural deformations of the NE and the NPC (reviewed in ref. 45). These range between the formation of herniations created by a double membrane seal over the cytoplasmic face of the NPCs (seen in *nup116* and *gle2* mutants) and NPC clustering (seen in *nup82* and *nup159* mutants).

The third stable yeast NPC subcomplex is the "Nup53-complex" composed of the core component Nup170p and two FG nucleoporins, Nup53p and Nup59p.[58] This complex has a symmetrical distribution with respect to the mid plane of the NPC, i.e., it is found on both the nuclear and cytoplasmic sides of the NPC and near its center. Despite the fact that none of these three nucleoporins is essential, distinct functions have been assigned to them. It appears that Nup53p serves as a primary and specific docking site for the karyorherin Kap121p, which is the main import factor for ribosomal proteins.[59] This is unlike other FG-nucleoporins, which have the property to bind more or less indiscriminately to various members of the β-karyopherin family. Kap121p, in turn, also appears to be required for nuclear import of Nup53 and its subsequent association with the NPC, which is mediated by Nup170p.

Nup170p also appears to be required for the NPC association of two further FG-nucleoporins Nup1p and Nup2p,[60] although these two nucleoporins are not part of the Nup53p complex and display a distinct localization, being detected only at the nuclear side of the NPC (see also below). Nup170p is homologous to the yeast Nup157 (which is also synthetic lethal to Nup170p) and to the abudant mammalian nucleoporin Nup155, which can complement *nup170* mutants in a remarkable display of functional conservation between yeast and mammals.[61] Nup53p has been reported to also physically interact with Nic96p, probably providing a link between the neighboring Nsp1p and Nup53p complexes.[62] Both Nup170p and Nup53p in collaboration with other genetically interacting nucleoporins are somehow involved in NE morphology. Depletion or overexpression of Nup170p in cells lacking Pom152 causes dramatic deformations in the NE.[61] The depletion results in enlarged and distorted NE containing massive extensions and invaginations and displaying long stretches without detectable NPCs. The overexpression, on the other hand, results in the appearance of membraneous structures that lie parallel and beneath the inner nuclear membrane, resembling intranuclear annulate lamellae.

Overexpression of Nup53p induces the formation of intranuclear, tubular membranes that fuse to form flattened double membrane lamellae resembling the nuclear membrane.[59] These lamellae displayed pores and contained two pore membrane proteins Pom152 and Ndc1p. However they lacked all other nucleoporins tested as well as visible NPCs. All these data suggest that the stoichiometric relationship between NPC components is required for the maintenance of normal nuclear structure and provide a link between the biogenesis of the NPC and the nuclear membrane (see also below). Intriguingly, a mutation in *NUP170* (which is supressed by *NUP157*) leads to defects in chromosome segragation and kinetochore function.[63]

The fourth well-characterized NPC subcomplex is the Nup84p-complex.[64] This symmetrically localized complex consists of exclusively non-FG nucleoporins and comprises seven proteins: Nup84p, Nup85p, Nup120p, Nup133p, Nup145-C, Seh1p and Sec13p. Mutations in the first five components severely impair growth and cause an inhibition of nuclear export of mRNA but not detectable defects in nuclear protein import [41,45]. In agreement with this, the mRNA export factors Mex67p-Mtr2p have been shown to associate physically with Nup85p.[65] Furthermore, yeast strains carrying mutations in these nucleoporins exhibit severe nuclear morphology defects such as NPC clustering (in *nup133, nup120* and *nup145-C* mutants) or NPC clustering combined with nuclear membrane abnormalities (e.g., long thin projections detached from the main body of the nucleus, "herniations", "blisters" and invaginations found in *nup84* and *nup85* mutants). The Nup145-C component corresponds to the carboxy-terminal part of Nup145p. This protein is proteolytically cleaved in vivo by a self-catalyzed mechanism to produce two functionally independent moieties, Nup145-N which contains FG-repeats and is not part of the Nup84p-complex and Nup145-C.[66,67]

The presence of Sec13p in the Nup84-complex is both surprising and intriguing. Sec13p is an essential protein, subunit of the COPII coat complex, with a well-established function in the budding of ER-derived vesicles.[68] However, a pool of Sec13p is found associated with the NPC and certain mutant alleles of Sec13p cause clustering and herniations of the NPCs.[69] Seh1p, which is non-essential, is related by sequence homology to Sec13p but, by localization criteria, behaves as a bona fide nucleoporin.

The Nup84p complex (lacking Nup133p) was purified from yeast and shown to exhibit a rough Y-shaped, triskelion like shape with a diameter of 25 nm and a molecular mass of 375 kDa.[69] The individual components were mapped onto this structure by an amazing series of sequential reconstitution experiments using proteins co-expressed in *E. coli*.[70] The two short arms of the Y-structure contain Nup120p and the Nup85p-Seh1p dimer, respectively. These connect to the Sec13p-Nup145-C dimer which together with the more distally placed Nup84p form the longer stalk. This hexameric complex can also finally connect in vitro to Nup133p via Nup84p to form a 0.5-MDa assembly. Taken together, the Nup84p-complex appears to be implicated in mRNA nuclear export, formation of the NPC and organization-biogenesis of the nuclear membrane. Its importance is reflected to the fact that an NPC subcomplex of similar composition and function has also been recently detected in vertebrates comprising Nup160, Nup96, Nup107, Nup133 and Sec13 (the homologues of yeast Nup120p, Nup145-C, Nup84p, Nup133p and Sec13p, respectively).[5,71]

The organization of nucleoporins in complexes discussed above together with their localization to particular substructures of the NPC provides a rough architectural map of the yeast NPC and offers clues for the possible mechanisms of nucleocytoplasmic transport and of NPC biogenesis and dynamics (Fig. 1). One conclusion that can be drawn from these studies is that three of the four main sub-complexes, i.e., the majority of the nucleoporins, are found on both sides of the NPC and at roughly the same distance from the central plane of the NE. This symmetrical distribution is most likely a reflection of the high level of symmetry displayed by the NPC as a whole and mostly by its central core. Furthermore, the distribution of the FG nucleoporins, which are thought to function as docking sites for soluble transport factors, all along and throughout the pathway across the NPC has important implications for the models of the translocation process.[13]

However, the directionality of the transport processes also creates the need for at least some elements of structural asymmetry which would corroborate the functional asymmetry imposed by Ran. These are indeed displayed by the peripheral structures of the NPC such as the cytoplasmic fibrils or the nuclear basket. A minority of nucleoporins with distinct functions in transport has been localized exclusively or predominantly to either of these structures. As already discussed, the Nup82-complex, involved in mRNA and probably also in rRNA export,[72] may constitute part of the cytoplasmic fibers of the NPC. Another set of nucleoporins such as Nup100p, Gle1p and Nup42p with a role in mRNA export[33,73] are also found mainly in the same location. On the contrary, a set of nucleoporins including Nup1p, Nup2p and Nup60p, which are though to play a specific role in the "classical" protein import pathway involving importin α (Kap60p, Srp1p), form part of the nuclear basket structure.[74-78] Therefore, the yeast nucleoporins localized on the peripheral NPC filaments may facilitate the rate limiting step of the translocation process which is the release of the transport substrates to their desired destinations, may it be nucleus or cytoplasm.

Apart of their functional implications, the nucleoporin interaction and localization studies also suggest that a NPC is built from a set of prefabricated subcomplexes which are recruited hierarchically to the construction site of a nuclear pore. As exemplified by the spectacular reconstitution of the Nup84p-complex, which is thought to constitute almost 15% of the mass of the NPC, the road has been paved for the ultimate goal of the in vitro assembly and subsequent detailed mapping of the structure and function of the entire yeast nuclear pore complex.

Biogenesis of the Yeast NPCs and Their Role in the Organization of the NE

Despite the wealth of data on the composition, structure and function of the yeast NPC, very little is known about how and where this complex organelle assembles. Since the yeast nuclear envelope remains intact throughout the cell cycle, NPC de novo biogenesis has to occur by insertion of the NPCs into the plane of a continuous nuclear membrane. In the case of vertebrate NPCs, monitoring the reassembly of NPCs in the reforming NE at the end of mitosis has allowed the observation of intermediates during NPC formation.[79] Such an analysis is, of course, not possible in yeast and no obvious intermediates in NPC formation have been detected in yeast cells, with one possible exception.

As already mentioned, overexpression of Nup53p leads to the formation of intranuclear double membrane lamellae interrupted by pores but lacking NPC structures.[59] It is therefore likely that these pores represent abortive intermediates in NPC assembly and that NPC formation is triggered by the binding of Nup53p and its interacting proteins at loci along the inner nuclear membrane. At these sites a regulated fusion may occur between the inner and the outer membrane leading to the formation of a pore of the correct diameter followed closely by the insertion of NPC subcomplexes in the NE.

The assembly process of a nascent NPC may be similar to the duplication of the SPB, the other major organelle of the yeast NE.[80] In this case, the formation of a pore across the NE and the insertion of the new SPB occur simultaneously. The fact that one of the proteins required for SPB duplication, Ndc1p, is also a component of the NPC suggests that there might be shared functions in the assembly of these two complexes into the yeast NE.[40] A link between NPC biogenesis and the formation of ER derived vesicles has been suggested by the presence of Sec13p at the NPC as a part of the Nup84p complex.[69] Sec13p may play a role in the membrane fusion step of NPC assembly or the stabilization of the bent nuclear pore membrane.

Analysis of the number and distribution of NPCs at distinct stages of the cell cycle has shown that the number of NPCs increases steadily and almost linearly, from approximately 85 per nucleus in G1-phase to 140 per nucleus at early mitosis.[81] This suggests that the assembly of the NPCs occurs continuously during the cell cycle. The NE surface area also rises throughout the cell cycle but not steadily and at a different rate than the number of NPCs, the greatest increase being observed between the S-phase and early mitosis. As a result, the density of the NPCs peaks in S-phase and drops in mitotic cells. The functional significance of these observations is not clear but they certainly indicate a timely coordination between assembly of NPCs and proliferation of the nuclear membrane. One might speculate that NPCs need to reach a certain density before the nuclear envelope can expand and extend into the growing bud.

The number of NPCs per nucleus falls significantly when cells carrying conditional mutations in Nic96p or Nup192p are shifted to the restrictive temperature, although the structure of the remaining NPCs appears normal.[49] It seems that these two core nucleoporins are required for early steps in NPC construction and in their absence assembly of new NPCs fails completely. As presented in detail in the previous section, mutations in other nucleoporin genes have a more dramatic effect on the structure of the NE membrane. One of the phenotypes observed in several different nucleoporin mutants is perturbations of the outer nuclear membrane at the NPC leading to the formation of double membrane seals over the cytoplasmic faces of the NPC. Electron-dense material often accumulates between these NPCs and the membrane seals resulting in "herniations" of the NE around individual NPCs.

A different phenomenon detected in the same as well as other nucleporin mutants, is a massive change in the shape of the nucleus causing thin finger-like projections, invaginations and extreme convolutions of the NE that appear to increase the total surface area of the nucleus. All these NE deformations are difficult to explain. One possibility is that the mutations affect the NPC function in nucleocytoplasmic transport, leading indirectly to the mislocalization or reduced synthesis of factors involved in nuclear membrane proliferation and maintenance. However, there is not always a correlation between a defect in NE structure and a defect in transport.

Another possibility is that formation of NPCs and growth of the nuclear membrane are coordinated processes, so that defects in NPC formation also destabilize the NE membrane and vice versa. This idea is supported by the fact that several integral membrane proteins of the NE (discussed in detail below) interact functionally with the NPC and when mutated can cause NE deformations similar to the ones already described. Additionally, defects in membrane lipid synthesis can also affect NPC function. A mutation in the yeast acetyl coenzyme A carboxylase (Acc1p), the enzyme responsible for fatty acid elongation, not only causes severe nuclear membrane perturbations but also impairs nuclear mRNA export.[82] It is thought that the lack of very-long-chain fatty acids observed in this mutant leads to destabilization of the curved nuclear pore membrane with detrimental effects on both NE structure and NPC assembly or function.

Finally, a third explanation for the NE deformities caused by mutations in nucleoporins is that NPCs provide attachements to cyto- or nucleo-skeletal structures that control the shape of the nuclear membrane. Although no direct link has been demonstrated between the NPC and the yeast actin- and microtubule-based networks, circumstantial evidence points to a connection between the NPC and cytoskeletal elements, the nature of which has not yet been fully characterized. First, a mutant allele of the yeast divergent actin gene *ACT2* has been identified which displays defects in NPC structure and nuclear import although the actin cytoskeleton remains normal.[83] In this mutant the NPCs are no longer embedded within, but rather lie on either side of, the nuclear envelope suggesting that Act2p may act as a scaffolding protein for the assembly of the NPC or the maintenance of its integrity.

Second, NPCs in normal yeast cells are neither evenly nor randomly distributed over the surface of the NE.[81] They rather display a modest level of clustering, creating regions of high and low local densities. One such region is the area of the NE in contact with the vacuole[84,85] which is devoid of NPCs. On the other hand, an area around the SPB becomes a preferential NPC clustering site in early mitotic nuclei. Interestingly, mutants in several yeast nucleoporins (see above) give rise to a "clustering "phenotype, wherein the NPCs are found to be concentrated in a single or very few patches of the NE, which sometimes aggregate to form grape-like structures. The most straightforward explanation for all these is that certain nucleoporins play a role in attaching NPC substructures to distinct cyto- or nucleo-skeletal elements. These attachments, however, have to be of dynamic nature.

Analysis of NPC movement in NPC clustering mutants has shown that NPCs can move laterally through the double nuclear membrane and redistribute fast to either form clusters (upon induction of the mutant phenotype) or break away from them (upon induction of the wild-type phenotype) to achieve a normal distribution.[84,86] Lateral diffusion of the NPCs in the plane of the NE was also observed during karyogamy, where fluorescently-labeled NPCs were shown to move from the donor section of the NE to that of the recipient nucleus. These results contrast to the recent observations that mammalian NPCs form immobile networks, most likely through their connections to the nuclear lamina.[87] Genetic experiments in *Droshophila* and *C. elegans* also indicate that the nuclear lamina is responsible for anchoring and evenly distributing the NPCs in the NE of higher eukaryotes.[88,89] Yeast do not contain a lamina and the nature of the postulated skeletal elements that regulate the distribution of the NPCs is still unknown. However, a connection between the yeast NPCs and the nuclear interior seems to be provided by filamentous structures formed by the proteins Mlp1 and Mlp2p, the yeast members of the Tpr family of proteins.[90,91]

It has been suggested that Mlp1p and Mlp2p form extensive filamentous structures radiating into the nucleoplasm from foci at the NPC periphery, perhaps attached at the distal ends of the NPC nuclear baskets.[90] In addition, a physical interaction has been observed between Mlp2p and nucleoporin Nic96p, which has been also localized at the distal ring of the nuclear basket.[91,92] Surprisingly, Mlp2p also physically interacts with Yku70, a component of the protein complex involved in telomeric localization near the NE, telomere maintenance and DNA double-strand break repair.[93,94] Furthermore, when Mlp1p and Mlp2p were deleted telomeres were no longer localized to the nuclear periphery, double-strand DNA break repair was impaired and telomeric gene silencing was reduced. The same defects plus nucleoplasmic mislocalization of Mlp2 were observed when Nup145-C was deleted. According to these data, it was suggested the Mlp proteins form NPC extensions (probably docked to either or both of Nic96p and Nup145-C) that tether chromosome ends to the nuclear periphery, thus implicating the NPCs into the subnuclear localization and functional organization of the chromatin. Alternatively, the Mlp proteins may be components of a nucleoskeletal structure functionally homologous to the nuclear lamina and responsible for the architectural arrangement of both NPCs and heterochromatin.

In conclusion, an emerging idea is that the yeast NPC is involved in cellular processes beyond its well-established function in nucleocytoplasmic transport. Perhaps, in the absence of a nuclear lamina and its associated proteins, yeast had no option but to entrust the NPC, the most abudant proteinaceous component of the NE, with a major role in the organization of the nuclear periphery. The few known yeast NE membrane proteins may also assist in this role as discussed in the following and last part.

Integral Membrane Proteins of the Yeast NE and Their Function

Despite the continuity of the inner and outer nuclear membranes, the two membranes are thought to perform distinct functions. The ONM is part of the rough ER, while the INM is considered to be responsible for the unique characteristics of the NE (reviewed in ref. 95). In higher eukaryotes the INM contains a characteristic set of integral membrane proteins which include the LBR (lamin B receptor), LAP1, LAP2, emerin, MAN1 and nurim. It is believed that at least some of these proteins contribute to the architecture of the NE by mediating interactions between the nuclear membrane and the nuclear lamina or the peripheral chromatin. None of these proteins, except LBR, have homologues in the yeast *S. cerevisiae*. Even in the case of LBR, the homology is restricted to the C-terminal membrane-spanning domain (also found in the yeast sterol reductase) while the N-terminal part of the LBR, which interacts with lamins and chromatin, has no apparent homologues in yeast.

In yeast, the inner nuclear membrane remains a completely unexplored territory and no proteins are known that exclusively localize in this domain. However, there is a handful of recently identified transmembrane proteins that can be classified as NE proteins mainly by functional criteria. These proteins are important for the morphology, organization and function of the NE and/or the NPC, although they do not directly associate with the NPC, and include Snl1p, Spo7p, Nem1p, Nvj1p and Brr6p. I will also briefly discuss Ire1p because of its putative inner nuclear membrane localization and because it may represent the only other means of transducing signals across the nuclear membrane apart of the NPC.

Snl1p is a small type I integral membrane protein (18.3 kDa) with a single putative transmembrane domain that localizes to both the nuclear and ER membranes.[96] Its gene was identified as a high-copy suppressor of the lethal phenotype caused by expression of the carboxy-terminal domain of Nup116p in the *nup116* null strain. Furthermore, high copy *SNL1* also suppressed the temperature sensitivity of cells carrying mutations in the nucleoporins Gle2p and Nic96p. Since at least two of the mutants suppressed by *SNL1* are characterized by NE deformations such as herniations, it has been suggested that Snl1p may function to stabilize the nuclear pore membrane or facilitate the fusion event that leads to the formation of new NPCs.

Both Spo7p and Nem1p were found to genetically interact with Nup84p as well as other nucleoporins.[97] Spo7p and Nem1p bind to each other forming a biochemically stable complex, localize to both the nuclear membrane and the ER and behave as integral membrane proteins. Their hydrophobic domains are likely to adopt an unusual loop-configuration causing the exposure of both N- and C-termini to the cytoplasm. Spo7p appears to be unique to yeasts while Nem1p belongs to a large conserved family of proteins, members of which are present at different cell locations and have phosphatase activity.[98] The most striking feature of Nem1p and Spo7p is that their absence causes a dramatic morphological change of the nucleus although cell viability is not impaired. Instead of being round the nuclei in *spo7* and *nem1* null strains are irregularly shaped and elongated, exhibiting long and thin projections that penetrate deeply into the cytoplasm. These elongations, which can take up the appearance of a second nucleus connected to the first one with a thin tubular double membrane, contain NPCs and intranuclear soluble proteins but, strikingly, not DNA.

Therefore, Nem1p and Spo7p are required for the normal spherical organization of the yeast nucleus. Their exact mechanism of action is not known but they may mediate or control the interactions of the nuclear membrane with the chromatin and underlying cyto- or nucleo-skeletal structures involved in maintaining the nuclear shape. Alternatively, they may negatively control the proliferation of the nuclear membrane by inhibiting its outgrowth at distinct sites. The phenotype of the *spo7* and *nem1* mutants may actually have a physiological counterpart. The only normal situation that the nuclear membrane detaches from the chromatin to form a protrusion is when the NE starts penetrating the mother-bud neck at the onset of mitosis.[99]

This physiological process of nuclear budding certainly requires structural rearrangements, the regulation of which may actually be defective in the *spo7* and *nem1* mutants.

SPO7 was originally identified as a gene essential for sporulation and more specifically premeiotic DNA replication.[100] Interestingly, also Nem1p as well as the NPC components that are functionally linked to the Spo7p/Nem1p complex are also essential for sporulation.[97] Therefore, not only the spherical nuclear organization but also the correct function/distribution of the NPCs is a prerequisite for meiotic nuclear division. The exact link between NE structure and the process of meiosis in yeast remains a very interesting, but still largely uninvestigated, question.

NE morphology is also affected in *brr6* mutants.[101] In this case, however, the structural defects consist of outer nuclear membrane bulges and the formation of clustered herniations, reminiscent of the deformations caused by mutations in Gle2p and Nup116p. In addition, depletion of Brr6p, which is an essential protein, also causes an altered distribution of NPCs and a defect in mRNA and protein nuclear export. Brr6p behaves as an integral membrane protein and localizes to the nuclear rim but appears not to be tightly associated with the NPC, although it genetically interacts with several nucleoporins. It has been suggested that Brr6p may be a constituent of a novel peri-pore NE domain, which would not only be required for NPC spacing but also for directing nuclear export cargoes to the gated channel of the NPC.

The role of Nvj1p appears to be quite different from the other membrane proteins described so far. Nvj1p was not identified by a genetic interaction with NPC components but rather by its physical association to the vacuolar membrane protein Vac8p.[85] Nvj1p appears to be concentrated in small patches or rafts at sites of close contact between the nucleus and the vacuole, an area from which NPCs are excluded. This localization of Nvj1p depends on Vac8p because in its absence Nvj1p re-localizes all around the nucleus. When Nvj1p is missing, nucleus-vacuole junctions are absent and upon its overexpression nucleus-vacuole contacts are multiplied. Overall, these results demonstrate the presence of a specialized NE domain, part of which is Nvj1p, involved in inter-organelle attachments.

Accumulation of unfolded proteins in the ER triggers a signaling cascade, known as the unfolded protein response (UPR), which leads to transcriptional up regulation of many target genes.[102] A key element of this pathway in both yeast and multicellular organisms is Ire1p, a transmembrane kinase, which is thought to transmit the UPR signal from the ER lumen to the nucleus.[103] Ire1p contains an N-terminal lumenal "stress sensing" domain, an ER transmembrane domain, and a cytoplasmically oriented part with kinase and endoribonuclease activities. In response to ER stress, Ire1p is activated and, along with the tRNA ligase Rlg1p, splices the mRNA of the yeast transcription factor Hac1p. The spliced mRNA can then produce active Hac1p, which can induce the transcription of the UPR target genes. In *C. elegans* and mammals the target of the endonucleolytic activity of Ire1p has been recently shown to be the mRNA of the transcription factor XBP1, also required for activation of the UPR.[104-107] The yeast tRNA ligase Rlg1p is a nuclear enzyme localized close to the inner side of the NE.[108] This, plus the fact that splicing reactions normally occur inside the nucleus, leads to the assumption that yeast Ire1p has to be also, at least transiently, located to the inner nuclear membrane. Indeed, this appears to be the case for the mammalian Ire1p, as recently suggested by biochemical fractionation experiments.[107]

The main factors of the UPR in yeast are also involved in the regulation of phospholipid synthesis, suggesting that the production of ER proteins is coordinated with the production of ER membrane lipids.[109] Indeed, this coordination appears to be a main feature of ER biogenesis. An example thereof, is the proliferation of the ER membrane upon overexpression of a number of different ER and non-ER integral membrane proteins.[99,109] This often results in the production of "karmellae", stacked membrane pairs wrapped around the nucleus. Nuclear envelope biogenesis probably occurs via membrane flow from the ER. It is not unlikely that the rules applied to ER biogenesis are also valid for the formation of the NE. A hypothesis would be that the suggested coordination between NPC and nuclear membrane biogenesis is ensured by a signaling pathway similar or identical to the UPR. This could lead to the integration of the biogenetic mechanisms for the two major membrane compartments and landmarks of the eukaryotic cell.

Acknowledgements

I would like to thank Dr. E. Hurt, Dr. S. Siniossoglou and Dr. E. Paraskeva for their critical reading of the manuscript and their useful comments.

References

1. Yang Q, Rout MP, Akey CW. Three-dimensional architecture of the isolated yeast nuclear pore complex: Functional and evolutionary implications. Mol Cell 1998; 1:223-234.
2. Stoffler D, Fahrenkrog B, Aebi U. The nuclear pore complex: From molecular architecture to functional dynamics. Curr Opin Cell Biol 1999; 11:391-401.
3. Ryan KJ, Wente SR. The nuclear pore complex: A protein machine bridging the nucleus and cytoplasm. Curr Opin Cell Biol 2000; 12:361-371.
4. Wente SR. Gatekeepers of the nucleus. Science 2000; 288:1374-1377.
5. Vasu SK, Forbes DJ. Nuclear pores and nuclear assembly. Curr Opin Cell Biol 2001; 13:363-375.
6. Strambio-de-Castillia C, Rout MP. The structure and composition of the yeast NPC. Results Probl Cell Differ 2002; 35:1-23.
7. Fahrenkrog B, Aebi U. The vertebrate nuclear pore complex: from structure to function. Results Probl Cell Differ 2002; 35:25-48.
8. Pante N, Kann M. Nuclear pore complex is able to transport macromolecules with diameters of ~39 nm. Mol Biol Cell 2002; 13:425-434.
9. Görlich D, Kutay U. Transport between the cell nucleus and the cytoplasm. Annu Rev Cell Dev Biol 1999; 15:607-660.
10. Conti E, Izaurralde E. Nucleocytoplasmic transport enters the atomic age. Curr Opin Cell Biol 2001; 13:310-319.
11. Quimby BB, Corbett AH. Nuclear transport mechanisms. Cell Mol Life Sci 2001; 58:1766-1773.
12. Kuersten S, Ohno M, Mattaj IW. Nucleocytoplasmic transport: Ran, beta and beyond. Trends Cell Biol 2001; 11:497-503.
13. Rabut G, Ellenberg J. Nucleocytoplasmic transport: Diffusion channel or phase transition? Curr Biol 2001; 11:R551-554.
14. Rout MP, Aitchison JD. The nuclear pore complex as a transport machine. J Biol Chem 2001; 276:16593-16596.
15. Strom AC, Weis K. Importin-beta-like nuclear transport receptors. Genome Biol 2001; 2:reviews 3008.1-3008.9.
16. Fornerod M, Ohno M. Exportin-mediated nuclear export of proteins and ribonucleoproteins. Results Probl Cell Differ 2002; 35:67-91.
17. Reed R, Hurt E. A conserved mRNA export machinery coupled to pre-mRNA splicing. Cell 2002; 108:523-531.
18. Simos G, Grosshans H, Hurt E. Nuclear export of tRNA. Results Probl Cell Differ 2002; 35:115-131.
19. Azuma Y, Dasso M. The role of Ran in nuclear function. Curr Opin Cell Biol 2000; 12:302-307.
20. Bischoff FR, Scheffzek K, Ponstingl H. How Ran is regulated. Results Probl Cell Differ 2002; 35:49-66.
21. Smith AE, Slepchenko BM, Schaff JC et al. Systems analysis of Ran transport. Science 2002; 295:488-491.
22. Dasso M. Running on Ran: Nuclear transport and the mitotic spindle. Cell 2001; 104:321-324.
23. Clarke PR, Zhang C. Ran GTPase: A master regulator of nuclear structure and function during the eukaryotic cell division cycle? Trends Cell Biol 2001; 11:366-371.
24. Fagotto F, Glück U, Gumbiner BM. Nuclear localization signal-independent and importin/karyopherin-independent nuclear import of β-catenin. Curr Biol 1998; 8:181-190.
25. Yokoya F, Imamoto N, Tachibana T et al. Beta-catenin can be transported into the nucleus in a Ran-unassisted manner. Mol Biol Cell 1999; 10:1119-1131.
26. Hetzer M, Mattaj IW. An ATP-dependent, Ran-independent mechanism for nuclear import of the U1A and U2B" spliceosome proteins. J Cell Biol 2000; 148:293-303.
27. Aitchison JD, Rout MP. The road to ribosomes. Filling potholes in the export pathway. J Cell Biol 2000; 151:F23-26.
28. Ho JH, Kallstrom G, Johnson AW. Nmd3p is a Crm1p-dependent adapter protein for nuclear export of the large ribosomal subunit. J Cell Biol 2000; 151:1057-1066.

29. Gadal O et al. Nuclear export of 60s ribosomal subunits depends on Xpo1p and requires a nuclear export sequence-containing factor, Nmd3p, that associates with the large subunit protein Rpl10p. Mol Cell Biol 2001; 21:3405-3415.

30. Milkereit P et al. Maturation and intranuclear transport of pre-ribosomes requires Noc proteins. Cell 2001; 105:499-509.

31. Saveanu C et al. Nog2p, a putative GTPase associated with pre-60S subunits and required for late 60S maturation steps. Embo J 2001; 20:6475-6484.

32. Senger B et al. The nucle(ol)ar Tif6p and Efl1p are required for a late cytoplasmic step of ribosome synthesis. Mol Cell 2001; 8:1363-1373.

33. Rout MP et al. The yeast nuclear pore complex: composition, architecture, and transport mechanism. J Cell Biol 2000; 148:635-651.

34. Miller BR, Forbes DJ. Purification of the vertebrate nuclear pore complex by biochemical criteria. Traffic 2000; 1:941-951.

35. Sträßer K, Baßler J, Hurt E. Binding of the Mex67p/Mtr2p heterodimer to FXFG, GLFG, and FG repeat nucleoporins is essential for nuclear mRNA export. J Cell Biol 2000; 150:695-706.

36. Chaillan-Huntington C, Braslavsky CV, Kuhlmann J et al. Dissecting the interactions between NTF2, RanGDP, and the nucleoporin XFXFG repeats. J Biol Chem 2000; 275:5874-5879.

37. Strawn LA, Shen T, Wente SR. The GLFG regions of Nup116p and Nup100p serve as binding sites for both Kap95p and Mex67p at the nuclear pore complex. J Biol Chem 2001; 276:6445-6452.

38. Fribourg S, Braun IC, Izaurralde E et al. Structural basis for the recognition of a nucleoporin FG repeat by the NTF2-like domain of the TAP/p15 mRNA nuclear export factor. Mol Cell 2001; 8:645-656.

39. Shulga N, Mosammaparast N, Wozniak R et alS. Yeast nucleoporins involved in passive nuclear envelope permeability. J Cell Biol 2000; 149:1027-1038.

40. Chial HJ, Rout MP, Giddings TH et al. *Saccharomyces cerevisiae* Ndc1p is a shared component of nuclear pore complexes and spindle pole bodies. J Cell Biol 1998; 143:1789-1800.

41. Doye V, Hurt EC. From nucleoporins to nuclear pore complexes. Curr Opin Cell Biol 1997; 9:401-411.

42. Fahrenkrog B, Hurt EC, Aebi U et al. Molecular architecture of the yeast nuclear pore complex: Localization of Nsp1p subcomplexes. J Cell Biol 1998; 143:577-588.

43. Grandi P, Doye V, Hurt EC. Purification of NSP1 reveals complex formation with 'GLFG' nucleoporins and a novel nuclear pore protein NIC96. EMBO J 1993; 12:3061-3071.

44. Grandi P, Schlaich N, Tekotte H et al. Functional interaction of Nic96p with a core nucleoporin complex consisting of Nsp1p, Nup49p and a novel protein Nup57p. EMBO J 1995; 14:76-87.

45. Fabre E, Hurt E. Yeast genetics to dissect the nuclear pore complex and nucleocytoplasmic trafficking. Annu Rev Genet 1997; 31:277-313.

46. Nehrbass U, Rout MP, Maguire S et al. The yeast nucleoporin Nup188p interacts genetically and physically with the core structures of the nuclear pore complex. Journal of Cell Biology 1996; 133:1153-1162.

47. Zabel U et al. Nic96p is required for nuclear pore formation and functionally interacts with a novel nucleoporin, Nup188p. J Cell Biol 1996; 133:1141-1152.

48. Kosova B, Pante N, Rollenhagen C et al. Nup192p is a conserved nucleoporin with a preferential location at the inner site of the nuclear membrane. J Biol Chem 1999; 274:22646-22651.

49. Gomez-Ospina N et al. Yeast nuclear pore complex assembly defects determined by nuclear envelope reconstruction. J Struct Biol 2000; 132:1-5.

50. Grandi P et al. Nup93, a vertebrate homologue of yeast Nic96p, forms a complex with a novel 205-kDa protein and is required for correct nuclear pore assembly. Mol Biol Cell 1997; 8:2017-2038.

51. Miller BR, Powers M, Park M et al. Identification of a new vertebrate nucleoporin, Nup188, with the use of a novel organelle trap assay. Mol Biol Cell 2000; 11:3381-3396.

52. Belgareh N et al. Functional characterization of a Nup159p-containing nuclear pore subcomplex. Mol Biol Cell 1998; 9:3475-3492.

53. Hurwitz ME, Strambio-de-Castillia C, Blobel G. Two yeast nuclear pore complex proteins involved in mRNA export form a cytoplasmically oriented subcomplex. Proc Natl Acad Sci USA 1998; 95:11241-11245.

54. Bailer SM et al. Nup116p and Nup100p are interchangeable through a conserved motif which constitutes a docking site for the mRNA transport factor Gle2p. EMBO J 1998; 17:1107-1119.

55. Bailer SM et al. Nup116p associates with the Nup82p-Nsp1p-Nup159p nucleoporin complex. J Biol Chem 2000; 275:23540-23548.

56. Ho AK et al. Assembly and preferential localization of Nup116p on the cytoplasmic face of the nuclear pore complex by interaction with Nup82p. Mol Cell Biol 2000; 20:5736-5748.

57. Bailer SM, Balduf C, Hurt E. The Nsp1p carboxy-terminal domain is organized into functionally distinct coiled-coil regions required for assembly of nucleoporin subcomplexes and nucleocytoplasmic transport. Mol Cell Biol 2001; 21:7944-7955.

58. Marelli M, Aitchison JD, Wozniak RW. Specific binding of the karyopherin Kap121p to a subunit of the nuclear pore complex containing Nup53p, Nup59p, and Nup170p. J Cell Biol 1998; 143:1813-1830.

59. Marelli M, Lusk CP, Chan H et al. A link between the synthesis of nucleoporins and the biogenesis of the nuclear envelope. J Cell Biol 2001; 153:709-724.

60. Kenna MA, Petranka JG, Reilly JL et al. Yeast Nle3p/Nup170p is required for normal stoichiometry of FG nucleoporins within the nuclear pore complex. Mol Cell Biol 1996; 16:2025-2036.

61. Aitchison JD, Rout MP, Marelli M et al. Two novel related yeast nucleoporins Nup170p and Nup157p: Complementation with the vertebrate homologue Nup155p and functional interactions with the yeast nuclear pore-membrane protein Pom152p. J Cell Biol 1995; 131:1133-1148.

62. Fahrenkrog B et al. The yeast nucleoporin Nup53p specifically interacts with Nic96p and is directly involved in nuclear protein import. Mol Biol Cell 2000; 11:3885-3896.

63. Kerscher O, Hieter P, Winey M et al. Novel role for a *Saccharomyces cerevisiae* nucleoporin, Nup170p, in chromosome segregation. Genetics 2001; 157:1543-1553.

64. Siniossoglou S et al. A novel complex of nucleoporins, which includes Sec13p and a Sec13p homolog, is essential for normal nuclear pores. Cell 1996; 84:265-275.

65. Santos-Rosa H et al. Nuclear mRNA export requires complex formation between Mex67p and Mtr2p at the nuclear pores. Mol Cell Biol 1997; 18:6826-6838.

66. Teixeira MT et al. Two functionaly distincts domains generated by in vivo cleavage of Nup145p: A novel biogenesis pathway for nucleoporins. EMBO J 1997; 16:5086-5097.

67. Teixeira MT, Fabre E, Dujon B. Self-catalyzed cleavage of the yeast nucleoporin Nup145p precursor. J Biol Chem 1999; 274:32439-32444.

68. Kaiser C, Ferro-Novick S. Transport from the endoplasmic reticulum to the Golgi. Curr Opin Cell Biol 1998; 10:477-482.

69. Siniossoglou S et al. Structure and assembly of the Nup84p complex. J Cell Biol 2000; 149:41-54.

70. Lutzmann M, Kunze R, Buerer A et al. Modular self-assembly of a Y-shaped multiprotein complex from seven nucleoporins. Embo J 2002; 21:387-397.

71. Belgareh N et al. An evolutionarily conserved NPC subcomplex, which redistributes in part to kinetochores in mammalian cells. J Cell Biol 2001; 154:1147-1160.

72. Gleizes PE et al. Ultrastructural localization of rRNA shows defective nuclear export of preribosomes in mutants of the Nup82p complex. J Cell Biol 2001; 155:923-936.

73. Strahm Y et al. The RNA export factor Gle1p is located on the cytoplasmic fibrils of the NPC and physically interacts with the FG-nucleoporin Rip1p, the DEAD-box protein Rat8p/Dbp5p and a new protein Ymr 255p. Embo J 1999; 18:5761-5777.

74. Denning D et al. The nucleoporin Nup60p functions as a Gsp1p-GTP-sensitive tether for Nup2p at the nuclear pore complex. J Cell Biol 2001; 154:937-950.

75. Dilworth DJ et al. Nup2p dynamically associates with the distal regions of the yeast nuclear pore complex. J Cell Biol 2001; 153:1465-1478.

76. Solsbacher J, Maurer P, Vogel F et al. Nup2p, a yeast nucleoporin, functions in bidirectional transport of importin alpha. Mol Cell Biol 2000; 20:8468-8479.

77. Hood JK, Casolari JM, Silver PA. Nup2p is located on the nuclear side of the nuclear pore complex and coordinates Srp1p/importin-alpha export. J Cell Sci 2000; 113:1471-1480.

78. Booth JW, Belanger KD, Sannella MI et al. The yeast nucleoporin Nup2p is involved in nuclear export of importin alpha/Srp1p. J Biol Chem 1999; 274:32360-32367.

79. Gant TM, Goldberg MW, Allen TD. Nuclear envelope and nuclear pore assembly: Analysis of assembly intermediates by electron microscopy. Curr Opin Cell Biol 1998; 10:409-415.

80. Adams IR, Kilmartin JV. Localization of core spindle pole body (SPB) components during SPB duplication in *Saccharomyces cerevisiae*. J Cell Biol 1999; 145:809-823.

81. Winey M, Yarar D, Giddings TH Jr, Mastronarde DN. Nuclear pore complex number and distribution throughout the *Saccharomyces cerevisiae* cell cycle by three-dimensional reconstruction from electron micrographs of nuclear envelopes. Mol Biol Cell 1997; 8:2119-2132.

82. Schneiter R et al. A yeast acetyl coenzyme a carboxylase mutant links very-long-chain fatty acid synthesis to the structure and function of the nuclear membrane-pore complex. Mol Cell Biol 1996; 16:7161-7172.

83. Yan C, Leibowitz N, Mélèse R. A role for the divergent actin gene, ACT2, in nuclear pore structure and function. EMBO J 1997; 16:3572-3586.

84. Belgareh N, Doye V. Dynamics of nuclear pore distribution in nucleoporin mutant yeast cells. J Cell Biol 1997; 136:747-759.

85. Pan X et al. Nucleus-vacuole junctions in *Saccharomyces cerevisiae* are formed through the direct interaction of Vac8p with Nvj1p. Mol Biol Cell 2000; 11:2445-2457.

86. Bucci M, Wente SR. In vivo dynamics of nuclear pore complexes in yeast. J Cell Biol 1997; 136:1185-1199.

87. Daigle N et al. Nuclear pore complexes form immobile networks and have a very low turnover in live mammalian cells. J Cell Biol 2001; 154:71-84.

88. Lenz-Bohme B et al. Insertional mutation of the *Drosophila* nuclear lamin Dm0 gene results in defective nuclear envelopes, clustering of nuclear pore complexes, and accumulation of annulate lamellae. J Cell Biol 1997; 137:1001-1016.

89. Liu J et al. Essential roles for *Caenorhabditis elegans* lamin gene in nuclear organization, cell cycle progression, and spatial organization of nuclear pore complexes. Mol Biol Cell 2000; 11:3937-3947.

90. Strambio-de-Castillia C, Blobel G, Rout MP. Proteins connecting the nuclear pore complex with the nuclear interior. J Cell Biol 1999; 144:839-855.

91. Kosova B et al. Mlp2p, a component of nuclear pore attached intranuclear filaments, associates with nic96p. J Biol Chem 2000; 275:343-350.

92. Fahrenkrog B, Aris JP, Hurt EC et al. Comparative spatial localization of protein-A-tagged and authentic yeast nuclear pore complex proteins by immunogold electron microscopy. J Struct Biol 2000; 129:295-305.

93. Galy V et al. Nuclear pore complexes in the organization of silent telomeric chromatin. Nature 2000; 403:108-112.

94. Tham WH, Zakian VA. Telomeric tethers. Nature 2000; 403:34-35.

95. Georgatos SD. The inner nuclear membrane: Simple, or very complex? Embo J 2001; 20:2989-2994.

96. Ho AK, Raczniak GA, Ives EB et al. The integral membrane protein Snl1p is genetically linked to yeast nuclear pore complex function. Mol Biol Cell 1998; 9:355-373.

97. Siniossoglou S, Santos-Rosa H, Rappsilber J et al. A novel complex of membrane proteins required for formation of a spherical nucleus. Embo J 1998; 17:6449-6464.

98. Siniossoglou S, Hurt EC, Pelham HR. Psr1p/Psr2p, two plasma membrane phosphatases with an essential DXDX(T/V) motif required for sodium stress response in yeast. J Biol Chem 2000; 275:19352-19360.

99. Wente S, Gasser S, Caplan A. The nucleus and nucleocytoplasmic transport in *Saccharomyces cerevisiae*. In: Pringle JR, Broach JR, Jones EW, eds. The Yeast *Saccharomyces*: Cell Cycle and Cell Biology. Cold Spring Harbor: CSHL Press, 1997.

100. Esposito MS, Esposito RE. Genes controlling meiosis and spore formation in yeast. Genetics 1974; 78:215-225.

101. de Bruyn Kops A, Guthrie C. An essential nuclear envelope integral membrane protein, Brr6p, required for nuclear transport. Embo J 2001; 20:4183-4193.

102. Patil C, Walter P. Intracellular signaling from the endoplasmic reticulum to the nucleus: the unfolded protein response in yeast and mammals. Curr Opin Cell Biol 2001; 13:349-355.

103. Urano F, Bertolotti A, Ron D. IRE1 and efferent signaling from the endoplasmic reticulum. J Cell Sci 2000; 113:3697-3702.

104. Shen X et al. Complementary signaling pathways regulate the unfolded protein response and are required for *C. elegans* development. Cell 2001; 107:893-903.

105. Yoshida H, Matsui T, Yamamoto A et al. XBP1 mRNA is induced by ATF6 and spliced by IRE1 in response to ER stress to produce a highly active transcription factor. Cell 2001; 107:881-891.

106. Calfon M et al. IRE1 couples endoplasmic reticulum load to secretory capacity by processing the XBP-1 mRNA. Nature 2002; 415:92-96.

107. Lee K et al. IRE1-mediated unconventional mRNA splicing and S2P-mediated ATF6 cleavage merge to regulate XBP1 in signaling the unfolded protein response. Genes Dev 2002; 16:452-466.

108. Clark MW, Abelson J. The subnuclear localization of tRNA ligase in yeast. J Cell Biol 1987; 105:1515-1526.

109. Powell KS, Latterich M. The making and breaking of the endoplasmic reticulum. Traffic 2000; 1:689-694.

CHAPTER 8

Nuclear Envelope Breakdown and Reassembly in *C. elegans*: Evolutionary Aspects of Lamina Structure and Function

Yonatan B. Tzur and Yosef Gruenbaum

Abstract

The *C. elegans* is an excellent model system to study the biological functions of the nuclear lamina. Though theoretically feasible in mice, the equivalent transgenic studies are far simpler more rigorous when done in *C. elegans*. In addition, the protein composition of the nuclear lamina in *C. elegans* is simpler as compared to that of higher eukaryotes. For example, there is only a single lamin gene and only three LEM domain genes in *C. elegans*, including homologs for emerin and MAN1. Furthermore, genetic analysis in *C. elegans* is revealing new lamina genes that are involved in nuclear migration, and lamina genes with a novel motif termed the SUN domain. The *C. elegans* nuclear envelope disassembles very late compared to vertebrates and *Drosophila*. The timing of nuclear envelope disassembly in *C. elegans* is novel; The pore complexes are absent only in metaphase, the nuclear membranes and lamina are absent only in mid-late anaphase and intranuclear mRNA splicing factors are still present in prometaphase. The dynamic of the nuclear envelope in *C. elegans* probably reflects its evolutionary position between unicellular and more complex eukaryotes.

The Structure and Protein Composition of the Nuclear Lamina in *C. elegans*

The structure of the nuclear envelope in *C. elegans* is similar to that of other higher eukaryotes. Like all metazoans, it is composed of an outer nuclear membrane, which is continuous with the endoplasmic reticulum and is covered with ribosomes. The outer nuclear membrane is separated from the inner nuclear membrane by a 20-40 nm wide perinuclear space. The two membranes fuse at the nuclear pore complexes. The nuclear lamina is a fibrous protein meshwork associated with the inner nuclear membrane. Abutting the nuclear lamina is the peripheral chromatin, which contains a large fraction of the heterochromatin(for recent reviews see refs. 1-3).

The nuclear envelope physically separates the nucleoplasm from the cytoplasm. This separation creates the cell nucleus in which DNA replication, RNA processing and ribosome assembly occur, while protein synthesis and cell signaling occur in the cytoplasm. The nuclear pore complexes regulate transport of macromolecules between the nucleoplasm and the cytoplasm.

The nuclear lamina includes lamins and lamin-associated proteins. A search for known nuclear lamina genes in the data bases of the nearly complete genome sequences of yeast, *C. elegans* and human reveals that the genetic composition of the nuclear lamina in *C. elegans* resembles that of *Drosophila* and human (Table 1). In contrast, yeast cells do not have any of

Nuclear Envelope Dynamics in Embryos and Somatic Cells
edited by Philippe Collas. ©2002 Eurekah.com and Kluwer Academic/Plenum Publishers.

the known metazoan nuclear lamina genes. However, in *C. elegans* the number of nuclear lamina genes and their splicing isoforms is smaller, as compared to *Drosophila* and human, indicating an evolutionarily increase in the complexity of the nuclear lamina. For example, the major protein constituents of the nuclear lamina are the lamin proteins, which are type V intermediate filament proteins (reviewed in ref. 4). The human genome contains three lamin genes: *LMNA, LMNB1* and *LMNB2,* which give at least seven splicing isoforms (lamins A, C, C2, A_10 and B1-3). The *Drosophila* genome has two lamin genes: lamin Dm_0 and lamin C, while the *C. elegans* genome contains only a single lamin gene, termed *lmn-1,* which encodes a single lamin protein, termed Ce-lamin.[5,6] Interestingly, most known disease-causing mutations in the human lamin A gene are conserved in *Drosophila* and *C. elegans* lamins (reviewed in ref.7). While most of the Ce-lamin protein is present at the nuclear periphery, a significant fraction of Ce-lamin is present in the nuclear interior.[5] This cellular distribution of Ce-lamin is similar to that of vertebrate lamins.[8]

The LEM-domain is a 43 amino-acids-long conserved motif composed of helical turn followed by two alpha helices connected by an 11 or 12-residue loop.[9-11] LEM-domain genes are found in all metazoan cells. The *C. elegans* genome contains only three LEM-domain genes, termed *emr-1, lem-2* and *lem-3* (Table 1). *Emr-1* encodes Ce-emerin and *lem-2* encodes Ce-MAN1. Ce-emerin and Ce-MAN1 are both integral proteins of the inner nuclear membrane associated with nuclear lamins.[12] *Lem-3* encodes the LEM-3 protein, which does not contain a transmembrane domain.

Emerin, MAN1 and a putative LEM-3 homolog are present in human. In contrast, a homolog to the vertebrate LAP2, which is a LEM-domain protein encoding at least six splicing isoforms, is not present in *C. elegans* or *Drosophila.* The human and the *Drosophila* genomes contain other LEM-domain genes that are not found in *C. elegans,* including the *Drosophila* otefin and several novel genes.

Another common motif in integral proteins associated with the nuclear lamina is the SUN domain, which is found in the *S. pombe* sad1 and the *C. elegans* UNC-84[13] and *mtf-1* genes (unpublished observation), and in several *Drosophila* and human genes. However, the cytological location of the SUN-domain proteins in human and *Drosophila* still needs to be determined. Sad1 is associated during meiosis and mitosis with both the nuclear envelope and the spindle pole body,[14] while UNC-84 is required for proper migration of nuclei in the hyp-7 cells, P-cell lineage and in the gut, and for nuclear anchoring in multi-nucleated cells. The Ring Finger Binding Protein (RFBP) was recently discovered in vertebrates to be an integral protein of the inner nuclear membrane.[15] It would be interesting to determine if the *C. elegans* RFBP homolog is also associated with the nuclear lamina.

Metazoan cells also contain non-membrane proteins that are associated with peripheral and internal nuclear lamins or lamin associated proteins (reviewed in ref. 1). These proteins include the hypophosphorylated form of the vertebrate retinoblastoma protein (pRb), the transcription factor Oct-1, barrier to autointegration factor (BAF), germ cell-less (GCL) and the *Drosophila* Young-Arrest (YA). Homologies for pRb, BAF and GCL are present in *C. elegans,* but their interaction with the nuclear lamina was not determined.

Possible Functions of the Nuclear Lamina in *C. elegans*

Genetic analysis of the *lmn-1* gene in *C. elegans* has the advantage of studying the consequences of complete depletion of lamins or a significant reduction in their amounts in the whole animal. The function of Ce-lamin *in vivo* was analyzed using the RNA interference technique (RNAi) by injecting *lmn-1* dsRNA into the gonads of adult hermaphrodites.[5] The *lmn-1* gene is an essential gene, since reduction in the amounts of Ce-lamin protein produce embryonic lethality. Interphase nuclei with reduced amounts of lamin lose their round shape and show a rapid change in their nuclear morphology, which demonstrates the critical role of intact nuclear lamina in determining the shape of the nucleus. Surprisingly, the abnormality in nuclear shape does not interfere with the ability of most nuclei to undergo mitosis and the timing of nuclear divisions appears similar to that of wild type embryos.

Table 1. *Comparison between known nuclear lamina genes in C. elegans and human. The numbers in the table represent the number of homologs found in "GenBank".*

	C. elegans	Human
Lamin A	0	1
Lamin B	1	2
LEM3	1	1
Man1	1	1
Emerin	1	1
LAP2	0	1
Other LEM's	0	Many
Unc 84	1	2-3
Matefin	1	0
RFBP	1	1
LBR	0	1
LAP1	0	1
Nurim	0	1

The condensation of the chromatin in the interphase nuclei of *lmn-1 (RNAi)* embryos is abnormal. This may be due to insufficient attachment of chromatin to the nuclear lamina. Given that lamins, as well as many nuclear membrane proteins, can bind directly to a chromatin partner (reviewed in ref. 3). it is possible that attachments between chromatin and lamina components are required to keep the chromatin in its normal condensation state. In addition, several inner nuclear membrane proteins, which also bind chromatin and may be involved in its spatial organization, require lamins for their nuclear envelope localization (see below).

One of the most common phenotypes in the *lmn-1 (RNAi)* embryos are "anaphase bridges", which are chromatin bridges connecting two sets of chromosomes unable to be separated from each other. "Anaphase bridges" are also observed in mammalian mitotic cells unable to degrade their cohesin proteins, which connect the two sister chromatids, via the anaphase promoting complex (APC)/cyclosome pathway.[16,17] The similarity between the phenotypes in lamin deficient embryos and mammalian cells with defective APC/cyclosome pathway suggests a link between nuclear lamina functions and the regulation of mitosis. Other common phenotypes in the *lmn-1 (RNAi)* are loss of chromosomes and unequal separation of chromosomes into daughter nuclei. These phenotypes indicate that functional lamina is also required for additional mitotic functions. In addition, the instability of nuclear morphology in the *lmn-1 (RNAi)* embryos may be a primary defect eventually leading to some defects in cell cycle progression and chromatin organization.

The Ce-lamin protein is required for the correct spacing of the nuclear pore complexes, since lack or reduced amounts of Ce-lamin cause clustering of the nuclear pore complexes. The role of the nuclear lamina in anchoring and spacing nuclear pore complexes is probably conserved in evolution, since similar phenotypes are seen in *Drosophila* cells with reduced lamin Dm_0 and in mouse cells lacking the *LMNA* gene.[18-20]

Loss of emerin or autosomal dominant mutations in the human *LMNA* gene cause Emery Dreifuss muscular dystrophy (EDMD), but the disease mechanism is not understood. The Ce-emerin protein is expressed and localized at the nuclear lamina in essentially all cell types except sperm, which lack Ce-emerin. Ce-emerin colocalizes with Ce-lamin and Ce-lamin is required for the nuclear lamina localization of emerin (Yosef Gruenbaum and Kathy Wilson, unpublished observations), similar to the mislocalization of emerin in certain cells in the *LMNA* knockout mouse.[20] In contrast, other known *C. elegans* lamina proteins do not depend on Ce-emerin for their localization. Under normal growth conditions, elimination of Ce-emerin expression by injection of *emr-1* dsRNA (*emr-1 (RNAi)*) cause no detectable phenotypes throughout development, similar to the loss of human emerin in most cells (Yosef Gruenbaum and Kathy Wilson unpublished observations).

Mutations in the *C. elegans unc-84* gene cause defects in nuclear migration and anchoring.[13] UNC-84 protein co-localizes at the nuclear periphery[13] and is requires intact nuclear lamina for its nuclear envelope localization.

There are other roles suggested for the nuclear lamina, including role in DNA replication.[21] disassembly and reassembly of the nuclear envelope during mitosis[22] and apoptosis.[1] Given the similarity in protein composition and cellular localization of lamina proteins between *C. elegans* and vertebrates, it is likely that the nuclear lamina in *C. elegans* also plays a role in these activities.

Nuclear Dynamics in *C. elegans* During Mitosis

In vertebrates, the nuclear envelope undergoes an open mitosis. It starts to break down at prophase. By late prophase/prometaphase transition all nuclear envelope components disassemble (Table 2). Nuclear pore complex subunits are dispersed into the cytoplasm,[23-25] nuclear membranes and membrane proteins merge into the ER network[26,27] and the lamina depolymerizes into both soluble and membrane-associate pools.[28,4] During the anaphase/telophase transition, the vertebrate nuclear envelope reassembles around the decondensing chromatin.[29] The nuclear membranes attach to the chromatin and fuse to give the inner and outer nuclear membranes, nuclear pore complexes assemble and nuclear proteins including a large fraction of the lamins are imported back into the nucleus (reviewed in refs. 30,22). In *Drosophila* early embryos, nuclear pore complexes break down during prophase, similar to vertebrates. However, until mid-late anaphase the nuclear membranes and a fraction of the lamina remain in a spindle envelope (Fig. 1B) and are supplemented by a temporary second layer of membranes.[31,32] In contrast, *S. cerevisiae* has a closed mitosis, wherein the nuclear envelope remains intact and tubulin proteins are imported to allow mitotic spindles to assemble inside the nucleus.[33]

The *C. elegans* nuclear envelope disassembles very late in mitosis compared to vertebrates and *Drosophila*. The nuclear membranes and nuclear lamins remain in the nuclear periphery during metaphase and early anaphase, disassembling completely only during mid-late anaphase (Fig. 1A). Nuclear pore complexes remain in the nuclear periphery during prometaphase and intranuclear mRNA splicing factors leave the nucleus only after prometaphase.[12] *C. elegans* cells achieve a fully open mitosis only during mid-late anaphase when the nuclear lamina and membranes are also disassembled. Thus, *C. elegans* has a fully open mitosis, similar to other metazoans, and different from the closed mitosis in single-cells eukaryotes such as *S. cerevisiae*.[34] The main difference between *C. elegans* and other complex eukaryotes is the stage at which mitosis becomes fully open.

The picture emerging from the above data is that metazoan evolution was accompanied by an increase in the ability of different nuclear envelope components to disassemble early in mitosis, while the timing of reassembly of nuclear envelope components at the end of mitosis remained similar between *C. elegans* and vertebrates.

Open mitosis probably co-evolved with the nuclear lamina. Lamina proteins probably conferred a selective advantage to metazoan creatures, perhaps related to improvements in chromatin organization, or improved nuclear signaling, DNA replication or gene expression. However, lamin filaments could interfere with chromosome segregation during mitosis and therefore lamina disassembly is probably required to allow chromosome segregation. Open mitosis has another advantage, which is the exposure of chromatin to cytosolic proteins, which might provide new means to regulate the cell cycle through access to cytosolic replication licensing factors[1]. Furthermore, the process of nuclear assembly itself might provide new mechanisms for regulating chromatin structure during development and differentiation.

Table 2. **The timing of nuclear envelope disassembly in S. cerevisiae, C. elegans, D. melanogaster, and vertebrates. (+) indicates that the component is mostly intact; (+/-) partial disassembly of the component and release to the cytoplasm, (+/--) almost complete disassembly of the component; (-) complete disassembly of the component. INT-interphase; PRO – prophase; PROM – prometaphase, MET – metaphase, E. ANA – early anaphase; L. ANA – late anaphase; TEL – telophase; E. G1- early G1**

		Yeast	*C. elegans*	*C. elegans* early	*Drosophila* late	*Drosophila* early	Vertebrates late
Lamina	INT		+	+	+	+	+
	PRO		+	+	+	+/-	+/-
	PROM		+	+	+	-	-
	MET		+	+/-	+/-	-	-
	E. ANA		+	+/-	+/--	-	-
	L. ANA		-	-	-	-	-
	TEL		+/-	+/-	+/-	+/-	+/-
	E. G1		+	+	+	+	+
Nuclear	INT	+	+	+	+	+	+
Pore	PRO	+	+	+	+/-	+/-	+/-
Complexes	PROM	+	+	+	-	-	-
	MET	+	+	-	-	-	-
	E. ANA	+	-	-	-	-	-
	L. ANA	+	-	-	-	-	-
	TEL	+	+	+	+	+/-	+/-
	E. G1	+	+	+	+	+	+
Membranes	INT	+	+	+	+	+	+
	PRO	+	+	+	+	+/-	+/-
	PROM	+	+	+	+	-	-
	MET	+	+	+	+	-	-
	E. ANA	+	+	+	+	-	-
	L. ANA	+	-	-	+/-	-	-
	TEL	+	+	+	+	+/-	+/-
	E. G1	+	+	+	+	+	+

In *C. elegans* the timing of disassembly of the nuclear pore complexes, and to a lesser extent the nuclear lamin depends on embryo age. In embryos with fewer than 24-26 cells, nucleoporins (detected by mAb414) are present in the nuclear periphery during interphase, prophase, prometaphase and to most extent in metaphase. Nucleoporins completely disassemble during early anaphase a reassemble during telophase.[12] In older embryos (more than 24-26 cells) pore complexes disintegrate in prometaphase and are absent during metaphase and anaphase. Nuclear pore complexes in *C. elegans* disassemble earlier than other nuclear envelope structures, similar to vertebrates and *Drosophila*.[35] However, the nuclear pore complexes remain until after prometaphase, strikingly later than their disassembly in early prophase in mammalian cells and *Drosophila* (reviewed in ref. 1). There is no obvious explanation why the disassembly of the nuclear pore complexes occurs at different stages in early and late embryos. The length of cell division is variable in both early and late embryos and cell divisions are not synchronous. In addition, in *C. elegans*, there is no obvious equivalent to the midblastula transition since

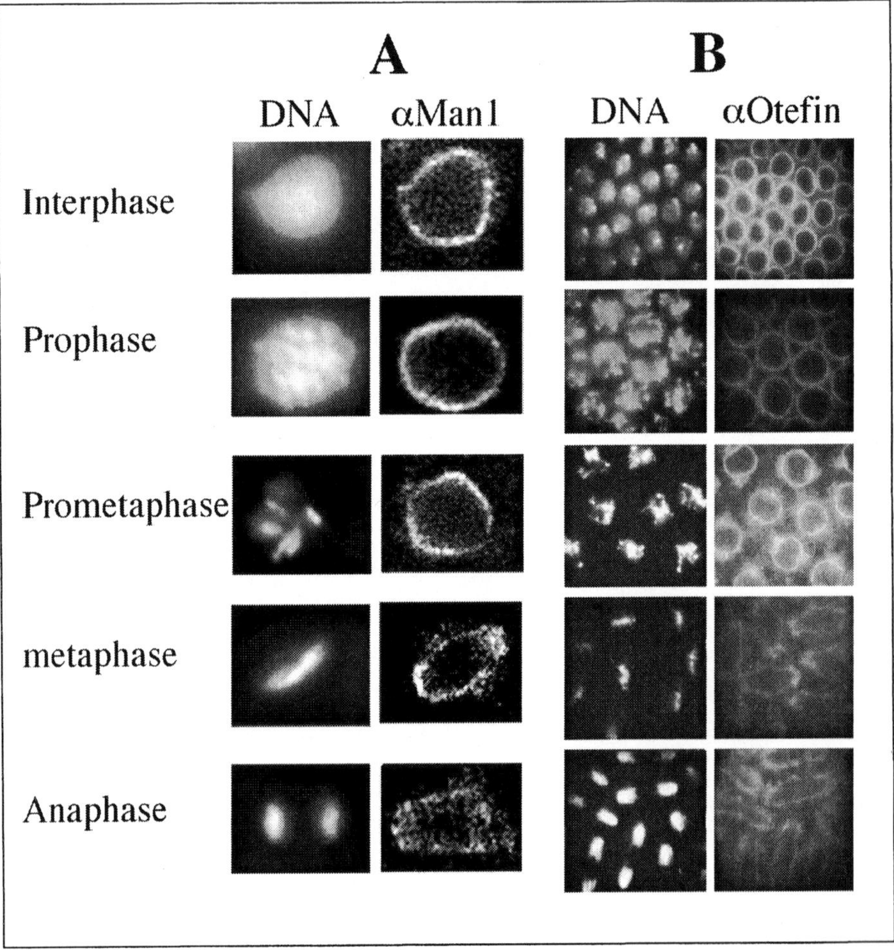

Figure 1. Nuclear membrane disassembly during mitosis in *C. elegans* and *Drosophila* early embryos. Immunofluorescence localization of the *C. elegans* Ce-MAN1 (A) and the *Drosophila* Otefin (B) at different stages of mitosis. Representative nuclei from most different stages in mitosis are shown.

transcription begins as early as the 3-4 cell stage[36,37] and is lineage-dependent. However, the breakpoint between 'early' and 'later' embryonic phenotypes occurs around the 24-cell stage, when the progenitor cells for all six major lineages (AB, MS, E, C, D, P4) have been created. It is also possible that there are differences in protein composition of the nuclear envelope in early and late embryos. If true, such differences could be important for selective transport of proteins and may also affect the timing of nuclear pore complex (and to some extent the nuclear lamina) disassembly during mitosis.

In *C. elegans* embryos, Ce-lamin is maintained in a spindle envelope structure during metaphase and early anaphase. The exception is near the spindle poles, where lamin staining becomes progressively weaker starting in prometaphase, with a large gap at both poles seen during early anaphase. This local disruption of lamina integrity is consistent with mechanical puncturing by spindle microtubules as seen in *C. elegans* and in other organisms.[31,32,38] During mid-late anaphase, the Ce-lamin filaments completely disassemble and the protein is dispersed in the cytoplasm. In later embryos (>24-26 cells), the lamina disassembles more extensively already during prometaphase and metaphase. Thus, in *C. elegans* embryos, the lamina structure persists much longer than the lamina in vertebrate cells. The timing of Ce-lamin assembly is similar to vertebrate and *Drosophila* lamins,[39,8] Ce-lamin reassociates with chromatin during telophase, but do not completely reassemble until G1-phase.

Analysis of integral membrane proteins in *C. elegans* reveals that nuclear membranes completely disassemble only during mid-late anaphase in both early and late *C. elegans* embryos[12] (Gruenbaum, unpublished observations). These proteins maintain a spindle envelope staining during prophase, prometaphase, metaphase and early anaphase and completely disassemble only during mid-late anaphase. Thus, in *C. elegans* embryos, the nuclear membranes and membrane proteins persist much longer than nuclear membranes of vertebrate cells.

The late disassembly of the nuclear envelope in *C. elegans* is in correlation with the finding that intranuclear proteins are released from nuclei during mitosis only at metaphase. For example, the endogenous SR (serine/arginine-rich) family of conserved small nuclear ribonucleoproteins (snRNPs) are involved in mRNA splicing, and can be detected with monoclonal antibody mAb104.[40] MAb104 recognizes phosphorylated snRNPs from a wide variety of vertebrate and invertebrate species, including *C. elegans*, during both interphase and mitosis. In mammalian cells, these splicing factors are released into the cytoplasm during early prophase.[40,41] In contrast, in *C. elegans*, these proteins have intranuclear localization during interphase, early prophase and late prophase, and dispersed in the cytoplasm only in metaphase, consistent with the disassembly of the nuclear pore complexes and invasion of microtubules into the nucleus.[12]

Acknowledgments

We thank Michael Brandeis and Merav Cohen for useful discussions and comments on the manuscript. This work was supported by grants from the USA-Israel Binational Science Foundation (BSF), the Israel Science Foundation (ISF) and the German-Israel Foundation GIF #1-573-036.13.

References

1. Cohen M, Lee KK, Wilson KL et al. Transcriptional repression, apoptosis, human disease and the functional evolution of the nuclear lamina. Trends Bioc Sci 2001; 26:41-47.
2. Hutchison CJ, Alvarez-Reyes M, Vaughan OA. Lamins in disease: why do ubiquitously expressed nuclear envelope proteins give rise to tissue-specific disease phenotypes? J Cell Sci 2000; 114:9-19.
3. Wilson KL, Zastrow MS, Lee KK. Lamins and disease: Insights into nuclear infrastructure. Cell 2001; 104:647-650.
4. Stuurman N, Heins S, Aebi U. Nuclear lamins: Their structure, assembly, and interactions. J Struct Biol 1998; 122:42-66.
5. Liu J, Rolef-Ben Shahar T, Riemer D et al. Essential roles for *Caenorhabditis elegans* lamin gene in nuclear organization, cell cycle progression, and spatial organization of nuclear pore complexes. Mol Biol Cell 2000; 11:3937–3947.
6. Riemer D, Dodemont H, Weber K. A nuclear lamin of the nematode *Caenorhabditis elegans* with unusual structural features; cDNA cloning and gene organization. Eur J Cell Biol 1993; 62:214-223.
7. Gruenbaum Y, Wilson KL, Harel A et al. Nuclear lamins—Structural proteins with fundamental functions. J Struct Biol 2000; 129:313-323.
8. Moir RD, Spann TP, Lopez-Soler RI et al. The dynamics of the nuclear lamins during the cell cycle—Relationship between structure and function. J Struct Biol 2000; 129:324-334.
9. Cai M, Huang Y, Ghirlando R et al. Solution structure of the constant region of nuclear envelope protein LAP2 reveals two LEM-domain structures: One binds BAF and the other binds DNA. EMBO J 2001; 20:4399-4407.
10. Laguri C, Gilquin B, Wolff N et al. Structural Characterization of the LEM Motif Common to Three Human Inner Nuclear Membrane Proteins. Structure 2001; 9:503-511.
11. Wolff N, Gilquin B, Courchay K et al. Structural analysis of emerin, an inner nuclear membrane protein mutated in X-linked Emery^Dreifuss muscular dystrophy. FEBS Lett 2001; 501:1-6.
12. Lee KK, Gruenbaum Y, Spann P et al. *C. elegans* nuclear envelope proteins emerin, MAN1, lamin, and nucleoporins reveal unique timing of nuclear envelope breakdown during mitosis. Mol Biol Cell 2000; 11:3089-3099.
13. Malone CJ, Fixsen WD, Horvitz HR et al. UNC-84 localizes to the nuclear envelope and is required for nuclear migration and anchoring during *C. elegans* development. Development 1999; 126:3171-3181.

14. Hagan I, Yanagida M. The product of the spindle formation gene sad1+ associates with the fission yeast spindle pole body and is essential for viability. J Cell Biol 1995; 129:1033-1047.

15. Mansharamani M, Hewetson A, Chilton BS. An atypical nuclear P-type ATPase is a RING-finger binding protein. J Biol Chem 2001; 276:3641–3649.

16. Zur A, Brandeis M. Securin degradation is mediated by fzy and fzr, and is required for complete chromatid separation but not for cytokinesis. EMBO J 2001; 20:792-801.

17. Hauf S, Waizenegger IC, Peters J-M. Cohesin cleavage by separase required for anaphase and cytokinesis in human cells. Science 2001; *293*:1320-1323.

18. Harel A, Goldberg M, Ulitzur N et al. Structural organization and biological roles of the nuclear lamina. In: Boulikas T, ed. Textbook of Gene Therapy and Molecular Biology: "From Basic Mechanism to Clinical Applications". 1998; 1:529-542.

19. Lenz-Bohme B, Wismar J, Fuchs S et al. Insertional mutation of the *Drosophila* nuclear lamin dm(0) gene results in defective nuclear envelopes, clustering of nuclear pore complexes, and accumulation of annulate lamellae. J Cell Biol 1997; 137:1001-1016.

20. Sullivan T, Escalente-Alcalde D, Bhatt H et al. Loss of A-type lamin expression compromises nuclear envelope integrity leading to muscular dystrophy. J Cell Biol 1999; 147:913-920.

21. Moir RD, Spann TP, Herrmann H, Goldman RD. Disruption of nuclear lamin organization blocks the elongation phase of DNA replication. J Cell Biol 2000; 149:1179-1192.

22. Gant TM, Wilson KL. Nuclear assembly. Annu Rev Cell Dev Biol 1997; 13:669-695.

23. Davis LI, Blobel G. Identification and characterization of a nuclear pore complex protein. Cell 1986; 45:699-709.

24. Gerace L, Blobel G. Nuclear lamina and the structural organization of the nuclear envelope. Cold Spring Harb Symp Quant Biol 1982; 46:967-978.

25. Snow CM, Senior A, Gerace L. Monoclonal antibodies identify a group of nuclear pore complex glycoproteins. J Cell Biol 1987; 104:1143-1156.

26. Ellenberg J, Lippincott-Schwartz J. Dynamics and mobility of nuclear envelope proteins in interphase and mitotic cells revealed by green fluorescent protein chimeras. Methods 1999; 19:362-372.

27. Yang Y, Smith HC. Multiple protein domains determine the cell type-specific nuclear distribution of the catalytic subunit required for apolipoprotein B mRNA editing. Proc Natl Acad Sci USA 1997; 94:13075-13080.

28. Moir RD, Spann TP, Goldman RD. The dynamic properties and possible functions of nuclear lamins. Int Rev Cytol 1995; 162b:141-182.

29. Haraguchi T, Koujin T, Hayakawa T et al. Live fluorescence imaging reveals early recruitment of emerin, LBR, RanBP2, and Nup153 to reforming functional nuclear envelopes. J Cell Sci 2000; 113:779-794.

30. Collas P, Poccia D. Membrane fusion events during nuclear envelope assembly. Subcell Biochem 2000; 34:273-302.

31. Harel A, Zlotkin E, Nainudel ES et al. Persistence of major nuclear envelope antigens in an envelope-like structure during mitosis in *Drosophila melanogaster* embryos. J Cell Sci 1989; 94:463-470.

32. Stafstrom JP, Staehelin AL. Dynamics of the nuclear envelope and of nuclear pore complexes during mitosis in the *Drosophila* embryo. Euro J Cell Biol 1984; 34:179-189.

33. Gerace L, Burke B. Functional organization of the nuclear envelope. Annu Rev Cell Biol 1988; 4:335-374.

34. Heath IB. Variant mitoses in lower eukaryotes: Indicators of the evolution of mitosis. Int Rev Cytol 1980; 64:1-80.

35. Terasaki M, Campagnola P, Rolls M et al. A new model for nuclear envelope breakdown. Mol Biol Cell 2001; 12:503-510.

36. Newman-Smith ED, Rothman JH. The maternal-to-zygotic transition in embryonic patterning of Caenorhabditis elegans. Curr Opin Genet Dev 1998; 8:472-480.

37. Seydoux G, Fire A. Soma-germline asymmetry in the distributions of embryonic RNAs in *Caenorhabditis elegans*. Development 1994; 120:2823-2834.

38. Terasaki M. Dynamics of the endoplasmic reticulum and golgi apparatus during early sea urchin development. Mol Biol Cell 2000; 11:897-914.

39. Hutchison CJ, Bridger JM, Cox LS et al. Weaving a pattern from disparate threads: Lamin function in nuclear assembly and DNA replication. J Cell Sci 1994; 107:3259-3269.

40. Roth MB, Murphy C, Gall JG. A monoclonal antibody that recognizes a phosphorylated epitope stains lampbrush chromosome loops and small granules in the amphibian germinal vesicle. J Cell Biol 1990; 111:2217-2223.

41. Zahler AM, Neugebauer KM, Lane WS, Roth MB. Distinct functions of SR proteins in alternative pre-mRNA splicing. Science 1993; 260:219-222.

Nuclear Envelope Assembly in Gametes and Pronuclei

D. Poccia, T. Barona, P. Collas and B. Larijani

Abstract

Studies using a cell-free system from fertilized sea urchin eggs have revealed several novel aspects of nuclear envelope formation unlikely to be restricted to either sea urchins or gametes. These include detergent-resistant lipophilic structures that target nuclear envelope precursor membrane vesicle binding and incorporate into the nuclear envelope upon membrane fusion, a highly phosphatidylinositol (PI)-enriched membrane fraction conferring special fusion properties on the system, PI-specific phospholipase C-induced membrane fusion, polarized binding of cytoplasmic vesicles to the nucleus, and a role for PI phosphorylated in the 3 position by PI-3 kinase.

Introduction

The nuclear envelope is disassembled and reassembled at each mitosis in typical animal cells. Mitotic NE dynamics are orchestrated by cell cycle control mechanisms and thus coordinated with other mitotic events. Multiple nuclei in a common cytoplasm typically undergo these processes synchronously.[1]

The processes of disassembly and reassembly may also occur in interphase and are usually but not always coordinate in nuclei sharing a common cytoplasm, for example in fertilized eggs.[2,3] Most eggs are fertilized during meiosis and female nuclei may be arrested at various stages of meiosis I or II. Male nuclear envelopes however are disassembled and reassembled in all cases. Although the timing may vary, most of the biochemical events underlying interphase, mitotic and meiotic nuclear envelope dynamics are likely to be similar.

This review will focus on recent biochemical understanding of the processes of male pronuclear membrane assembly and disassembly in fertilized sea urchin eggs investigated with cell-free extracts. These studies have revealed several novel features of nuclear membrane dynamics which will be discussed in relation to mitotic or meiotic transitions in other organisms.

Background

Sea urchin eggs are arrested in G1 (G0) and activated by the fertilizing sperm. Electron microscope studies show that the sperm nucleus enters the egg with a nuclear envelope lacking pores.[4,5] The envelope is rapidly disassembled in vivo, its membranes vesiculating as the sperm chromatin decondenses from its compact conoid shape to a uniformly euchromatic spherical mass. During this process, remnants of the sperm nuclear envelope at the tip and base of the conical nucleus, which line two cup-shaped cavities (the acrosomal and centriolar fossae), are retained. Membrane vesicles from the egg cytoplasm accumulate along the sides of the nucleus and fuse to form a nuclear envelope with pores. The sperm remnants are incorporated into the new male pronuclear envelope. Swelling of the nucleus follows as the male pronucleus migrates

Nuclear Envelope Dynamics in Embryos and Somatic Cells
edited by Philippe Collas. ©2002 Eurekah.com and Kluwer Academic/Plenum Publishers.

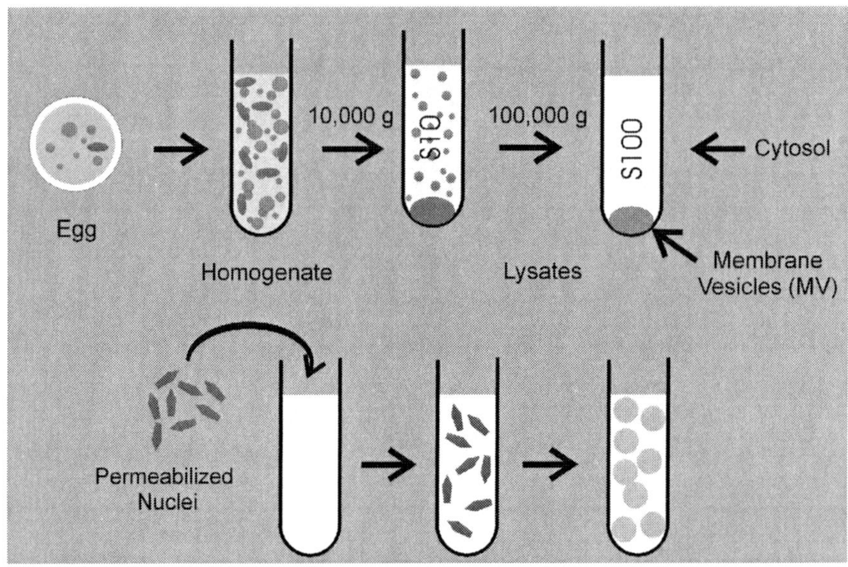

Figure1. Preparation and use of a cell-free system for nuclear envelope formation. Fertilized eggs are homogenized at 10 min post-fertilization, centrifuged at 10,000 g for 10 min and the clear supernatant (S10) removed. S10 cytoplasm may be fractionated into a membrane fraction (MV) and a cytosolic extract (S100 or S150) by centrifugation for 1-2 hr at 100,000 g or 150,000 g. Isolated sperm nuclei are extracted with 0.1% Triton X-100, washed and added to cytoplasm or cytosol and incubated for appropriate lengths of time with additional ATP (ATP generating system) and/or GTP. Please see http://www.eurekah.com/chapter.php?chapid=697&bookid=56&catid=15 for the color version of the figure.

through the cytoplasm and encounters the female pronucleus whose envelope remains intact throughout these processes. A fusion of their two nuclear envelopes results in formation of a single zygotic nucleus prior to initiation of replication and transcription in the first cell cycle.

Most of these in vivo transitions have been mimicked in a cell-free system (Fig. 1).[6] Below we review the details of sea urchin sperm male pronuclear envelope dynamics as revealed by in vitro experiments. Many comprehensive reviews of nuclear envelope transitions in meiotic cell extracts are available for comparison.[7-11]

Sperm Nuclear Envelope Disassembly

At the completion of spermatogenesis, the sperm nuclear envelope is closely apposed to the highly condensed sperm chromatin and lacks pores.[4] This morphology suggests a nuclear envelope non-permissive for transport. Since transcription has ceased prior to the last stages of spermatogenesis and does not begin again until the sperm nuclear envelope has been replaced following fertilization,[12,13] there is no need to export RNA.

Presumably nuclear import of proteins is also unnecessary for an inert nucleus which is inactive in replication and transcription at late spermiogenic stages. However, immediately following fertilization, cytoplasmic proteins including sperm histone kinases need access to the chromatin. This is provided by disassembly of the envelope. As noted, disassembly takes place everywhere but at the tip and base of the sperm head and appears to proceed via vesiculation of the sperm nuclear envelope membranes.[14] This vesiculation has not been reproduced in vitro. Instead nuclear membranes are solubilized with a non-ionic detergent prior to exposure to egg cytoplasmic extracts.[6] The resulting nuclei retain membranes at the tip and base[15] and a lamina as revealed by immunofluorescence with a variety of anti-lamin antibodies.[15-17]

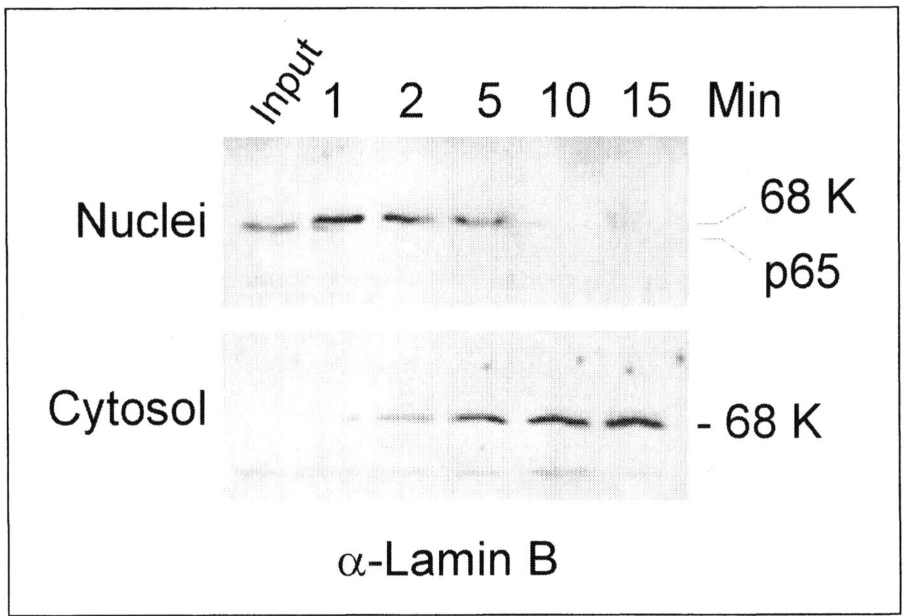

Figure 2. Solubilization of sperm B-type lamins by the cell-free system. Nuclei were incubated in egg cytosol and aliquots removed at various times. Nuclear and cytosolic proteins were separated on SDS gels, blotted and reacted with anti-lamin B antibodies. The 68 K protein is phosphorylated lamin B (p65). Reproduced with permission from Collas et al, J Biol Chem 1997; 272:21274-21280.

The sperm nuclear lamina is immediately phosphorylated and is solubilized within minutes by egg cytosolic extract (Fig. 2). A protein kinase C (PKC) activity, requiring Ca^{++} and inhibited by the PKC inhibitor chelerythrine, is necessary and sufficient to remove the lamina.[17] Immunodepletion of sea urchin PKC from the cytosol inhibits B-type lamin phosphorylation and its solubilization. Purified rat αβγPKC phosphorylates the B-type lamin at the same sites as the sea urchin cytosolic enzyme.

Only if the lamina is removed does the chromatin decondense and change shape from conical to spherical suggesting a role for the sperm lamina in maintaining the conoid shape of the nucleus or as a physical barrier to chromatin decondensation.[16] However removal of the lamina with purified rat αβγPKC in the absence of cytosol does not result in decondensation, indicating that a second step is necessary to permit decondensation (Fig. 3).[17] This step requires sperm-specific histone H1 and H2B phosphorylation by a CDK1-like kinase also present in the cytosol.[18] The two enzymes are sufficient to decondense the nuclei in cytosol-free physiological buffers.

Thus the disassembly of the sperm nuclear membranes permits access of cytosolic factors to the sperm chromatin, and the removal of the lamina allows chromatin decondensation. These two events are likely to be required for full male pronuclear chromatin activation and a failure of either would interfere with male pronuclear development and paternal contribution to the zygote.

Membrane Vesicle Fractions Contributing to the Nuclear Envelope

Nuclear envelope disassembly leaves the sperm nucleus devoid of paternal nuclear envelope except at the tips. The egg provides the necessary membranes and lamins to complete a new nuclear envelope. There are at least three egg membrane vesicle (MV) populations contributing

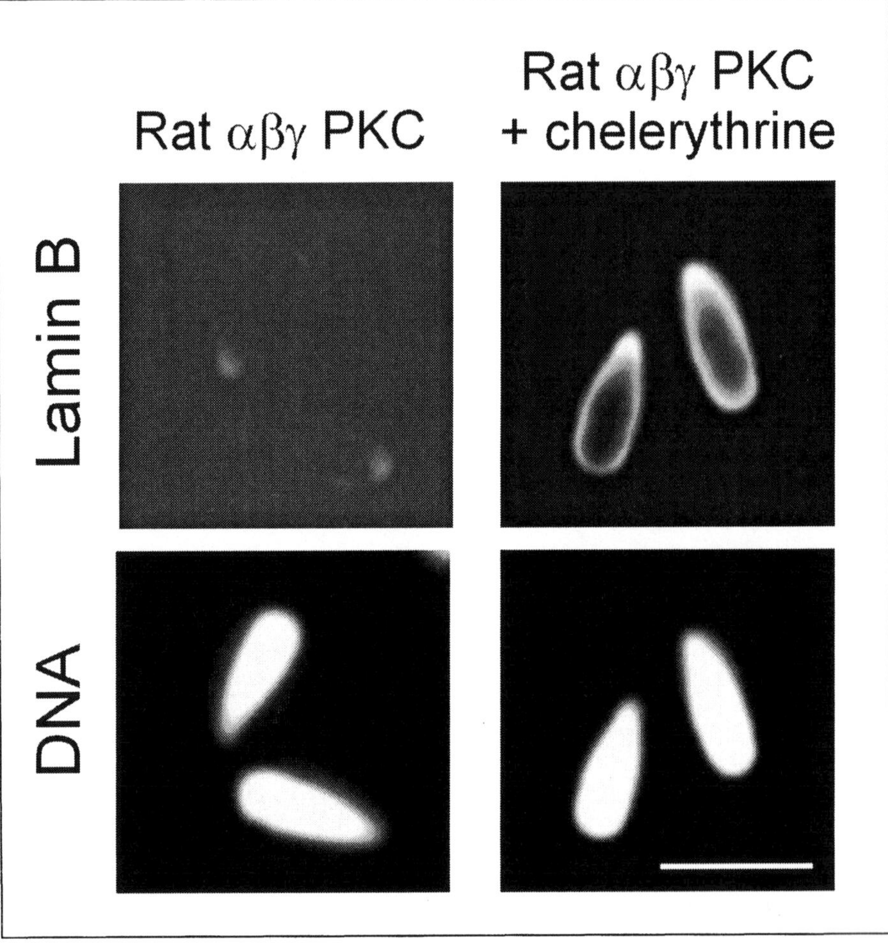

Figure 3. PKC is not sufficient to decondense sperm nuclei. Nuclei were treated with purified rat αβγ protein kinase C without cytosol and stained with anti-lamin B antibodies and Hoechst 33342. Although B-type lamins are removed, nuclei retain their condensed conical shape in the absence of other cytosolic factors. Bar = 5 mm. Reproduced with permission from Collas et al, J Biol Chem 1997; 272:21274-21280. Please see http://www.eurekah.com/chapter.php?chapid=697&bookid=56&catid=15 for the color version of the figure.

uniquely to the new nuclear envelope.[19] These can be fractionated from 10,000 g supernatants of cytoplasmic extracts of eggs on the basis of buoyant density.

Each has distinct biochemical properties. MV1 is the least dense. It has an unusual phospholipid composition: ca. 90% phosphatidylinositol (PI)-10% phosphatidylcholine as determined by 2-dimensional NMR (Fig. 4).[20] MV2α is enriched in an enzymatic marker of the Golgi apparatus: α-D-mannosidase. MV2β is enriched in an enzymatic marker of the endoplasmic reticulum: α-D-glucosidase.[19] MV2β contributes most of the membrane for the new nuclear envelope consistent with in vivo observations that endoplasmic reticulum is the main source of nuclear envelope membranes.[21] It contains B-type lamins and lamin B receptor in separate vesicles, thus defining at least two subpopulations.[22] Lamin B receptor (LBR) is an

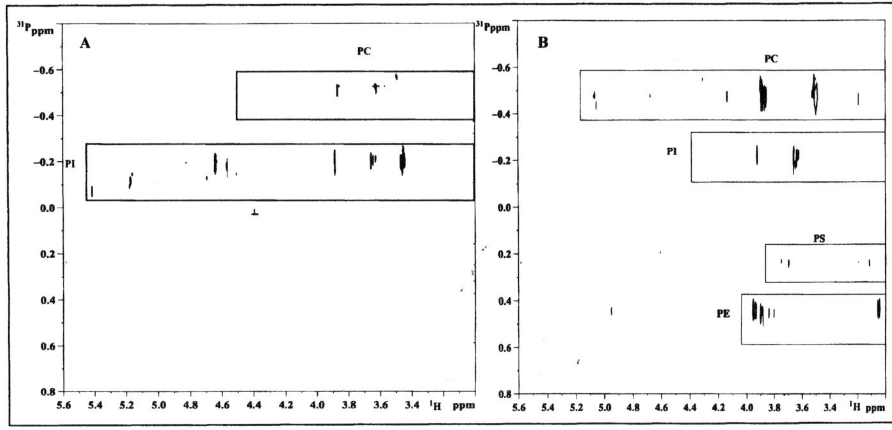

Figure 4. Two-dimensional NMR analyses of phospholipids extracted from MV1 and MV2 ($\alpha+\beta$). Protons of the lipid head group and backbone are edited by the phosphorous resonances to which they are coupled. Diagnostic peaks for phospholipids are boxed. Quantification of peaks indicates that MV1 (A) is almost 90% PI. MV2 (B) has a composition similar to typical cytoplasmic membranes. Reproduced with permission from Larijani et al, Lipids. 2000; 35:1289-1297.

integral membrane protein of the inner nuclear membrane which binds B-type lamins and heterochromatin[23,24] and is discussed in detail elsewhere in this volume.

The in vivo status of MV fractions is not clear since they are recovered following homogenization of eggs, but it seems unlikely that such fractions differing in marker enzymes or lipid composition could be artifactually created by random shearing. In unfertilized egg extracts prepared using the same homogenization protocols, co-precipitation and density gradient experiments show that most of the B-type lamins and LBR are found in the same membrane vesicles, but by 12 minutes post-fertilization they are mostly found in separate vesicles, reflecting an underlying alteration of cytoplasmic structure (Fig. 5).[25] It is known that the endoplasmic reticulum (ER) in vivo changes at fertilization, becoming fragmented,[26,27] which may sort different sets of vesicles. Unactivated but fertilized eggs fail to support nuclear envelope formation, but activation by cytoplasmic alkalinization reestablishes nuclear envelope assembly[28] and this appears to be at least in part due to differences in the membranes.[25] Whether alkalinization which normally follows fertilization is related to reticular reorganization and competence to support in vitro nuclear envelope assembly remains to be elucidated.

Binding of Egg Cytoplasmic Vesicles to Sperm Chromatin and Nuclear Envelope Remnants

All egg membrane vesicle binding to sperm nuclei in vitro requires materials located at the tip and base, the sites of sperm nuclear envelope remnants. We have called these lipophilic structures or LSs since they can be labeled with fluorescent lipophilic dyes (Fig. 6).[15,19] The remnants in the electron microscope consist of cup-shaped membranes lining the acrosomal and centriolar fossae separated from the chromatin by dense osmiophilic cups that form during spermatogenesis and may constitute a special region of the heterochromatin[14] (Fig. 7). Almost nothing is known about the composition of this region of the nucleus except that lamin B receptor (LBR) is exclusively localized there in mature sperm[22] and B-type lamins associated with this region are the last lamins removed from the nucleus by cytosol in vitro (Fig 8).[17]

The LSs define two poles of the spherical chromatin mass created by decondensation. Their membranous elements are not extracted in low concentrations of Triton-X100 in the cold,[15] a property shared by lipid "rafts" and other detergent-resistant membranes (DRGs).[29] They can

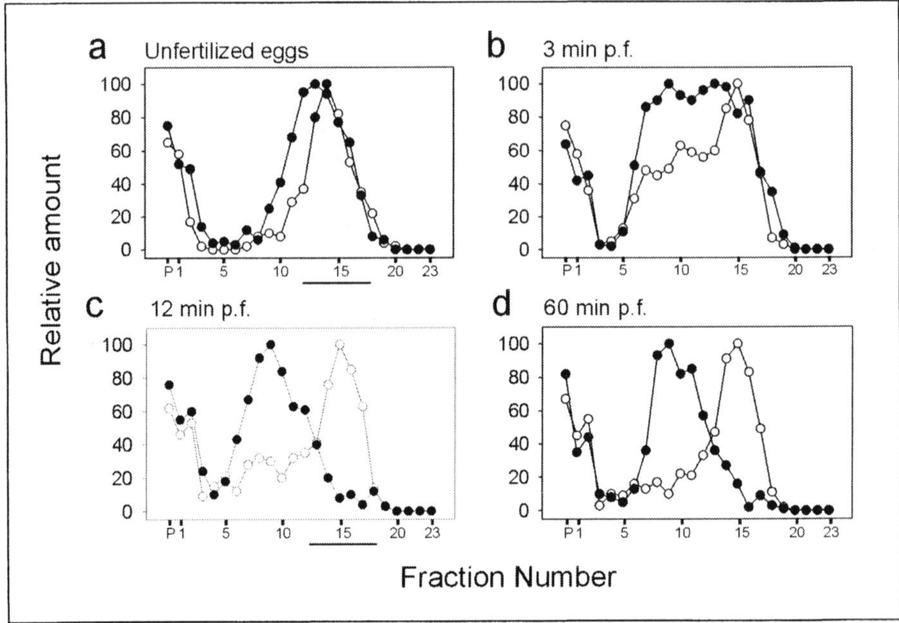

Figure 5. Redistribution of membrane-bound B-type lamins and lamin B receptor following fertilization. Vesicles were floated to density equilibrium in sucrose gradients. Fractions were collected at times indicated and immunoblotted. After densitometry, proportions were normalized to the fraction with the highest amount. Open circles = lamin B receptor. Closed circles = B-type lamins. Co-localization was evaluated by co-immunoprecipitation and blotting. Reproduced with permission from Collas et al, Eur J Cell Biol 2000; 79:10-16.

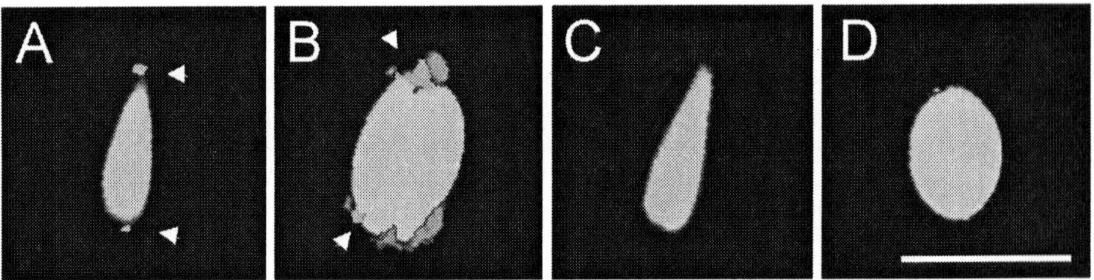

Figure 6. Loss of MV binding to nuclei upon removal of LSs. LSs were labeled with fluorescent 3,3'-dihexyloxacarbocyanine iodide (DiO-C_6 green) and removed with 1% Triton X-100. Nuclei were decondensed briefly in cytoplasm containing MVs labeled with fluorescent 1,1'-dioctadecyl-3,3,3',3'-tetramethylindocarbocyanine (DiI-C_{18} red). A. Control nucleus. B. MV binding to control nucleus. C. Extracted nucleus. D. Lack of binding to extracted nucleus. Nuclei were labeled blue with Hoechst 33342. Bar = 5 mm. Reproduced with permission from Collas and Poccia, Dev Biol 1995; 169:123-135. Please see http://www.eurekah.com/chapter.php?chapid=697&bookid=56&catid=15 for the color version of the figure.

be solubilized in high detergent concentrations and reconstituted exclusively to the tips of the sperm by lowering the concentration of detergent.[15]

By contrast, B-type lamins and LBR at the tips are resistant to even 1% detergent extraction, requiring high salt concentrations (0.4 M NaCl) as well. However both proteins are removed under physiological conditions in vitro by egg cytosol prior to reformation of the nuclear envelope (Fig 8).[22] Thus the membrane and LBR/lamin B have different solubility

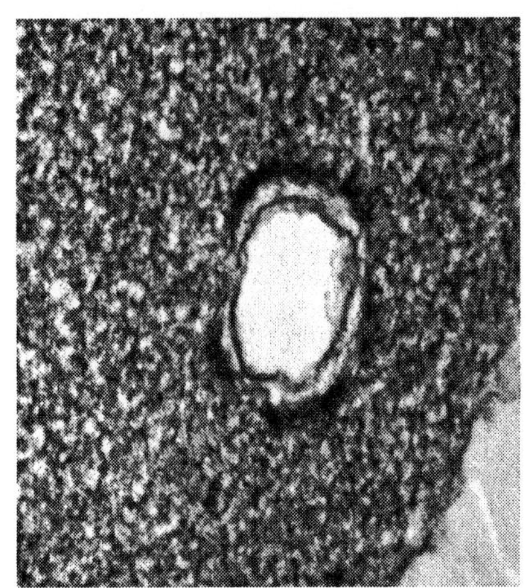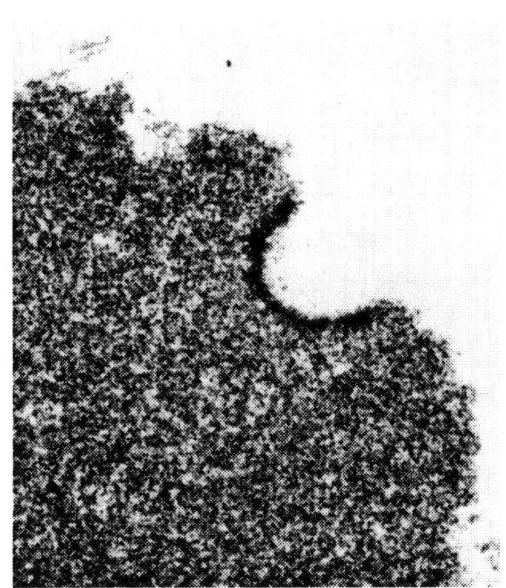

Figure 7. Electron microscopy of the LS region of sperm nuclei. Osmiophilic cup and membrane of LSs in tangential and longitudinal views. Left: cup is lined with membranous remnant of sperm nuclear envelope in 0.1% Triton X-100 extracted nucleus. Right: membrane is extracted by 1% Triton X-100 but cup remains. Bar = 0.5 μm. Reproduced with permission from Collas and Poccia, Dev Biol 1995; 169:123-135.

properties in the LS region, and unlike the membrane, those proteins are not directly incorporated into the male pronuclear envelope.

Binding of egg MVs to the sperm nucleus requires cytosol and ATP but not ATP hydrolysis. Stable complexes of membrane and nucleus can be demonstrated by centrifugation of the nuclei through 1 M sucrose.[15] Experiments with reconstitutes show that egg cytoplasmic MVs will only bind to the nuclei if LSs are present. They will only bind to one pole if LSs only reconstitute to one pole. Binding to both poles is required for full nuclear envelope formation (Fig. 9).

Binding experiments with separate MV fractions indicate that MV1 and MV2α vesicles bind exclusively to the polar regions of the male pronucleus, even in great membrane excess.[19] By contrast, MV2β vesicles bind around the entire surface of the nuclei (Fig. 10). MV1 and MV2α require peripheral membrane proteins to bind; MV2β does not. An integral membrane protein necessary for MV2β to bind is the lamin B receptor, since immunodepletion of cytoplasmic extracts with anti-LBR abolishes all binding activity of this fraction but has no effect on binding of MV1 and MV2α vesicles.[22] Therefore not all MVs require LBR to bind to chromatin and the LBR-containing MV2β vesicles need LSs at least for initiating binding at the poles, which then spreads equatorially (Fig. 11).[15]

Fusion of Nuclear Envelope Precursor Vesicles

Fusion of bound vesicles results in a continuous nuclear envelope surrounding the male chromatin (Fig. 12). An intact nuclear envelope is likely to be needed for DNA synthesis as it is in amphibian egg extracts[30,50] and for export of RNAs whose synthesis begins in sea urchins at about the time of replication, 30 minutes post-fertilization.[13]

Fusion incorporates the lipophilic material of LSs with the MV fractions[15] as predicted from in vivo observations.[14] Therefore LSs have two roles in nuclear envelope formation, serving as initial sites of vesicle docking and contributing a fraction of the new nuclear envelope. MV1 is apparently the only fraction which can fuse directly to LSs, suggesting a special fusigenic and mediating role of this fraction (Fig. 13).[19]

MV1 is required for LS fusion to any of the other fractions and for fusion of MV2α with MV2β. MV2α is needed for MV2β vesicles to fuse with LSs and MV1. These experiments

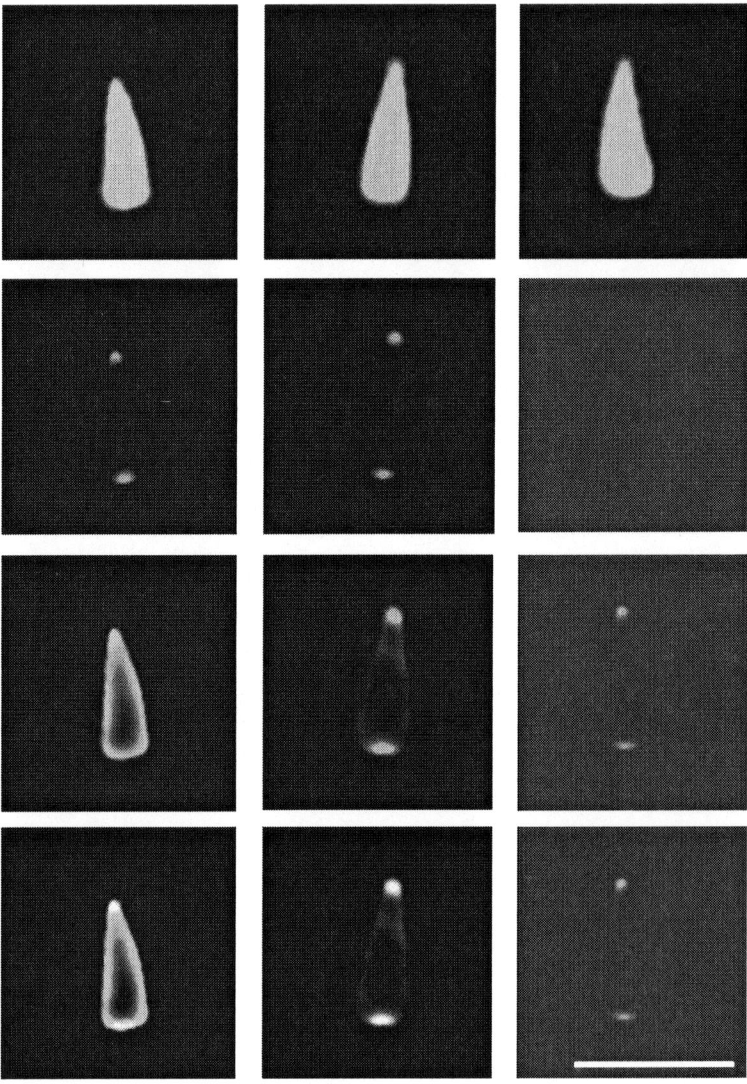

Figure 8. Fertilized egg extracts remove lamin B receptor and lamin B from chromatin within minutes. Nuclei were incubated in egg cytosol and aliquots removed at 0, 5 and 10 min (left to right) and stained with Hoechst 33342, anti-lamin B receptor (red) and anti-lamin B (green). The bottom panel is a merge of the middle two panels. Reproduced with permission from Poccia and Collas, Devel Growth Differ 1997; 39:541-550. Please see http://www.eurekah.com/chapter.php?chapid=697&bookid=56&catid=15 for the color version of the figure.

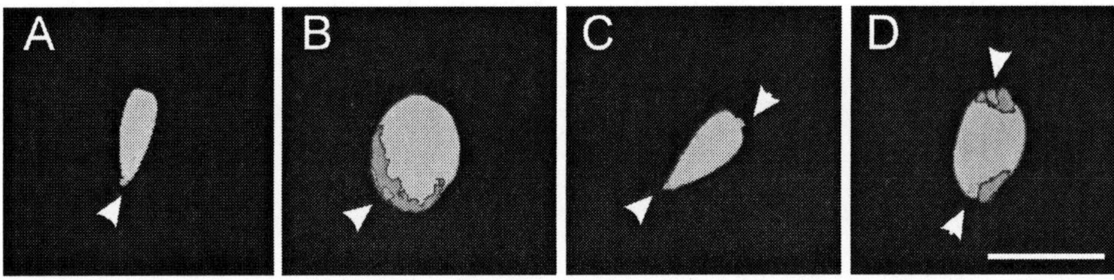

Figure 9. Reconstitution of LSs to nuclear poles targets MV vesicle binding. LSs stripped from nuclei with 1% Triton X-100 were reconstituted by dilution of the Triton to 0.1%. Reconstituted nuclei were incubated with cytoplasm containing MVs labeled red with DiI-C$_{18}$. LSs were labeled green with DiO-C$_6$. A, B. Unipolar reconstitute before and after incubation. C, D. Bipolar reconstitute before and after incubation. MVs only bind to poles reconstituted with LSs (arrows). Bar = 5 μm. Reproduced with permission from Collas and Poccia, Dev Biol 1995; 169:123-135. Please see http://www.eurekah.com/chapter.php?chapid=697&bookid=56&catid=15 for the color version of the figure.

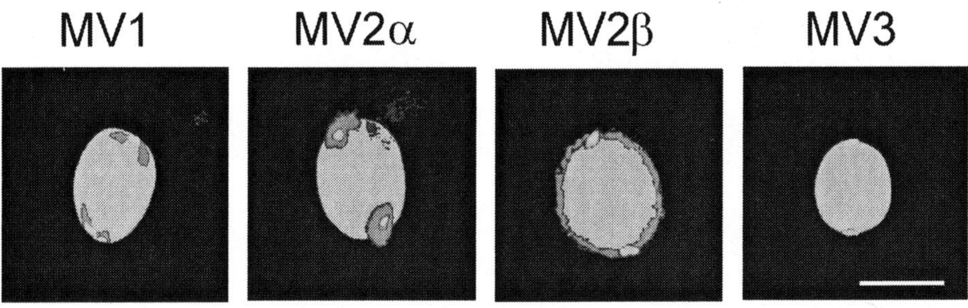

Figure 10. Binding sites of MV fractions on decondensed sperm nuclei. Decondensed chromatin labeled with Hoechst 33342 (blue) and associated LSs labeled with DiO-C$_6$ (green) were incubated with individual MV fractions labeled with DiI-C$_{18}$ (red) in cytosol. Bar = 5 μm. Reproduced with permission from Collas and Poccia, J Cell Sci 1996; 109:1275-1283. Please see http://www.eurekah.com/chapter.php?chapid=697&bookid=56&catid=15 for the color version of the figure.

indicate a complex hierarchy of interactions among the membrane precursors to the nuclear envelope and possibly a differentially regulated recruitment. In this regard, MV2β can form nuclear envelopes in the absence of MV1[31] suggesting that the roles of MV1 and MV2α may be regulatory, that they may provide novel binding sites to the chromatin, or that they may provide a mechanism to incorporate the LSs into the nuclear envelope to serve some unknown function.

Fusion of MVs to form a nuclear envelope occurs in cell-free extracts from 10-15 min fertilized eggs. If however either the cytosol or MV fractions are derived from unfertilized eggs, nuclear envelope formation fails.[25] Microinjection and electron microscopy experiments support the notion that immature or unactivated, unfertilized eggs do not support male pronuclear envelope formation.[28,32,33] Since in vitro assembly assays are all carried out at alkaline pH, the failure to form male pronuclear envelopes in unactivated egg extracts cannot be only due to pH, but pH may lead to membrane reorganization or changes in soluble components.

In fertilized activated cytoplasm, once MVs are assembled on the surface of the sperm nucleus, membrane fusion can be initiated by hydrolysis of GTP,[34] which is also required for nuclear envelope formation in cell-free *Xenopus* extracts[35] and many other types of membrane fusion reactions in vitro.[36,37] Fusion is inhibited by the non-hydrolyzable analogs GTPγS or GTP-PNP. This effect is probably mediated by monomeric G proteins such as compartment-specific Rab proteins required for membrane fusion events that occur in the endocytic or secretory

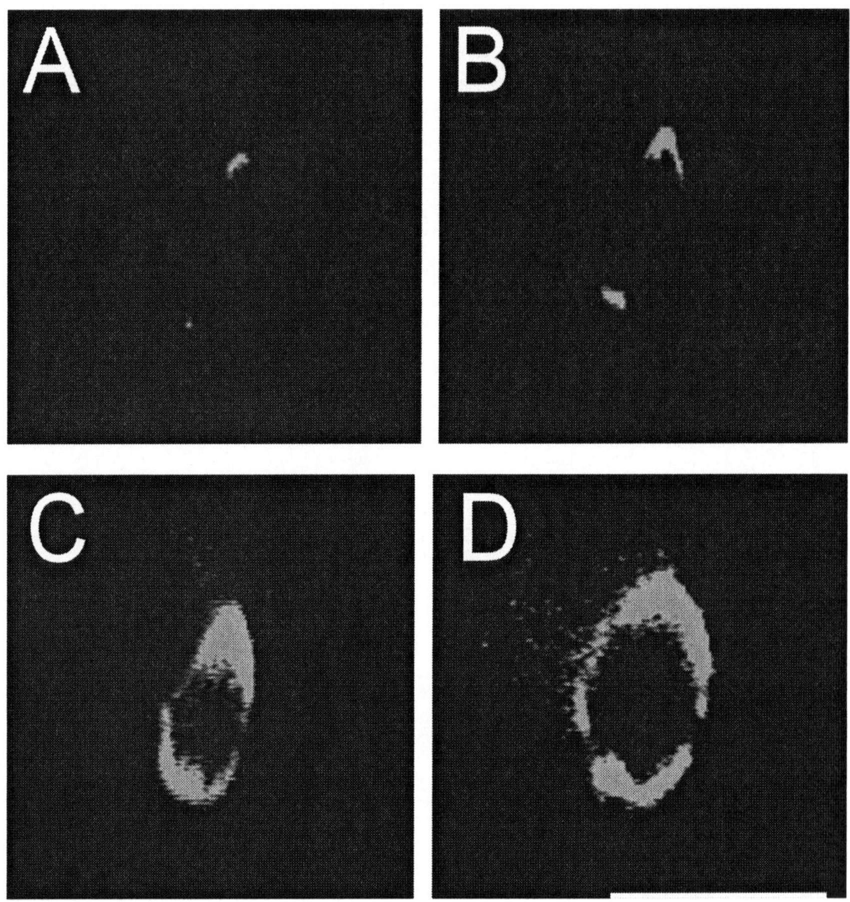

Figure 11. Polarized MV binding to sperm nuclei in vitro. Total MVs labeled with DiI-C$_{18}$. The same nucleus photographed at A 2, B 10, C 15, D 20 min after addition of nuclei to cytoplasm. Binding initiates at the acrosomal and centriolar fossa regions, sites of the LSs (compare with Figure 9). Bar = 5 μm. Reproduced with permission from Collas and Poccia, Dev Biol 1995; 169:123-135. Please see http:// www.eurekah.com/chapter.php?chapid=697&bookid=56&catid=15 for the color version of the figure.

pathways.[36,37] The relevant GTPase molecules for nuclear envelope formation have not been characterized.

A surprising second initiator of fusion is exogenous phosphatidylinositol (PI)-specific phospholipase C (PI-PLC).[31] A bacterial PI-PLC promotes fusion in the absence of added GTP or in the presence of GTPγS (Fig. 14). Unlike mammalian PI-PLCs, this enzyme prefers unphosphorylated PI and so does not produce inositol triphosphate, but instead D-*myo*-inositol 1,2-cyclic phosphate, which is then hydrolyzed to D-*myo*-inositol 1-phosphate.[38] However, as is the case for eukaryotic PI-PLCs, diacylglycerol (DAG) is produced, leading to the suggestion that MV1 might contain high concentrations of PI as a source of diacylglycerol. DAG is known to destabilize artificial protein-free lipid bilayers leading to fusion.[39,40,41] Exogenous PI-PLC does not lead to fusion when MV1 is removed from the system, suggesting that quantities in

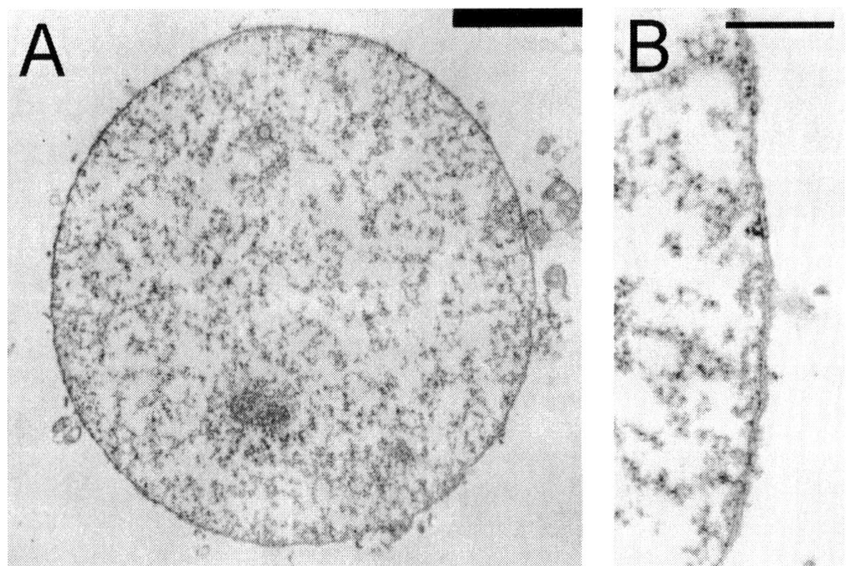

Figure 12. Electron micrographs of swollen sea urchin male pronucleus formed in vitro. A. Whole nucleus. B. Close up of nuclear envelope showing double membrane and pores (arrows). Bar = 0.1 μm. Reproduced with permission from Collas and Poccia, Meth Cell Biol 1998; 53: 417-452.

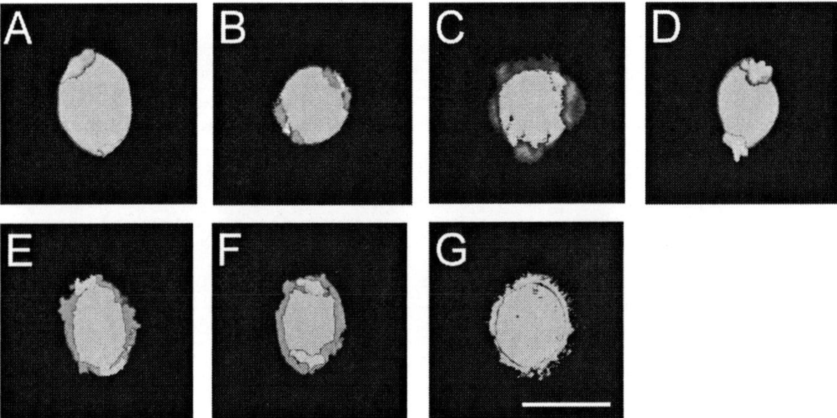

Figure 13. Fusion of MV fractions with LSs and each other. MVs or LSs were labeled with DiO-C$_6$ (green) or DiI-C$_{18}$ (red). Purified MVs with S100 were added in various combinations to nuclei. Orange indicates GTP-induced fusion from mixing of lipophilic dyes in the same membranes. A. Fusion of red MV1 with green LSs. B. Lack of fusion of green LSs with red MV2α. C. Lack of fusion of green LSs with red MV2β. D. Fusion of green MV1 with red MV2α. E. Lack of fusion of green MV1 with red MV2β. F. Lack of fusion of green MV2α with red MV2β. G. Fusion of green MV1 with red MB2β in the presence of unlabelled MV2α. Blue is Hoechst 33342 stained chromatin. Bar = 5 μm. Reproduced with permission from Collas and Poccia, J Cell Sci 1996; 109:1275-1283. Please see http://www.eurekah.com/chapter.php?chapid=697&bookid=56&catid=15 for the color version of the figure.

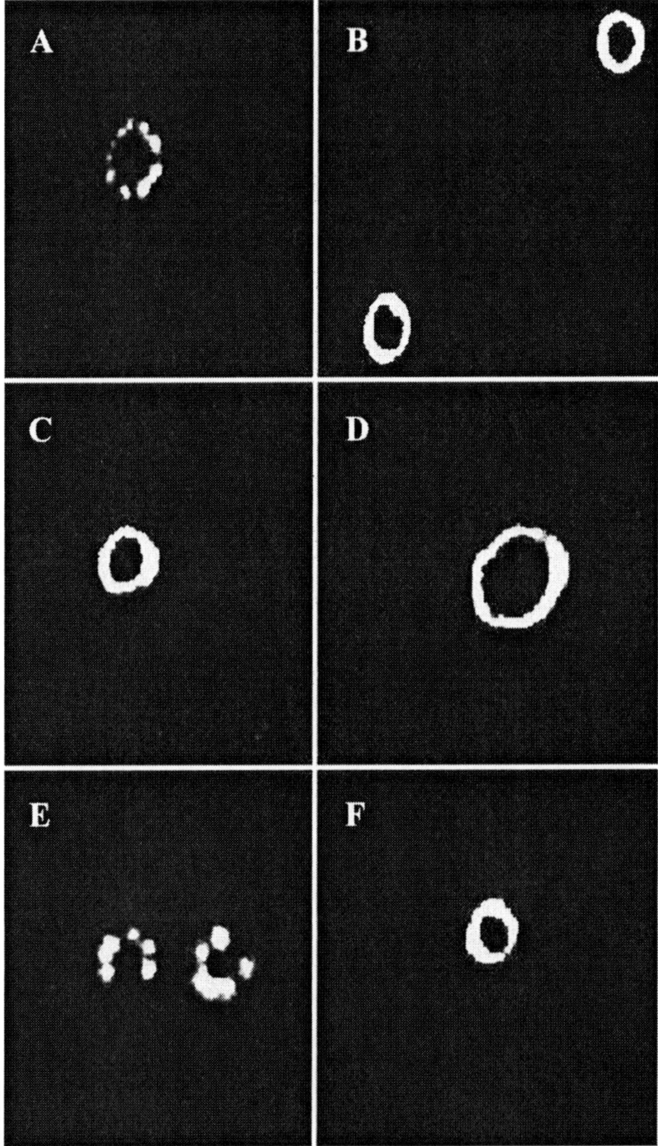

Figure 14. MV fusion is induced by a bacterial phosphatidyl inositol specific phospholipase C (PI-PLC). A. ATP-dependent binding of MVs. B. GTP-dependent fusion. C. PI-PLC dependent MV fusion in the absence of GTP. D. Swelling of complete nuclei treated as in C. E. Inhibition of PI-PLC fusion by the PI-PLC inhibitor 1-O-octadecyl-2-O-methyl-sn-glycero-3-phosphorylcholine. (ET-18-OCH$_3$). F. GTP induced fusion is not inhibited by ET-18-OCH$_3$. MVs labeled with DiO-C$_6$. Reproduced with permission from Larijani et al, Biochem J. 2001; 356:495-501. Please see http://www.eurekah.com/chapter.php?chapid=697&bookid=56&catid=15 for the color version of the figure.

excess of the 5-10% PI typical of most cellular membranes are required to produce fusion by this mechanism.[31] The physiological relevance of this inducer of fusion remains to be established.

GTP however initiates significant amounts of fusion in MV1-depleted fractions, presumably by MV2α homotypic fusion. This and GTP-initiated fusion of unfractionated MVs are subject to blocking by the phosphatidylinositol 3 kinase (PI-3 kinase) inhibitors wortmannin and LY 294002 near their IC_{50}'s, suggesting a role for PI phosphorylated in the 3 position of the inositol ring.[31] The roles of different phosphorylated PIs containing phosphates at the 3 position (i.e., PI-3 P; PI-3,4 P_2; PI-3,4,5 P_3) are not known, nor is their relative abundance in the cell-free system.

Completion of Male Pronuclear Envelope Formation

The nuclear envelope that forms immediately following fusion in vitro is devoid of lamins but contains lamin B receptor (Fig. 15).[22] The nuclei remain as spheres of 4 μm in diameter until provided with additional ATP. The ATP leads to import of lamins from the cytosol, and further membrane fusions as the nucleus almost doubles in diameter.[16] The swollen nuclei contain lamins in association with the lamin B receptor, and if B-type lamin uptake is blocked (by immunodepletion or inhibiting pore function) the nuclei fail to swell, raising the possibility that lamin polymerization might drive swelling of the nucleus.[16]

Comparison with Other Systems and Speculations

Novel aspects of the sea urchin system discussed above include: 1) LSs, remnants of the sperm nuclear envelope exhibiting detergent-resistance and serving as required sites of egg MV binding to chromatin, 2) multiple MV fractions contributing uniquely and in an ordered fashion to the nuclear envelope, 3) a membrane vesicle subfraction MV1 highly enriched in phosphatidylinositol and required for PI-PLC-induced fusion, and 4) a role for PI-3 kinase in GTP-induced nuclear envelope formation. Below we discuss each of these in turn.

LS s in Other Cell Types

LS regions serve as required binding sites for egg cytoplasmic MVs and their lipophilic materials are incorporated into the nuclear envelope upon fusion. In the sea urchin, two LSs are needed for a complete nuclear envelope and they define opposite poles of the nucleus. One of these poles is the region of the centriolar fossa whose centrosome serves to organize an astral array of microtubules during male pronuclear migration. The monopolar sperm aster contrasts with the bipolar LSs. In vivo, egg cytoplasmic endoplasmic reticulum derivatives are organized as a spherical halo around the male pronucleus during its migration.[26] Given the geometry of the LSs vs. the centrosome, and the typical continuity of the outer nuclear membrane with the endoplasmic reticulum, it is possible that LSs have a role in cytoplasmic membrane reorganization different from that of microtubules.

LSs are not restricted to sea urchins, but can be demonstrated in sperm of vertebrates including fish, frogs, rabbits, foxes, bulls and mice (Fig. 16).[42] They appear as one or two regions of detergent-resistant lipophilic material, one of which is always associated with the implantation fossa (centriolar fossa). Vertebrate LSs can bind echinoderm or mammalian MVs and echinoderm LSs can bind mammalian MVs. Although binding is not species-specific, membrane fusion induced by GTP hydrolysis is. This suggests that common mechanisms might act to initiate male pronuclear envelopes in a variety of species, but that some elements of the fusion machinery may differ.

It is not known if detergent-resistant membrane fractions initiating MV binding are typical of mitotic somatic chromosomes, but polarization of envelope formation has been observed in many cell types. For example, nuclear envelope formation in HeLa cells appears to derive from ER, initiating at the chromosomal regions away from the spindle-chromosome fibers and proceeding towards the kinetochores.[43] In rat epithelial cells, envelope formation was also observed to initiate at the region away from the fibers in continuity with the ER, and to complete near the kinetochores.[44] Although these observations are also consistent with the spindle fibers

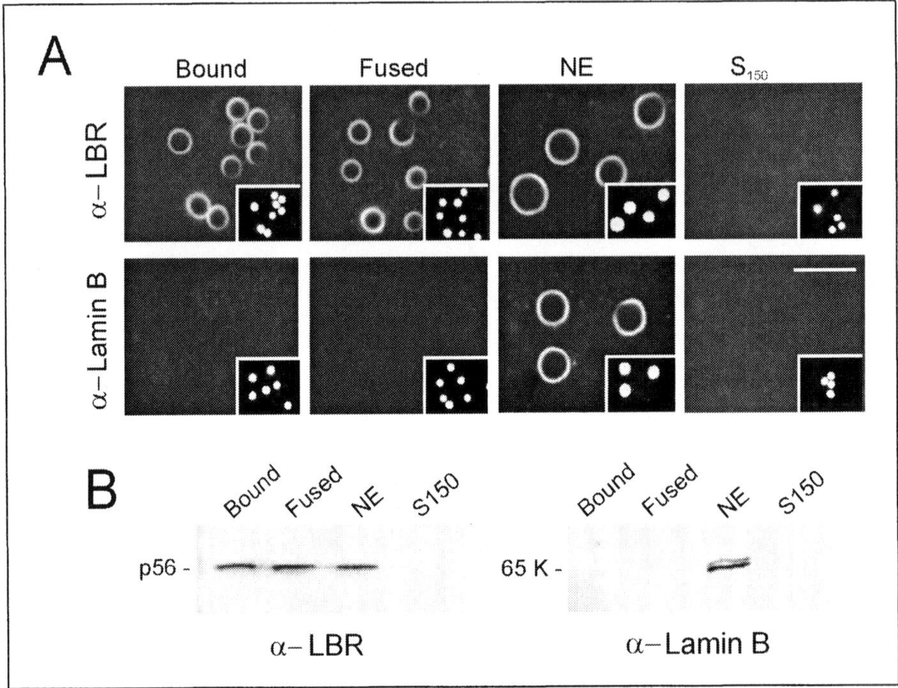

Figure 15. Independent recruitment of lamin B and lamin B receptor during nuclear envelope formation in vitro. A. Immunofluorescence of nuclei with bound MVs, fused MVs forming a nuclear envelope, swollen nuclei (NE) and S150 control (no membranes). B. Immunoblots corresponding to time points shown in A. Insets show blue Hoechst 33342 staining, same fields reduced. Lamin B is absent from initial nuclear envelopes but is imported and required for swelling. Reproduced with permission from Collas and Poccia, J Cell Biol 1996; 135:1715-1725. Please see http://www.eurekah.com/ chapter.php?chapid=697&bookid=56&catid=15 for the color version of the figure.

sterically hindering approach of MVs, they are also consistent with a spatially ordered assembly of vesicles. In this regard, binding of nuclear envelope components in certain cell types appears to begin at discrete sites associated with individual chromosomes.[45,46]

Cultured mitotic cells contain vesicles differing in protein markers of inner membrane and pores differentially assembled at mitosis.[45] An LBR containing vesicle fraction from mammalian or avian cells is required for binding to chromosomes.[47] Non-uniform binding of LBR to mitotic chromosomes (resembling Q or G banding patterns) suggests receptors for LBR may be non-uniformly distributed on chromosomes.[47] The chromosomal caps or initial regions of binding may contain lamins, pore complexes, nuclear membrane proteins and heterochromatin protein 1 (see ref. 48 for discussion).

Early embryos may exhibit more atypical or accelerated patterns of nuclear envelope formation required by the rapid replication cycles of these cells. In some cells, individual chromosomes become fully surrounded by membranes forming karyomeres prior to fusion of each of these to form the nuclear envelope encircling the chromosomal set.[49-52, 63-65] For example, in sea urchin blastomeres, individual telophase chromosomes have well developed attached nuclear envelopes that appear to subsequently fuse.[51] Karyomeres can function as mini-nuclei, replicating independently.[50] In such cases, it is imperative to ensure that all chromosomes are included in the new nucleus, but how this is regulated is not known. In *Drosophila* embryos, NE breakdown during mitosis is incomplete, and only the region where the spindle fibers attach

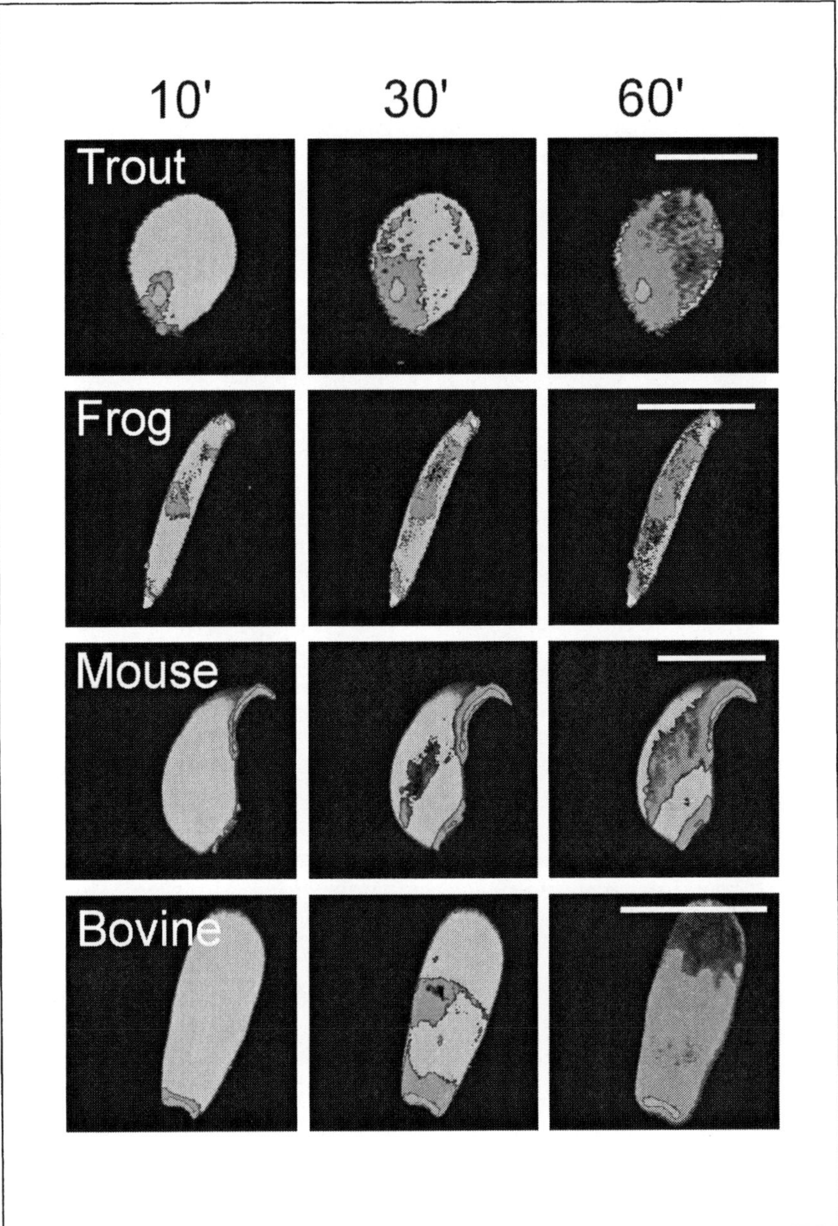

Figure 16. Binding of LSs to MVs is conserved from echinoderms to mammals. Various sperm nuclei were demembranated with 0.1% TX-100 and their LSs were labeled green with DiO-C$_6$. They were incubated in sea urchin egg cytoplasm whose membrane vesicles were labeled red with DiI-C$_{18}$. Binding of membranes was followed at successive times of incubation. Reproduced with permission from Collas and Poccia, Eur J Cell Biol 1996; 71:22-32. Please see http://www.eurekah.com/chapter.php?chapid=697&bookid=56&catid=15 for the color version of the figure.

breaks down.[52] This differential nuclear envelope breakdown is reminiscent of sea urchin sperm nuclear envelope breakdown, though more membrane is retained in the former case.

In any event, mechanisms involving nuclear envelope organizing centers associated with individual chromosomes might be typical of most or all somatic cells. LSs could ensure that all chromosomes become enclosed in an envelope, ready for incorporation into the full diploid nucleus, which would prevent aneuploidy. The distribution of LSs in one or two regions of various sperm nuclei might be explained if such sites on individual chromosomes were clustered together during spermatogenesis as are, for example, telomeres or centromeres in mammalian sperm.[53]

Multiple MV Fractions

Studies from several groups on assembly of nuclear envelopes in amphibian egg extracts have also shown that multiple MVs with distinct markers and ordered timing of chromosomal binding contribute to the nuclear envelope. The major precursor vesicle population for the male pronuclear envelope in *Xenopus* is enriched in an ER marker enzyme.[54] Differential timing of assembly of two distinct membrane populations containing different markers of ER/ outer nuclear membrane and inner nuclear membrane proteins leads to nuclear envelope assembly in *Xenopus* eggs and cultured cells.[55] Other studies show that two vesicle populations are necessary for the assembly of normal sized nuclei. One contains chromatin targeting molecules and membrane fusion machinery and the other contains chromatin targeting molecules and a molecule necessary for nuclear pore complex assembly. The first assembles a minimal nuclear membrane, and the second is needed for assembly of nuclear pore complexes and swelling.[56,57] Although the relationship between the MVs reported by these groups and the sea urchin MVs is not clear, in all cases assembly seems to require more than a single homogeneous population of precursor vesicles, ordered timing of membrane binding to chromatin, and a segregation of some element of the fusion machinery similar to the sea urchin system.

PI-Containing Membrane Vesicles and PI-PLC Induction of Nuclear Envelope Formation

A possible role of PI-rich MV1 is to provide a large source of DAG after hydrolysis. We are not aware of any reports of naturally occurring membranes with such high levels of PI as sea urchin egg MV1. Consistent with experiments showing that MV1 is needed for PI-PLC induced nuclear envelope formation, several in vitro studies indicate that large unilammellar vesicles in which diacylglycerol is produced by the action of phospholipase C (PI- or PC-specific) lead to membrane fusion.[39,40,41]

PI derivatives have been reported to have a role in fusion of endosomes and secretory granules.[58, 59] PIs phosphorylated in the 3, 4 and 5 positions have been implicated in various steps of endocytosis (reviewed in ref. 60). But it is still not clear which if any phosphorylated derivatives of the PI in MV1 is produced in vitro or whether nuclear envelope formation by PI-PLC in vitro is an artifact of non-physiological levels of the enzyme.

Role of PI-3K in Nuclear Envelope Formation

The wortmannin and LY 294002 sensitivity of GTP-induced fusion of MVs in forming a nuclear envelope implicates a PI-3 kinase activity in this event. Three classes of PI-3 kinases are known:[61] Class I which functions predominantly in tyrosine kinase surface receptor activated signalling pathways, Class II of unknown function and Class III which is involved in vacuolar protein sorting in yeast and lysosomal sorting and endosomal fusion reactions in mammals. The latter may hint at a possible reason for the sea urchin PI-3 kinase requirement for fusion. In the endosomal pathway, PI-3 P serves to recruit the small GTPase Rab5 and a protein EEA1 to the membrane.[37] These proteins are potent effectors of membrane fusion and similar molecules might have a role in nuclear envelope formation.

Highly curved small unilamellar vesicles containing PI are enhanced substrates for PI-3 kinase.[62] This offers a possible mechanism for recruitment of the enzyme to regions of high curvature such as "stalks", hypothesized to be intermediates in membrane fusion. In such a scenario, clustering of PI might stimulate local PI-3 P formation and assemble fusion machinery at the correct sites.

Issues for Future Investigation

Observations on nuclear envelope formation in sea urchin egg cytoplasmic extracts raise new issues concerning nuclear envelope formation and suggest several possible avenues for further investigations. What if any compositional or physical characteristics do lipophilic structures have in common with rafts and other DRGs? Is initiation of binding of membrane precursors to the nuclear envelope orchestrated by similar membranous elements in somatic cells? Do LSs have a role in organization of cytoplasmic membranes or insuring against aneuploidy? Do MVs enriched in PI contribute to somatic nuclear envelope formation? Are there classes of MVs which catalyze or confer directionality on fusion reactions? What are the minimum lipid and protein requirements for nuclear envelope precursor vesicles? Does production of high levels of DAG promote fusion of nuclear envelope precursor vesicles? Does PI-3 P serve to assemble parts of the fusion machinery for nuclear envelope formation as it does for endosomal vesicles? Is there a Rab protein unique to the nuclear envelope pathway?

Use of cell-free systems, mutated recombinant proteins and reconstitution of minimal vesicle populations of defined lipid compositions should lead to a deeper understanding of the assembly of membranes delimiting the nucleus.

References

1. Gurdon JB, Woodland HR. The cytoplasmic control of nuclear activity in animal development. Biol Rev 1968; 43:233-267.
2. Poccia D, Collas P. Transforming sperm nuclei into male pronuclei in vivo and In vitro. Curr Topics Devel Biol 1996; 34:25-88.
3. Longo FJ. Fertilization: A comparative ultrastructural review. Biol Reprod 1973; 9:149-215.
4. Longo FJ, Anderson E. Sperm differentiation in the sea urchins *Arbacia punctulata* and *Strongylocentrotus purpuratus*. J Ultrastruct Res 1969; 27:486-499.
5. Longo F. Regulation of pronuclear development. In: G Jagiello G, Vogel HJ, eds. Bioregulators of Reproduction. New York: Academic Press, 1981:529-557
6. Cameron LA, Poccia DL. In vitro development of the sea urchin male pronucleus. Dev Biol 1994; 162:568-578.
7. Cox LS, Hutchison CJ. Nuclear envelope assembly and disassembly. Subcellular Biochem 1994; 22: 263-325.
8. Marshall ICB, Wilson KL. Nuclear envelope assembly after mitosis. Trends Cell Biol 1997; 7:69-74.
9. Wilson KL, Wiese C. Reconstituting the nuclear envelope and endoplasmic reticulum. Seminars Cell Devel Biol 1996; 7:487-496.
10. Newport J, Spann T. Disassembly of the nucleus in mitotic extracts: Membrane vesicularization, lamin disassemby, and chromosome condensation are independent processes. Cell 1987; 48:219-230.
11. Burke B, Gerace L. A cell free system to study reassembly of the nuclear envelope at the end of mitosis. Cell 1986; 44:639-652.
12. Poccia D. Remodeling of nucleoproteins during gametogenesis, fertilization, and early development. Int Rev Cytol 1986; 105:1-65.
13. Poccia D, Wolff R, Kragh S et al. RNA synthesis in male pronuclei of the sea urchin. Biochim Biophys Acta 1985; 824:349-356.
14. Longo FJ, Anderson E. The fine structure of pronuclear development and fusion in the sea urchin, *Arbacia punctulata*. J Cell Biol 1968; 39:339-368.
15. Collas P, Poccia D. Lipophilic organizing structures of sperm nuclei target membrane vesicle binding and are incorporated into the nuclear envelope. Dev Biol 1995; 169:123-135.
16. Collas P, Pinto-Correia C, Poccia DL. Lamin dynamics during sea urchin male pronuclear formation In vitro. Exp Cell Res 1995; 219:687-698.
17. Collas P, Thompson L, Fields AP et al. PKC-mediated interphase lamin B phosphorylation and solubilization. J Biol Chem 1997; 272:21274-21280.
18. Stephens S, Beyer B, Balthazar-Stablein U et al. Two kinase activities are sufficient for sea urchin sperm nuclear decondensation in vitro. Molec Reprod Devel 2002; 62:496-503.

19. Collas P, Poccia D. Distinct egg cytoplasmic membrane vesicles differing in binding and fusion properties contribute to sea urchin male pronuclear envelopes formed in vitro. J Cell Sci 1996; 109:1275-1283.

20. Larijani B, Poccia DL, Dickinson LC. Phospholipid identification and quantification of membrane vesicle subfractions by two-dimensional ^{31}P-^{1}H nuclear magnetic resonance. Lipids 2000; 35:1289-1297.

21. Longo FJ. Derivation of the membrane comprising the male pronuclear envelope in inseminated sea urchin eggs. Dev Biol 1976; 49:347-368.

22. Collas P, Courvalin J-C, Poccia D. Targeting of membranes to sea urchin sperm chromatin is mediated by an LBR-like integral membrane protein. J Cell Biol 1996; 135:1715-1725.

23. Worman HJ, Yuan J, Blobel G et al. A lamin B receptor in the nuclear envelope. Proc Natl Acad Sci USA 1988; 85:8531-8534.

24. Ye Q, Callebaut I, Pezhman A et al. Domain-specific interactions of human HP1-type chromodomain proteins and inner nuclear membrane protein LBR. J Biol Chem 1997; 271:14983-14989.

25. Collas P, Barona T, Poccia DL. Rearrangements of sea urchin egg cytoplasmic membrane domains at fertilization. Eur J Cell Biol 2000; 79:10-16.

26. Terasaki M, Jaffe LA. Organization of the sea urchin egg endoplasmic reticulum and its reorganization at fertilization. J Cell Biol 1991; 114:929-940.

27. Terasaki M, Jaffe LA, Hunnicutt GR *et al.* Structural change of the endoplasmic reticulum during fertilization: Evidence for loss of membrane continuity using the green fluorescent protein. Dev Biol 1996; 179:320-328.

28. Carron CP, Longo FJ. Relation of intracellular pH and pronuclear development in the sea urchin, *Arbacia punctulata*. Dev Biol 1980; 79:478-487.

29. Brown D, London E. Functions of lipid rafts in biological membranes. Annu Rev Cell Dev Biol 1998; 14:111-136.

30. Spann TP, Moir RD, Goldman AE et al. Disruption of nuclear lamin organization alters the distribution of replication factors and inhibits DNA synthesis. J Cell Biol 1997; 136:1201-1212.

31. Larijani B, Barona TM, Poccia DL. Role for phosphatidylinositol in nuclear envelope formation. Biochem J 2001; 356:495-501.

32. Cothren CC, Poccia DL. Two steps required for male pronucleus formation in the sea urchin egg. Exp Cell Res 1993; 205:126-133.

33. Longo F. Insemination of immature sea urchin (*Arbacia punctulata*) eggs. Dev Biol 1978; 62:271-291.

34. Collas P, Poccia D. Formation of the sea urchin male pronucleus in vitro: Membrane-independent chromatin decondensation and nuclear-envelope dependent nuclear swelling. Molec Reprod Dev 1995; 42:106-113.

35. Boman AL, Delannoy MR, Wilson KL. GTP hydrolysis is required for vesicle fusion during nuclear envelope assembly in vitro. J. Cell. Biol 1992; 115:281-294.

36. Novick P, Zerial M. The diversity of Rab proteins in vesicle transport. Curr Opin Cell Biol. 1997; 9:496-504.

37. Zerial M, McBride H. Rab proteins as membrane organizers. Nat Rev Mol Cell Biol 2001; 2:107-117.

38. Griffith OH, Ryan M. Bacterial phosphatidylinositol-specific phospholipase C: structure, function, and interaction with lipids. Biochim Biophys Acta 1999; 1441:237-254.

39. Nieva JL, Goni FM, Alonso A. Liposome fusion catalytically induced by phospholipase C. Biochemistry 1989; 28:7364-7367.

40. Basanez G, Ruiz-Arguello MB, Alonso A et al. Morphological changes induced by phospholipase C and by sphingomyelinase on large unilamellar vesicles: A cryo-transmission electron microscopy study of liposome fusion. Biophys J 1997; 72:2630-2637.

41. Villar AV, Alonso A, Goni FM. Leaky vesicle fusion induced by phosphatidylinositol-specific phospholipase C: Observation of mixing of vesicular inner monolayers. Biochemistry 2000; 39:14012-14018.

42. Collas P, Poccia D. Conserved binding recognition elements of sperm chromatin, sperm lipophilic structures and nuclear envelope precursor vesicles. Eur J Cell Biol. 1996; 71:22-32.

43. Robbins E, Gonatas N. The ultrastructure of a mammalian cell during the mitotic cycle. J Cell Biol 1964; 21:429-463.

44. Zeligs JD, Wollman SH. Mitosis in rat thyroid epithelial cells in vivo. I. Ultrastructural changes in cytoplasmic organelles during the mitotic cycle. J Ultrastruct Res 1979; 66:53-77.

45. Buendia B, Courvalin JC. Domain-specific disassembly and reassembly of nuclear membranes during mitosis. Exp Cell Res 1997; 230:133-144.

46. Haraguchi T, Koujin T, Hayakawa T et al. Live fluorescence imaging reveals early recruitment of emerin, LBR, RanBP2, and Nup153 to reforming functional nuclear envelopes. J Cell Sci 2000; 113:779-794.
47. Pyrpasopoulou A, Meier J, Maison C et al. The lamin B receptor (LBR) provides essential chromatin docking sites at the nuclear envelope. EMBO J 1996; 15:7108-7119.
48. Kourmouli N, Theodoropoulos PA, Dialynas G et al. Dynamic associations of heterochromatin protein 1 with the nuclear envelope. EMBO J 2000; 19:6558-6568.
49. Fuchs J-P, Giloh H, Kuo C-H et al. Nuclear structure determination of the fate of the nuclear envelope in *Drosophila* during mitosis using monoclonal antibodies. J Cell Sci 1983; 64:331-349.
50. Lemaitre JM, Geraud G, Mechali M. Dynamics of the genome during early *Xenopus laevis* development: karyomeres as independent units of replication. J Cell Biol 1998; 142:1159-1166.
51. Harris P. Electron microscope study of mitosis in sea urchin blastomeres. J Biochem Biophys Cytol 11: 1961;419-431.
52. Harel A, Zlotkin E, Nainudel-Epszteyn S et al. Persistence of major nuclear envelope antigens in an envelope-like structure during mitosis in *Drosophila melanogaster* embryos. J Cell Sci 1989; 94:463-470.
53. Haaf T, Ward DC. Higher order nuclear structure in mammalian sperm revealed by *in situ* hybridization and extended chromatin fibers. Exp Cell Res 1995; 219:604-611.
54. Vigers GPA, Lohka MJ. A distinct vesicle population targets membranes and pore complexes to the nuclear envelope in *Xenopus* eggs. J Cell Biol 1991; 112:545-556.
55. Drummond S, Ferrigno P, Lyon C et al. Temporal differences in the appearance of NEP-B78 and an LBR-like protein during *Xenopus* nuclear envelope reassembly reflect the ordered recruitment of functionally discrete vesicle types. J Cell Biol 1999; 144:225-240.
56. Imai N, Sasagawa S, Yamamoto A et al. Characterization of the binding of nuclear envelope precursor vesicles and chromatin, and purification of the vesicles. J Biochem 1997; 122:1024-1033.
57. Sasagawa S, Yamamoto A, Ichimura T et al. In vitro nuclear assembly with affinity-purified nuclear envelope precursor vesicle fractions, PV1 and PV2. Eur J Cell Biol 1999; 78:593-600.
58. Roth MG, Sternweis PC. The role of lipid signalling in constitutive membrane traffic. Curr Opin Cell Biol 1997; 9:519-526.
59. De Camilli P, Emr SD, McPherson PS et al. Phosphoinositides as regulators in membrane traffic. Science 1996; 271:1533-1539.
60. D'Hondt K, Heese-Peck A, Riezman H. Protein and lipid requirements for endocytosis. Annu Rev Genet 2000; 34:255-295.
61. Wymann MP, Pirola L. Structure and function of phosphoinositide 3-kinases. Biochim Biophys Acta 1998; 1436:127-150.
62. Hubner S, Couvillon AD, Kas JA et al. Enhancement of phosphoinositide 3-kinase (PI 3-kinase) activity by membrane curvature and inositol-phospholipid-binding peptides. Eur J Biochem 1998; 258:846-853.
63. Longo FJ. An ultrastructural analysis of mitosis and cytokinesis in the zygote of the sea urchin, *Arbacia punctulata*. J Morphol 1972; 138:207-238.
64. Montag M, Spring H, Trendelenburg MF. Structural analysis of the mitotic cycle in pre-gastrula *Xenopus* embryos. Chromosoma 1988; 96:187-196.
65. Emanuelsson H. Karyomeres in early cleavage embyros of *Ophryotrocha labronica*. LaGreca and Bacci Wilhelm Roux' Archiv 1973; 173:27-45.

CHAPTER 10

Nuclear Envelope Dynamics in *Drosophila* Pronuclear Formation and in Embryos

Mariana F. Wolfner

The essential role of the nuclear envelope in basic cellular events such as DNA replication, nuclear organization, gene expression and the cell cycle is well known in many organisms. Recently it has become clear that the nuclear envelope plays an important role in development and in particular cell types (e.g., 1-4). The stages surrounding the formation of pronuclei in fertilized eggs involve dramatic changes in nuclear envelope composition, organization and function, making them of particular interest for investigating nuclear envelope dynamics. Knowing the basis of the changes in nuclear envelopes as one goes from gamete to differentiating embryo is instructive both to our understanding of nuclear assembly/disassembly and function, and to our understanding of this critical developmental stage. This article focuses on pronuclear and embryo nuclear envelope formation and function in *Drosophila melanogaster*. I will first describe the overall dynamics and composition of *Drosophila* nuclear envelopes, and then their developmental dynamics in the stages surrounding pronuclear formation and early embryonic development.

Drosophila Nuclear Envelopes

Composition

As in other animals, the nuclear envelope in *Drosophila* has three basic sub-parts: the nuclear membranes, a lamina just beneath the membrane layer, and nuclear pores that rivet the nuclear membranes and permit nucleo-cytoplasmic transport.

There are two nuclear membranes, each a lipid bilayer containing characteristic proteins. The outer nuclear membrane is continuous with the endoplasmic reticulum. The inner nuclear membrane contains proteins that can associate with lamin and/or chromatin. The *Drosophila* genome predicts genes for relatives of several proteins known to be in the inner membranes of vertebrates or nematodes. These include otefin, a LEM protein,[5-7] which has been shown to be a constituent of the Drosophila inner nuclear membrane, and predicted counterparts of two other LEM proteins (emerin and MAN-1) as well as nurim, the lamin B receptor (LBR), and unc-84 (see ref. 7); we still await reports of the subcellular localization of these predicted proteins.

The Drosophila nuclear lamina contains proteins related to those in other animal nuclear laminas. In particular, its major constituent is intermediate-filament proteins of the lamin family. As in most higher animals, Drosophila has lamins of two subtypes: A-type and B-type. However, its suite of lamins is simpler than that in vertebrates. *Drosophila* has a single gene for a B-type lamin called lamin Dm_0.[8] Lamin Dm_0 derivatives (called, collectively, lamin Dm, below) are present in the nuclear envelopes of nearly all *Drosophila* cells[9-14] and, this protein is essential for full viability.[15] Fertility of *Drosophila* females also demands sufficient levels of lamin Dm_0 derivatives in the female germ line.[15] The B-type lamin undergoes post-translational

Nuclear Envelope Dynamics in Embryos and Somatic Cells
edited by Philippe Collas. ©2002 Eurekah.com and Kluwer Academic/Plenum Publishers.

modifications, including phosphorylations that convert it from a soluble form (Dm_{mit}) to nuclear forms that can make polymeric fibers in the lamina (Dm_1, Dm_2.[11,16] *Drosophila*'s single A-type lamin ("lamin C", [17-19]), analogous to the vertebrate situation, is found only in certain cell types; in *Drosophila* these are a subset of differentiated cells. *Drosophila* nuclear laminas in early embryos include at least one protein without a vertebrate counterpart—the Young Arrest ("YA") protein,[1,20] a novel hydrophilic nuclear lamina component that is essential for embryonic development to initiate (see below). Made only during oogenesis (though excluded from nuclei at this time[21,22]), YA is found only in the nuclei of young (cleavage stage) embryos;[1, 20] after this time it is not detected in any *Drosophila* cells, except for developing oocytes.

Drosophila nuclear envelopes also contain pore complexes whose components are relatives of those of vertebrate nuclear pores:[23] including nup154,[24] gp210,[25] a Tpr homologue,[26] myosin-like proteins[27,28] and a predicted nup153 homologue with RNA binding activity.[29] Components of the nuclear import/export system, including importin family members and an exportin homologue, have also been found in *Drosophila* (e.g., see refs. 30-39).

Cell Cycle Dynamics

In at least some, and possibly all, *Drosophila* cells, nuclear envelopes do not completely break down during mitosis. This was first shown for cleavage stage *Drosophila* embryos, in an elegant electron microscopy study;[40] subsequent investigators extended this finding by documenting the behaviors of specific nuclear envelope, or envelope-associated, proteins during the cell cycles in embryos.[1,5,26] The envelopes of interphase nuclei in embryos are continuous around the nuclei. By prometaphase, the envelopes are seen to be open at the spindle poles. A second membrane layer, paralleling the nuclear membranes and, like them, open at the spindle poles, begins to form around the original nuclear envelope. Nuclear pores dissociate from the nuclear membranes beginning in prophase.[26,40] In metaphase, the nuclear lamina largely dissociates: most of its lamin Dm disappears from the nuclear periphery[5] and YA also becomes undetectable at the nuclear periphery.[1] However, the fenestrated, double "spindle envelope" remains, still open at the poles. Chromosomes segregate on the spindle within this envelope. At telophase, lamin Dm, otefin and nuclear pore proteins begin to be detected in the nuclear envelopes, and the second membrane layer disappears.[1,5,26,40] At the end of telophase in early embryos, the YA protein becomes detectable at the nuclear periphery, suggesting that it assembles into the nuclear envelope after lamin Dm.[1] These nuclear envelope dynamics may be characteristic of all *Drosophila* nuclei: spindle envelopes of similar appearance to those reported by Stafstrom and Staehelin[40] have also been seen in spermatogenic cells[41] and in cultured *Drosophila* cells.[42]

Developmental Changes in Nuclear Envelopes Around the Time of Fertilization

Dramatic changes in nuclear envelopes occur around the time of fertilization in animals. At these times, highly differentiated gametes, with unique nuclear structures, combine to form a zygote. The gamete nuclei are converted to pronuclei that then participate in a special series of events that generates the first zygotic nuclei of the embryo. In many animals, including *Drosophila*, the zygotic nuclei then go through a very rapid mitotic stage (cleavage), using maternal proteins stored in the egg. When the maternal "dowry" has been depleted, the zygotic genome takes control of cell division dynamics, and usually at about the same time the zygotic cells begin to undertake differentiative pathways.

The next sections of this article follow these events in *Drosophila*. I briefly describe the formation of gametes, pronuclei, zygotic nuclei, cleavage nuclei and nuclei committed to differentiative pathways. I then focus on the composition and changes in nuclear envelopes that accompany these events. If the roles of nuclear envelopes or their constituents are known, these are included. As illustrated in the Figure, some nuclear envelope components are nearly ubiquitous throughout these stages, others are present in, and essential for, the functions of particular nuclei, and still others appear or disappear when nuclei have certain characteristics or fates.

Gamete Nuclei

Oocytes

The development of an oocyte begins with 4 mitotic divisions by the daughter of a germ line stem cell (see ref. 43 for review); these divisions have incomplete cytokinesis. Fifteen of the cells produced in these divisions (the "nurse cells") then cease dividing. They endoreplicate their genomes and begin to synthesize RNAs and proteins that will be transferred to the oocyte to provision it for subsequent embryo development. The 16th cell, the oocyte, initiates meiosis, arresting in metaphase of meiosis I at the end of oogenesis. This arrest is maintained until the oocyte is ovulated.[44] Ovulation activates the oocyte to complete meiosis; though meiotic progression occurs at the time of fertilization, it is independent of sperm penetration.[44-47] Meiosis I and, immediately thereafter, meiosis II occur (for reviews see refs. 47-50). The process is extremely rapid, taking as little as 17 min to complete. It occurs without cytokinesis, and results in the presence of 4 haploid meiotic products in the activated egg.

Several nuclear envelope proteins have unique behaviors, or biochemical properties, in developing or activating oocytes. Oocyte nuclei (and nurse cell nuclei) initially have standard nuclear envelope compositions, including lamin Dm,[11] otefin[51] and nup154.[24] By late oogenesis, the distribution of these proteins in the oocyte nucleus appears to be different from that in typical nuclei (including the nurse cell nuclei): higher levels of lamin Dm and otefin are detected within the oocyte nucleus (relative to levels at its periphery) than in other cell types.[11,51] Whether this reflects a difference in nuclear structure, leading to a greater accessibility of internal lamin Dm to staining or is due to a redistribution of lamin Dm at this stage is not known. In late-stage oocytes, there is also a soluble form of lamin Dm, that remains detectable until the end of the maternally-driven cleavage-cycle phase of embryonic development.[11] This lamin isoform, which is presumably the lamin Dm donated to the oocyte by the nurse cells and stored for use in embryo mitosis, is thought to be the same as the mitotic isoform of lamin, Dm_{mit}, that is observed in tissue culture cells.[11] Finally, the YA protein is detected in oocytes; this is the first time it is found post-embryonically. However, this YA is in a highly phosphorylated form that is excluded from the nuclei, apparently by associating with a cytoplasmic retention complex.[21,22]

At the very end of oogenesis, the metaphase-arrested nucleus of the oocyte appears to disassemble: lamin Dm and otefin become undetectable at the oocyte's nuclear periphery, though the precise structure of any remaining nuclear envelope is not known.[11,51] The dynamics of the nuclear envelope during the rapid meiotic divisions that follow are also unknown. The interphase between meiosis I and meiosis II is extremely short and hard to "catch", and there may be no typical interphase at all. In preliminary experiments we have been unable to detect assembly of a nuclear lamina around the meiotic nuclei during the time between telophase of meiosis I and metaphase of meiosis II.[52]

The nuclear envelope plays important roles in oogenesis. Females homozygous for a partial loss-of-function allele of the lamin Dm_0 gene are sterile, indicating the essential function of this lamin in oogenesis.[15] At least part of the function of lamin Dm may relate to its role in polarity generation in the oocyte. In normal oocytes, the nucleus is located at the anterior end of the cell, and is closer to one side of the oocyte (see ref. 53 for review). That side is fated to become the dorsal side of the egg and embryo. mRNA from the gurken (grk) gene (a TGF-alpha family member) accumulates near the oocyte nucleus (see ref. 53 for review). This results in higher levels of GRK protein in the vicinity of the oocyte nucleus, which in turn activates a signaling system that confers a dorsal identity on the side of the oocyte nearest the nucleus. In oocytes carrying *misguided* mutations of the lamin Dm_0 gene, grk RNA is mis-localized, and

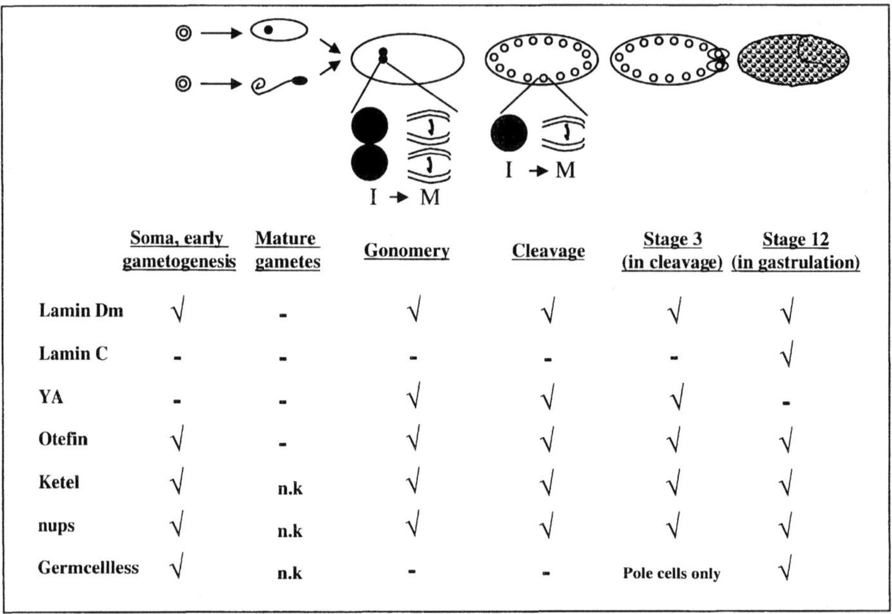

	Soma, early gametogenesis	Mature gametes	Gonomery	Cleavage	Stage 3 (in cleavage)	Stage 12 (in gastrulation)
Lamin Dm	√	-	√	√	√	√
Lamin C	-	-	-	-	-	√
YA	-	-	√	√	√	-
Otefin	√	-	√	√	√	√
Ketel	√	n.k	√	√	√	√
nups	√	n.k	√	√	√	√
Germcellless	√	n.k	-	-	Pole cells only	√

Figure 1. Nuclear envelope components in the nuclear periphery of gametes, pronuclei and early *Drosophila* embryos. The upper panel shows the dynamics of the nuclear cycles between gametogenesis and mid-gastrulation. DNA of the oocyte nucleus and female pronucleus (upper of the two nuclei in the Gonomery panel) is shown in red. DNA of the sperm nucleus and male pronucleus (lower of the two nuclei in the Gonomery panel) is in blue at http://www.eurekah.com/chapter.php?chapid=711&bookid=56&catid=15. Nuclei in the cleavage, Stage 3 and Stage 12 panels are zygotic nuclei. Their DNA is shown in purple at http://www.eurekah.com/chapter.php?chapid=711&bookid=56&catid=15. I = interphase, M = mitotic metaphase. The nuclear envelopes of gametogenic cells, oocytes and cleavage stage nuclei in interphase are shown as single black lines. Spindle envelopes (metaphase of the gonomeric and cleavage divisions) are shown as double black lines; the inner ones represent the nuclear membranes and the outer ones represent the additional membrane layer that forms at this stage. Below the drawings are listed the proteins or complexes known to be present in nuclear envelopes at the indicated stages (see references in the text; "nups" refers to nuclear pores and to the pore-associated antigen Bx34 [see ref. 26]): √ = detectable, — = not detectable, n.k. = not known. Stage 3 = the cleavage stage at which pole cells form (~1.5 hr of development). Stage 12 = the germ band retraction stage during gastrulation, at which lamin C is first detected (~10 hr of development).

the oocyte's dorsoventral polarity is consequently disrupted.[54] Thus, the nuclear envelope contributes to proper positioning of dorsoventral determinants in the developing oocyte.

Sperm

The development of sperm begins with mitotic divisions by the daughter of a stem cell in the male germline (see ref. 55 for review). After 4 mitotic divisions with incomplete cytokinesis, the nuclei undertake meiosis. After meiosis has completed, spermiogenesis occurs. During this process, the sperm nucleus becomes elongated and bounded by microtubules and the sperm itself elongates and loses most of its cytoplasm.

There are dramatic changes in the nuclear envelope of spermatogenic cells during this process. The nuclei in the cells undergoing mitosis have nuclear envelopes that contain lamin Dm,[13,56] otefin[51] and nup154,[24] and can accommodate YA if it is ectopically expressed in these cells.[13] The distribution of other nuclear envelope components in spermatogenic cells has not been reported. The nuclear envelopes of spermatogenic cells have spindle envelopes, and cell cycle dynamics of the type described above.[41] During spermiogenesis, when the sperm nucleus' shape changes, the composition of its nuclear envelope also changes. All detectable nuclear envelope antigens tested (lamin Dm, otefin, and, upon ectopic expression, YA[13,51,56]) become undetectable around the sperm nuclei. Membranes are reported to remain around the sperm nucleus at least until late in this stage. The nucleus of a mature *Drosophila* sperm either does not have a nuclear envelope or is bounded by an atypical envelope whose protein components are not cross-reactive with any somatic nuclear envelope protein tested thus far.

Pronuclei

Female

After the oocyte nucleus has completed meiosis, nuclear envelopes form around the four female meiotic products. All four meiotic products are within the egg cytoplasm and are indistinguishable in terms of nuclear envelope antigens at this time, even though their fates will be different. The haploid nucleus closest to the center of the egg will usually become the female pronucleus and contribute to the embryo's genome. The other three meiotic products will migrate to the egg periphery, where these "polar bodies" will enter a metaphase-like state but then degenerate (see ref. 48 for review).

The envelopes of the four products of oocyte meiosis contain lamin Dm and the YA protein. The presence of lamin Dm in the envelopes of these nuclei likely is simply a reflection of the return to standard nuclear envelope structure. YA's presence in the envelopes of the post-meiotic nuclei marks, however, the initial appearance of this protein at the nuclear periphery—YA was produced, but excluded from nuclei, during oogenesis.[21] YA's phosphorylation level drops during egg activation.[21] We hypothesize that this releases YA from a cytoplasmic retention complex, and allows it to enter nuclei for the first time when nuclei form after meiosis.[22]

Male

If the egg is penetrated by a sperm, the sperm's nucleus will have to be converted to a male pronucleus in order to participate in development. As in systems that are presently better characterized (e.g., see refs. 57,58 for review), the sperm nucleus' chromatin must decondense (presumably involving replacement of some of its chromatin-packaging proteins with proteins available in the egg), change shape,[13,14] and acquire a new nuclear envelope made from maternally-provided constituents including lamin Dm and YA.[13] Since the entire *Drosophila* sperm enters the egg, still bounded by its plasma membrane,[59] the first step in converting the sperm nucleus to a male pronucleus is removal of the sperm's plasma membrane. The product of the *sneaky* (*snky*) gene appears to be necessary for this membrane removal; evidence suggests that sperm from males mutant in *snky* fail to demembranate when they enter eggs.[60] In eggs fertilized by *snky* sperm, the sperm nucleus fails to decondense, change shape or migrate, and it does not become surrounded by lamin Dm or YA.[13,60]

Once the sperm nucleus has been exposed to the cytoplasm of the fertilized egg, it begins to swell and decondense its DNA. Partway through the decondensation process, the sperm nucleus acquires an envelope made from components present in the egg.[13,14,61] Analysis of subsequent steps in *Drosophila* male pronuclear formation have been most informed by genetic analysis, since existing in vitro nuclear assembly systems[62-64] have not been ideal for detailed examination of nuclear assembly: these systems are too inefficient, perhaps because they are made from embryos at mixed stages of the cell cycle.[63] The in vitro assembly systems have, however, been useful for identifying chromatin decondensation factors that may act in pronuclear formation.[65,66]

Phenotypic analysis of mutants in the maternal-effect gene *sésame* (*ssm*)[14] suggest that *Drosophila* male pronuclear formation follows the paradigm elegantly established in sea urchins, in which decondensation of the sperm nucleus occurs in two phases: an initial decondensation independent of acquisition of a new nuclear envelope, and a later decondensation dependent on a nuclear envelope derived from maternal components (see refs. 57, 58, 67, 68 for review).

In eggs lacking functional *ssm* gene product, sperm nuclei decondense only partially. The partially decondensed sperm nuclei have acquired nuclear envelopes that include lamin Dm and YA (provided by the egg).[14] The phenotype of the sperm nuclei in *ssm* embryos suggests that *ssm* may encode a maternal factor needed for the second, nuclear envelope-dependent, phase of male pronuclear formation. The molecular nature of the *ssm* gene product has not yet been reported.

Combining the Pronuclear Genomes

Once the sperm nucleus has acquired a maternally-provided nuclear envelope, microtubules from its associated centriole capture one of the female meiotic products (the female pronucleus). The female and male pronuclei, each bounded by a lamin Dm- and YA-containing nuclear envelope, migrate into the anterior/center of the embryo and become closely apposed, but do not fuse (1, 61, 69; see ref. 48 for review). The apposed pronuclei then initiate the first cell cycle of the early embryo. Pronuclear DNA replicates. A single spindle forms, using centrosomes that derived from a centriole donated by the sperm. A metaphase occurs, with the maternal chromosomes remain on one side of the spindle and the paternal ones on the other. The two chromosome complements then undergo anaphase, in coordination but again each on their own side of the spindle.[1,61,69] These characteristics lead to the designation of this division as "gonomeric". In late telophase of the gonomeric division, the parental chromosome complements finally mix; thus the first zygotic nuclei form at the end of the first cell cycle. Two independent lines of evidence confirm the light-microscopy determination of gonomery. In eggs of normal females fertilized by sperm from Wolbachia-infected male flies, development arrests soon after fertilization.[70,71] Many such zygotes show abnormalities during the gonomeric division. Their male-derived chromosomes are abnormally condensed and confined to one side of the gonomeric spindle; the maternal genome segregates, as normal, on the other side of the spindle.[72] The phenotype of the *ssm* mutant also shows that maternal and paternal chromosomes segregate separately during the gonomeric division. In fertilized ssm eggs, one side of the spindle is occupied by condensed male chromatin that cannot participate in the division, but maternally-derived chromatids still segregate as normal on the other side of the spindle.[14]

In these developmental events as well, the nuclear envelope and its components play important roles. Function of the YA protein is essential for the nuclei to enter the gonomeric cell cycle. In the absence of YA function, the apposed pronuclei appear to arrest prior to S phase of this cell cycle[1,73,74] Our data lead us to hypothesize that YA function is necessary for nuclei to pass a checkpoint that monitors completion of meiosis and creation of mitotically-competent nuclei. YA is capable of binding to chromatin, via interaction with DNA and histone H2B,[63,75] and also interacts with lamin Dm_0.[76] We hypothesize that its function is to confer a mitotically-competent state on nuclei.[74] Another nuclear envelope-related protein whose function is apparent at this early time in development is Ketel, a *Drosophila* homologue of importin-β.[34] Ketel protein is made during oogenesis and placed into eggs; it continues to be made from the zygotic genome in embryos.[34] Embryos lacking maternal Ketel product, or embryos produced by females carrying apparent dominant-negative Ketel mutations arrest development during the gonomeric cell cycle, with nuclei that appear abnormal. The arrest phenotype of these embryos indicates that Ketel is required for proper nuclear assembly following mitosis.[34,35] The mechanism by which Ketel exerts this function is not known. The simplest possible model—that Ketel mutations prevent all nuclear import and hence assembly of nuclei with proper nuclear envelopes—is ruled out since at least one substrate (cNLS-phycoerythrin) can enter nuclei in the mutant embryos. This suggests that Ketel has either a unique function specific to nuclear assembly at the end of mitosis[77,78] or that the import of at least one molecule essential for nuclear assembly is completely dependent on Ketel, with no alternative pathway available for its import.[34,35]

The nuclear envelope also confers the "gonomeric" character on this first cell cycle. Immuno-fluorescence studies showed that the separation of the parental genomes during the gonomeric division is due to the presence of a spindle envelope around each of the pronuclei.[61] The spindle envelopes, open at the poles, allow microtubules of the single spindle to penetrate, but keep the parental chromosome separate until they reach the spindle poles late in telophase.[61]

Cleavage Nuclei

The gonomeric division is followed by 12 very rapid mitotic divisions (reviewed in ref. 48), leading to several thousand nuclei. These divisions occur within a syncytium, are roughly synchronous, and are driven by maternal products that were stored in the egg during oogenesis. After about 7 nuclear cycles (including the gonomeric division) the nuclei begin to migrate to the periphery of the egg, forming a "syncytial blastoderm" (see ref. 48 for review). Towards the end of the syncytial phase of development, the division cycles slow from ~9 min to ~21 min, as maternal products are depleted. In the final cleavage cell cycle, the maternally-provided mRNA for *string* (*stg*), which encodes a cdc25-family phosphatase, is degraded. Further progress through mitotic cycles requires transcription of *stg* from the zygotic genome, and accumulation of STG protein to sufficient levels.[79-81] At this time also, membranes grow down into the egg, separating each nucleus into its own cell (initiating the cellular blastoderm stage). In such cellularized embryos, mitotic cycles become more independent, losing the synchrony seen in syncytial blastoderm embryos; however nuclei of cells with similar fate often share similar cell division behaviors (mitotic domains; see ref. 82).

During the pre-cellularization nuclear division cycles, nuclear envelopes show typical Drosophila nuclear envelope dynamics,[40] involving cyclic loss, and then restoration of nuclear pores and their associated proteins, otefin and lamin Dm (in telophase),[1,5,26,40] YA (later in telophase),[1] and presence of fenestrated membranes and a spindle envelope during mitosis. Lamin Dm_0 derivatives show changes in phosphorylation status that are thought to correspond to the appearance of nuclear, polymerized forms (Dm1, Dm2) in addition to the soluble form (thought to be Dm_{mit}) that was already present in early embryos.[11]

The nuclear envelope has several essential developmental functions during the cleavage stages. First, the timing of nuclear envelope reassembly plays an important role in cell cycle dynamics at this stage. A checkpoint, identified through the phenotype of the *grapes* (*grp*) mutation, involves the role of the nuclear envelope.[83,84] *grp* mutants arrest because their nuclei undertake cell cycles with improperly condensed chromatin. Careful examination of the dynamics of a variety of nuclear structures in *grp* vs. normal embryos in the presence and absence of aphidicolin, indicated that the defect in *grp* embryos results in premature breakdown of nuclear envelopes in those embryos. This premature breakdown occurs before the chromosomes are fully condensed or fully replicated, resulting in abnormal chromosome behaviors during mitosis.[84] The GRP protein is normally located in nuclei, but is released when nuclear envelopes break down; it is suggested to be part of a checkpoint that prevents nuclear envelope breakdown until S-phase is completed,[84] also allowing for attainment of the proper condensation state by chromosomes. A second role for the nuclear envelope is suggested by the tight parallel of lamin's, and its position and dynamics to those of mitotic spindles.[85] It was proposed that the nuclear lamina might play a role in facilitating spindle formation or dynamics during mitosis.[85] A third role for the nuclear envelope is more indirect, in the generation of some embryo polarities during this developmental stage. This role results from the nuclear envelope acting as a barrier that separates nuclear contents from the cytoplasm: dorsoventral polarity, for example, is controlled by the regulated access of the DL transcription factor to the nuclear interior. On the dorsal side of the embryo, a cytoplasmic complex keeps DL out of nuclei, causing them to take on a dorsal fate (e.g., see ref. 86 for review). Disruption of this complex on the ventral side of the embryo due to a signaling pathway (e.g., see ref. 87 for review) allows DL to enter nuclei on this side, conferring on them a ventral fate. Though roles, if any, for specific nuclear envelope proteins in the regulation of dorsoventral polarity are not known, this example serves to illustrate the critical role of a functional nuclear envelope in developmental choices.

Nuclei of Differentiating Cells in Embryos

The first overt differentiative change in the embryo is detected in the behavior of ~8 nuclei in an otherwise cleavage-stage embryo. During cycle 9 (counting the gonomeric division) these most posterior nuclei of the embryo bud out into individual cells. These "pole cells" are fated to form the germ line. Their divisional timing differs from the synchronous syncytial divisions of the remaining somatic nuclei in the embryo. The nuclear pores of pole cell nuclear envelopes acquire an associated protein that is the product of the *germcellless* (*gcl*) gene.[88,89] Function of gcl is essential for pole cell nuclei to take on a germ cell fate. *gcl* is also expressed at other times in development, but its function at those times is not known. A vertebrate counterpart of *gcl* has recently been reported to be involved in regulating transcription,[90] suggesting that modulation of transcription by GCL protein in pole cells may be important in specifying their fate.

Other nuclei and cells in Drosophila embryos only show signs of commitment to a differentiated fate after the cellular blastoderm stage. As nuclei take on differentiated fates, the composition of their envelopes change from that of cleavage nuclei: YA protein disappears[1,20] by the end of the blastoderm stage and later, during gastrulation, lamin C becomes detectable in nuclear envelopes of differentiated cells (initially in the nuclear envelopes of cells that will become oenocytes, hindgut and posterior spiracle, and later in additional cell types[19]).

The nuclear envelope is again seen to play important roles in the differentiative stages. For example, *misguided* mutant alleles of the lamin Dm_0 gene have a phenotype in differentiating cells: these mutations specifically disrupt the branching growth of trachae that occurs late in embryogenesis.[54] The aberrant growth and targeting by the tracheal branches suggests that lamin Dm plays a role in cell polarity.

Conclusion

Nuclear envelopes undergo dramatic changes in composition and dynamics during the developmental transition from gametes to differentiating embryos. These changes include modification and partial dissolution in gametogenesis, complete re-formation from maternal components in early embryos, and modulation of composition during embryogenesis as nuclei take on differentiation competence. In *Drosophila*, we now know much about the dynamics and functions of nuclear envelopes at these critical times, and several of the likely important molecular players. Continued exploitation of the genetics, genomics and molecular developmental biology in this model system will provide a detailed picture of the mechanism by which pronuclear envelopes form, break down, and regulate the start of development.

Acknowledgements

I thank my YA-colleagues past and present for many stimulating discussions, Drs. M. Goldberg, J. Liu and J. Yu for many helpful comments on this manuscript, and the NIH (R01-GM44659) for supporting our work in this area.

References

1. Lin H, Wolfner MF. The *Drosophila* maternal-effect gene *fs(1)Ya* encodes a cell cycle-dependent nuclear envelope component required for embryonic mitosis. Cell 1991; 64:49-62.
2. Furukawa K, Hotta Y. cDNA cloning of a germ cell specific lamin B_3 from mouse spermatocytes and analysis of its function by ectopic expression in somatic cells. EMBO J 1993; 12:97-106.
3. Manilal S, Man NT, Sewry CA et al. The Emery-Dreifuss muscular dystrophy protein, emerin, is a nuclear membrane protein. Hu Mol Genet 1996; 5:801-808.
4. Bonné G, Di Barletta MR, Varnous S et al. Mutations in the gene encoding lamin A/C cause autosomal dominant Emery-Dreifuss muscular dystrophy. Nat Genet 1999; 21:285-288.
5. Harel A, Zlotkin E, Nainudel-Epszteyn S et al. Persistence of major nuclear envelope antigens in an envelope-like structure during mitosis in *Drosophila melanogaster* embryos. J Cell Sci 1989; 94:463-470.

6. Padan R, Nainudel-Epszteyn S, Goitein R et al. Isolation and characterization of the *Drosophila* nuclear envelope otefin cDNA. J Biol Chem, 1990; 265:7808-7813.

7. Cohen M, Lee KK, Wilson KL et al. Transciptional repression, apoptosis, human disease and the functional evolution of the nuclear lamina. Trends Biochem Sci 2001; 26:41-47.

8. Gruenbaum Y, Landesman Y, Drees B et al. *Drosophila* nuclear lamin precursor Dmo is translated from either of two developmentally regulated mRNA species apparently encoded by a single gene. J Cell Biol 1988; 106:585-596.

9. Smith DE, Fisher PA. Identification, developmental regulation, and response to heat shock of two antigenically related forms of a major nuclear envelope protein in *Drosophila* embryos: Application of an improved method for affinity purification of antibodies using polypeptides immobilized on nitrocellulose blots. J Cell Biol 1984; 99:20-28.

10. Smith DE, Gruenbaum Y, Berrios M et al. Biosynthesis and interconversion of *Drosophila* nuclear lamin isoforms during normal growth and in response to heat shock. J Cell Biol 1987; 105:771-790.

11. Smith DE, Fisher PA. Interconversion of *Drosophila* nuclear lamin isoforms during oogenesis, early embryogenesis, and upon entry of cultured cells into mitosis. J Cell Biol 1989; 108:255-265.

12. Stuurman N, Maus N, Fisher PA. Interphase phosphorylation of the *Drosophila* nuclear lamin: site-mapping using a monoclonal antibody. J Cell Sci 1995; 108:3137-3144.

13. Liu J, Lin H Lopez JM et al. Formation of the male pronuclear lamina in *Drosophila*. Dev Biol 1997; 184:187-196.

14. Loppin B, Docquier M, Bonneton F et al. The maternal-effect mutation *sésame* affects the formation of the male pronucleus in *Drosophila melanogaster*. Dev Biol 2000; 222:392-404.

15. Lenz-Böhme B, Wismar J, Fuchs S et al. Insertional mutagenesis of the *Drosophila* nuclear lamin Dmo gene results in defective nuclear envelopes, clustering of nuclear pore complexes, and accumulation of annulate lamellae. J Cell Biol, 1997; 137:1001-1016.

16. Schneider U, Mini T, Jeno P et al. Phosphorylation of the major *Drosophila* lamin in vivo: site identification during both M-phase (meiosis) and interphase by electrospray ionization tandem mass spectrometry. Biochemistry 1999;3 8:4620-4632.

17. Bossie CA, Sanders MM. A cDNA from *Drosophila melanogaster* encodes a lamin C-like intermediate filament protein. J Cell Sci 1993; 104:1263-1272.

18. Riemer D, Weber K. The organization of the gene for *Drosophila* lamin C: limited homology with vertebrate lamin genes and lack of homology versus the *Drosophila* lamin Dmo gene. Euro J Cell Biol 1994; 63:299-306.

19. Riemer D, Stuurman N, Berrios M et al. Expression of *Drosophila* lamin C is developmentally regulated: analogies with vertebrate A-type lamins. J Cell Sci, 1995; 108:3189-3198.

20. Lopez J, Song K, Hirshfeld A et al. The *Drosophila* fs(1)Ya protein, which is needed for the first mitotic division, is in the nuclear lamina and in the envelopes of cleavage nuclei, pronuclei and nonmitotic nuclei. Dev Biol 1994; 163:202-211.

21. Yu J, Liu J, Song K et al. Nuclear entry of the *Drosophila melanogaster* nuclear lamina protein YA correlates with developmentally regulated changes in its phosphorylation state. Dev Biol 1999; 210:124-134.

22. Yu J, Garfinkel A, Wolfner MF. Interaction of the essential *Drosophila* nuclear protein YA with P0/AP3 in the cytoplasm: Implications for developmental regulation of YA's subcellular location. Dev Biol 2002; 244:429-441.

23. Filson AJ, Lewis A, Blobel G et al. Monoclonal antibodies prepared against the major *Drosophila* nuclear matrix-pore complex-lamina glycoprotein bind specifically to the nuclear envelope in situ. J Biol Chem 1985; 260:3164-3172.

24. Gigliotti S, Callaini, G, Andone S et al. Nup154, a new *Drosophila* gene essential for male and female gametogenesis is related to the nup155 vertebrate nucleoporin gene. J Cell Biol 1998; 142:1195-1207.

25. Berrios M, Meller VH, McConnell M et al. *Drosophila* gp210, an invertebrate nuclear pore complex glycoprotein. Eur J Cell Biol 1995; 67:1-7.

26. Zimowska G, Aris JP, Paddy MR. A *Drosophila* Tpr protein homolog is localized both in the extrachromosomal channel network and to nuclear pore complexes. J Cell Sci 1997; 110:927-944.

27. Berrios M, Fisher PA, Matz, EC. Localization of a myosin heavy chain-like polypeptide to *Drosophila* nuclear pore complexes. Proc Natl Acad Sci USA 1991; 88:219-223.

28. Strambio-De-Castillia C, Blobel G, Rout MP. Proteins connecting the nuclear pore complex with the nuclear interior. J Cell Biol 1999; 144:839-855.

29. Dimaano C, Ball JR, Prunuske AJ et al. RNA association defines a functionally conserved domain in the nuclear pore protein Nup153. J Biol Chem 2001; 276:45349-45357.

30. Torok I, Strand D, Schmitt R et al. The *overgrown hematopoietic organs-31* tumor suppressor gene of Drosophila encodes an importin-like protein accumulating in the nucleus at the onset of mitosis. J Cell Biol 1995; 129:1473-1489.

31. Siomi MC, Fromont M, Rain JC et al. Functional conservation of the transportin nuclear import pathway in divergent organisms. Mol Cell Biol 1998; 18:4141-4148.

32. Dockendorff TC, Tang Z, Jongens TA. Cloning of karyopherin-alpha3 from *Drosophila* through its interaction with the nuclear localization sequence of germcell-less protein. Biol Chem, 1999; 380:1263-1272.

33. Herold A, Suyama M, Rodrigues JP et al. TAP (NXF1) belongs to a multigene family of putative RNA export factors with a conserved modular architecture. Mol Cell Biol 2000; 20:8996-9008.

34. Lippai M, Tirian L, Boros I et al. The Ketel gene encodes a *Drosophila* homologue of importin-beta. Genetics 2000; 156:1889-1900.

35. Tirian L, Puro J, Erdelyi M et al. The Ketel(D) dominant-negative mutations identify maternal function of the *Drosophila* importin-beta gene required for cleavage nuclei formation. Genetics 2000; 156:1901-1912.

36. Fasken MB, Saunders R, Rosenberg M et al. A leptomycin B-sensitive homologue of human CRM1 promotes nuclear export of nuclear export sequence-containing proteins in *Drosophila* cells. J Biol Chem 2000; 275:1878-1886.

37. Mathe E, Bates H, Huikeshoven H et al. Importin-alpha3 is required at multiple stages of *Drosophila* development and has a role in the completion of oogenesis. Dev Biol 2000; 223:307-322.

38. Collier S, Chan HY, Toda T et al. The *Drosophila embargoed* gene is required for larval progression and encodes the functional homolog of *Schizosaccharomyces* Crm1. Genetics 2000; 155:1799-1807.

39. Lorenzen JA, Baker SE, Denhez F et al. Nuclear import of activated D-ERK by DIM-7, an importin family member encoded by the gene *moleskin*. Development 2001; 128:1403-1414.

40. Stafstrom JP, Staehelin LA. Dynamics of the nuclear envelope and of nuclear pore complexes during mitosis in the *Drosophila* embryo. Euro J Cell Biol 1984; 34:179-189.

41. Church K, Lin HPP. Meiosis in *Drosophila melanogaster*: II. The prometaphase-1 kinetochore microtubule bundle and kinetochore orientation in males. J Cell Biol 1982; 93:365-373.

42. Debec A, Marcaillou C, Structural alterations of the mitotic apparatus induced by the heat shock response in *Drosophila* cells. Biol Cell 1997; 89:67-78.

43. Spradling AC. Developmental genetics of oogenesis. In: Bate M, Martinez Arias A., eds.The Development of *Drosophila melanogaster*. Plainview: Cold Spring Harbor Laboratory Press, 1993:1-70.

44. Heifetz Y, Yu J, Wolfner MF. Ovulation triggers activation of *Drosophila* oocytes. Dev Biol 2001; 234:416-424.

45. Doane WW. Completion of meiosis in uninseminated eggs of *Drosophila melanogaster*. Science 1960; 132:677-678.

46. Mahowald AP, Goralski TJ, Caulton JH. In vitro activation of *Drosophila* eggs. Dev Biol 1983; 98:437-445.

47. Page AW; Orr-Weaver TL. Activation of the meiotic divisions in *Drosophila* oocytes. Dev Biol 1997; 183:195-207.

48. Foe VE, Odell GM, Edgar BA. Mitosis and morphogenesis in the *Drosophila* embryo: Point and counterpoint. In: Bate M, Martinez Arias A., eds. The Development of *Drosophila melanogaster*. Plainview: Cold Spring Harbor Laboratory Press, 1993:149-300.

49. Orr-Weaver TL. Meiosis in *Drosophila*: Seeing is believing. Proc Natl Acad Sci USA 1995; 92:10443-10449.

50. Endow SA, Komma DJ. Spindle dynamics during meiosis in *Drosophila* oocytes. J Cell Biol 1997; 137:1321-1336.

51. Ashery-Padan R, Weiss AM, Feinstein N et al. Distinct regions specify the targeting of otefin to the nucleoplasmic side of the nuclear envelope. J Biol Chem 1997; 272:2493-2499.

52. Berman CL. The role of YA protein in *Drosophila* female meiosis. 2000. MS Thesis, Cornell University: Ithaca, NY.

53. Nilson LA, Schupbach T. EGF receptor signaling in *Drosophila* oogenesis. Curr Top Dev Biol 1999; 44:203-243.

54. Guillemin K, Williams T, Krasnow MA. A nuclear lamin is required for cytoplasmic organization and egg polarity in *Drosophila*. Nat Cell Biol 2001; 3:848-851.

55. Fuller MT. Spermatogenesis. In: Bate M, Martinez Arias A., eds. The Development of *Drosophila melanogaster*. Plainview: Cold Spring Harbor Laboratory Press, 1993:71-147.

56. Whalen AM, McConnell M, Fisher PA. Developmental regulation of *Drosophila* DNA topoisomerase II. J Cell Biol, 1991; 112:203-213.

57. Poccia D, Collas P. Transforming sperm nuclei into male pronuclei in vivo and in vitro. In: Pederson RA, Schatten G, eds. Current Topics in Developmental Biology. San Diego: Academic Press, 1996:25-88.
58. Collas P, Poccia D. Remodeling the sperm nucleus into a male pronucleus at fertilization. Theriogenology 1998; 49:67-81.
59. Perotti ME. Ultrastructural aspects of fertilization in *Drosophila*. In: Afzelins, ed. The Functional Anatomy of the Spermatozoan. Proceedings of the Second International Symposium. Oxford: Pergamon Press, 1975:57-68.
60. Fitch KR, Wakimoto BT. The paternal effect gene *ms(3)sneaky* is required for sperm activation and the initiation of embryogenesis in *Drosophila melanogaster*. Dev Biol 1998; 197:270-282.
61. Callaini G, Riparbelli MG. Fertilization in *Drosophila melanogaster*: centrosome inheritance and organization of the first mitotic spindle. Dev Biol, 1996;176:199-208.
62. Crevel G, Cotterill S, DNA replication in cell-free extracts from *Drosophila melanogaster*. EMBO J 1991; 10:4361-4369.
63. Lopez JM, Wolfner MF. The developmentally regulated *Drosophila* embryonic nuclear lamina protein 'Young Arrest' (fs(1)Ya) is capable of associating with chromatin. J Cell Sci 1997; 110:643-651.
64. Ulitzur N, Harel A, Goldberg M et al. Nuclear membrane vesicle targeting to chromatin in a *Drosophila* embryo cell-free system. Mol Biol Cell 1997; 8:1439-1448.
65. Kawasaki K, Philpott A, Avilion AA et al. Chromatin decondensation in *Drosophila* embryo extracts. J Biol Chem 1994; 269:10169-76.
66. Crevel G, Huikeshoven H, Cotterill S et al. Molecular and cellular characterization of CRP1, a *Drosophila* chromatin decondensation protein. J Struct Biol 1997; 118:9-22.
67. Collas P, Poccia DL. Formation of the sea urchin male pronucleus in vitro: membrane-independent chromatin decondensation and nuclear envelope-dependent nuclear swelling. Mol Reprod Dev 1995; 42:106-113.
68. Poccia D, Collas P. Nuclear envelope dynamics during male pronuclear development. Dev Growth Differ 1997; 39:541-550.
69. Sonnenblick BD. The early embryology of *Drosophila melanogaster*. In: Demerec M, ed. The Biology of *Drosophila*. New York: John Wiley and Sons, 1950:62-167.
70. O'Neill SL, Karr TL. Bidirectional incompatibility between conspecific populations of *Drosophila simulans*. Nature 1990; 348:178-180.
71. Lassy CW, Karr TL. Cytological analysis of fertilization and early embryonic development in incompatible crosses of *Drosophila simulans*. Mech Dev 1996; 57:47-58.
72. Callaini G, Dallai R, Riparbelli MG. Wolbachia-induced delay of paternal chromatin condensation does not prevent maternal chromosomes from entering anaphase in incompatible crosses of *Drosophila simulans*. J Cell Sci 1997; 110:271-80.
73. Liu J, Song K., Wolfner MF. Mutational analyses of fs(1)Ya, an essential, developmentally regulated, nuclear envelope protein in *Drosophila*. Genetics 1995; 141:1473-1481.
74. Lopez JL, Berman CL, Yu J et al. The YA mutation defines a control point between meiosis and mitosis in *Drosophila*. In preparation.
75. Yu J, Wolfner MF. *Drosophila* YA binds to DNA and histone H2B with four domains. Mol Biol Cell 2002; 13:558-569.
76. Goldberg M, Lu H, Stuurman N et al. Interactions among *Drosophila* nuclear envelope proteins lamin, otefin, and YA. Mol Cell Biol 1998; 18:4315-23.
77. Tirian L, Timinszky G, Zhang C et al. Importin-beta is required for nuclear envelope assembly, submitted.
78. Timinszky G, Tirian L, Nagy FT et al. The importin-beta P445L dominant negative mutant protein loses RanGTP binding ability and causes arrest during the exit from mitosis in *Drosophila*. J Cell Sci 2002; 115:1675-1687.
79. Edgar BA, Datar SA. Zygotic degradation of two maternal Cdc25 mRNAs terminates *Drosophila*'s early cell cycle program. Genes Dev 1996; 10:1966-1977.
80. Edgar BA, Lehman DA, O'Farrell PH. Transcriptional regulation of string (cdc25): a link between developmental programming and the cell cycle. Development 1994; 120:3131-3143.
81. Edgar BA, Sprenger F, Duronio RJ et al. Distinct molecular mechanisms regulate cell cycle timing at successive stages of Drosophila embryogenesis. Genes Devel 1994; 8:440-452.
82. Foe VE, Alberts BM. Studies of nuclear and cytoplasmic behaviour during the five mitotic cycles that precede gastrulation in *Drosophila* embryogenesis. J Cell Sci 1983; 61:31-70.
83. Fogarty P, Campbell SD, Abu-Shumays R et al. The *Drosophila grapes* gene is related to checkpoint gene chk1/rad27 and is required for late syncytial division fidelity. Curr Biol 1997; 7:418-426.
84. Yu KR, Saint RB, Sullivan W. The Grapes checkpoint coordinates nuclear envelope breakdown and chromosome condensation. Nat Cell Biol 2000; 2:609-615.

85. Paddy MR, Saumweber H, Agard DA et al. Time-resolved, in vivo studies of mitotic spindle formation and nuclear lamina breakdown in *Drosophila* early embryos. J Cell Sci 1996; 109:591-607.
86. Steward R, Govind S. Dorsal-ventral polarity in the *Drosophila* embryo. Curr Opin Genet Dev 1993; 3:556-561.
87. Lemosy EK, Hong CC, Hashimoto C. Signal transduction by a protease cascade. Trends Cell Biol 1999; 9:102-107.
88. Jongens TA, Hay B, Jan LY et al. The *germcell-less* gene product: A posteriorly localized component necessary for germ cell development in *Drosophila*. Cell 1992; 70:569-584.
89. Jongens TA, Ackerman LD, Swedlow JR et al. Germcell-less encodes a cell type-specific nuclear pore-associated protein and functions early in the germ-cell specification pathway of *Drosophila*. Genes Devel 1994; 8:2123-2136.
90. Nili E, Cojocaru GS, Kalma Y et al. Nuclear membrane protein LAP2beta mediates transcriptional repression alone and together with its binding partner GCL (germcell-less). J Cell Sci 2001; 114:3297-307.

CHAPTER 11

The Distribution of Emerin and Lamins in X-Linked Emery-Dreifuss Muscular Dystrophy

G. E. Morris, S. Manilal, I. Holt, D. Tunnah, L. Clements, F.L. Wilkinson, C.A. Sewry and Nguyen thi Man

Introduction

Basic research into nuclear structure and function received a significant boost with the discovery in 1996 that an X-linked human genetic disease is caused by absence of a novel nuclear membrane protein, emerin.[1,2] When it emerged in 1999 that an autosomal-dominant form of this disease is caused by mutations in lamin A/C,[3] a well-known component of the nuclear lamina, interest was stimulated yet further among the large body of researchers concerned with nuclear structure in general and the lamina in particular.

Demonstration of a direct interaction between emerin and lamin A[4] confirmed the implication that both forms of Emery-Dreifuss muscular dystrophy (EDMD) are likely caused by loss of function of an emerin-lamin A/C complex, though this function remains unknown at the present time.

Emerin is a member of a small, but growing, family of integral nuclear membrane proteins that interact with the nuclear lamina. They attach to the inner nuclear membrane through a short hydrophobic transmembrane sequence and project into the nucleoplasm where they bind to lamins and chromatin (Fig. 1). The most studied member of this family is LAP2β (lamina-associated protein 2β), one of three splicing isoforms from the LAP2 gene.[5] Other family members in higher organisms include LAP1(see ref. 6), MAN1 (with two transmembrane sequences)[7] and lamin B receptor (LBR, with seven transmembrane sequences).[8] Nurim may also be considered a member of this family though it consists of little other than five transmembrane domains.[9] Distinct, though related, inner nuclear membrane proteins in *drosophila melanogaster* (otefin) and *c. elegans* (UNC84) have been reviewed by Gruenbaum et al.[10]

A Brief History of EDMD

Emery-Dreifuss muscular dystrophy (EDMD) was first described in 1966 in a large family in Virginia with eight affected males.[11] Although the age of onset is similar to Duchenne/Becker MD, EDMD has little else in common with these diseases. In particular, there is no calf hypertrophy in EDMD, little elevation of serum creatine kinase and no intellectual impairment.[12] It was not until 1986, however, that location of the EDMD gene to the Xq28 region[13,14] confirmed that the affected gene was distinct from Duchenne/Becker at Xp21. The name "emerin" was coined by Toniolo and coworkers[15] who identified the gene by positional cloning in 1994. The distinctive features of EDMD are (a) preferential weakness/wasting of muscles in the upper arm, shoulders and lower leg, (b) early appearance of stiffness

Nuclear Envelope Dynamics in Embryos and Somatic Cells
edited by Philippe Collas. ©2002 Eurekah.com and Kluwer Academic/Plenum Publishers.

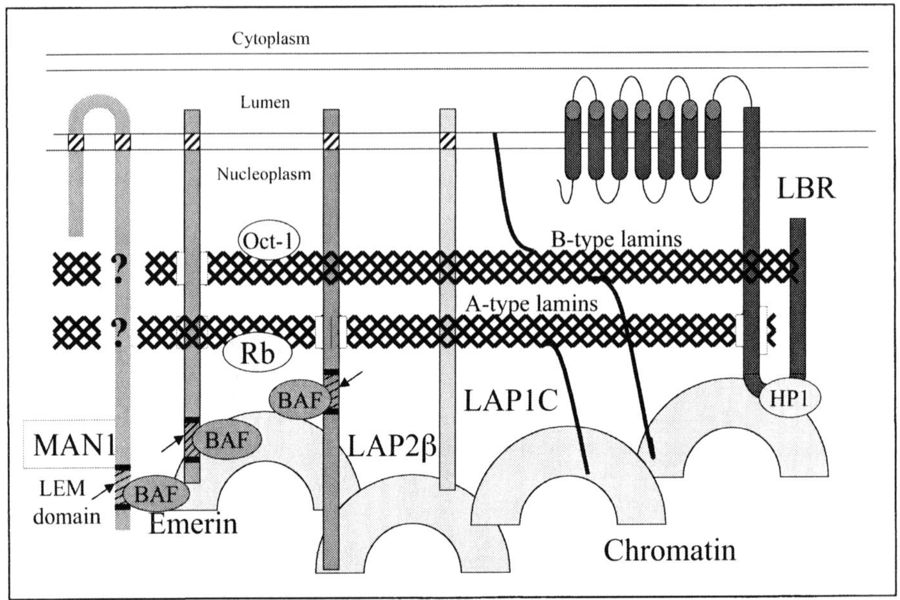

Figure 1. Integral proteins of the inner nuclear membrane and interacting partners of the nuclear lamina and chromatin.

("contractures") at the ankles, elbows and/or neck, and (c) very frequent cardiac conduction defects requiring insertion of a cardiac pacemaker.[12] Distinctive as these features now seem, when we consider the phenotypic variability of many neuromuscular disorders and the adeptness of many clinicians at fitting square pegs into round holes, it is a tribute to Alan Emery that EDMD was classified as a separate disorder so early.

The vast majority of emerin mutations in X-EDMD cause loss of the C-terminal transmembrane region and consequent absence of emerin (because truncated forms are unstable).[16] Such mutations include deletions of the entire emerin gene, partial deletions that remove the transmembrane-encoding DNA, smaller deletions that cause frameshifting and nonsense mutations causing early stop codons. Only 4 or 5 pathogenic missense or in-frame deletions are known (Fig. 2), though these include the five amino-acid YEESY deletion of the original Emery and Dreifuss (1966) X-EDMD family.

It once seemed that the autosomal dominant form of EDMD was clinically indistinguishable from the X-linked form, but it is now clear that lamin A/C missense mutations may cause more severe cardiomyopathy, often requiring heart transplant, and also more variable symptoms than absence of emerin.[17] The notion that lamin A/C mutations cause a variety of different diseases is largely a result of this variability of symptomatic expression. Thus, lamin A/C mutations appear to cause dilated cardiomyopathy with conduction defects (CMD1A; OMIM:115200) in families that do not display skeletal muscle problems,[18] a limb-girdle muscular dystrophy (LGMD1B; OMIM:159001) when muscle wasting is the main clinical feature[19] and AD-EDMD when all features are present. It was first thought that cardiomyopathy and EDMD were caused by mutations affecting different domains of the lamin A/C molecule,[18] but studies with larger patient numbers have not supported this view.[20] A rare example of autosomal-recessive EDMD, resulting from a consanguineous marriage, was found to be due to a lamin A/C mutation (H222Y) that was asymptomatic in the parents who had only one affected allele but produced a severe EDMD in the homozygote.[21] In contrast, a rather similar mutation (H222P) in a different family caused typical AD-EDMD in heterozygotes.[20]

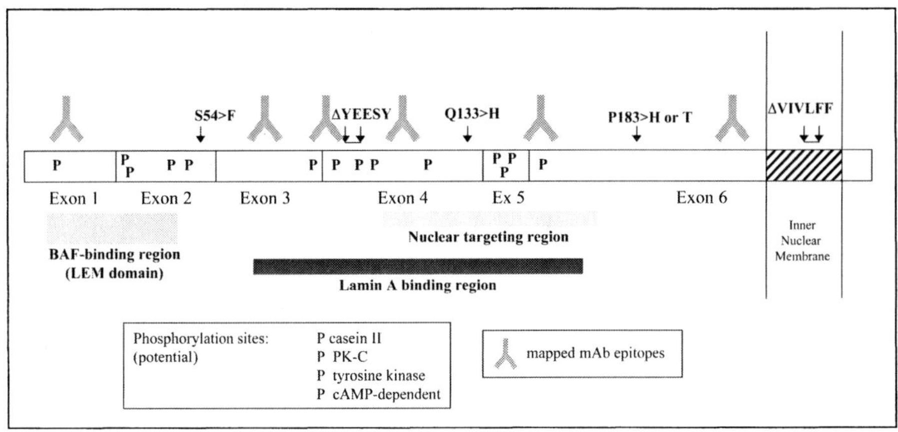

Figure 2. Primary structure of emerin and mutations in X-linked Emery-Dreifuss muscular dystrophy. Most mutations (over 95%) in X-EDMD are null mutations leading to complete deletion of early stop due to nonsense mutations or frameshift deletions. There are only five missense mutations or small in-frame deletions. Known functional domains of emerin are indicated. Each "P" represents a potential phosphorylation site.

Some variability is also observed in X-EDMD,[22,23] suggesting that lack of emerin and lamin A/C mutations cause related defects at the nuclear rim but other factors, such as genetic background (modifying genes), may determine the clinical consequences for different families or different individuals within a family. Involvement of other unidentified genes is emphasized by the fact that Wehnert and coworkers have identified an AD-EDMD family that has no mutations in either the lamin A/C gene or the emerin gene (M. Wehnert, personal communication). Dunnigan-type familial partial lipodystrophy (FPLD; OMIM: 151660) is a fourth disease caused by lamin A/C mutations and appears to be caused by a few specific amino-acid changes in the tail region of lamin A/C.[24] The principal clinical features of FPLD, insulin resistance, redistribution of fat deposits and onset at puberty, are quite different from EDMD. Recently, however, a patient with an R527P mutation in lamin A/C was found to display clinical features of both FPLD and EDMD[25] whereas this mutation causes typical EDMD in other families. One interpretation of this is that EDMD/CMD mutations and FPLD mutations affect different functions. Some mutations, like R527P, may alter lamin A/C structure so that both functions are affected (though only when genetic background permits). It has been suggested that that the FPLD mutations may specifically alter lamin A/C interaction with SREBP-1c, a transcription factor that regulates adipogenesis and locates to the nuclear rim.[24]

The Normal Distribution of Emerin and Lamins

The earliest work with anti-emerin antibodies showed that this protein is concentrated at the nuclear rim.[1,2] Emerin was proposed to be a type II integral protein of the inner nuclear membrane because of its sequence and structural homology with a known nuclear membrane protein, LAP2β.[2] When the emerin gene was first sequenced, an homology with thymopoietin was noted,[15] but the function of thymopoietin was not known at that time. It was not until the publication of the rat LAP2 sequence by the Gerace laboratory[25] that the identity of LAP2β and β-thymopoietin was first appreciated.[26] Biochemical and EM studies later confirmed the orientation of emerin with most of the molecule projecting into the nucleoplasm from the C-terminal transmembrane domain.[27,28]

The current hypothesis for targeting of inner nuclear membrane proteins is that the proteins are first inserted via their transmembrane sequences into the endoplasmic reticulum system (ER, which includes the Golgi and nuclear membranes). They then diffuse freely until they are trapped in the nuclear membrane by interactions of nucleoplasmic domains with, for example, the nuclear lamina and/or chromatin. This has been demonstrated for emerin by photobleaching studies.[29] Transfection studies with emerin fragments at first seemed inconsistent with this view, since the transmembrane domain alone seemed capable of directing emerin to the nuclear rim[30], but more recent studies suggest that the isolated transmembrane domain locates throughout the ER system, as expected.[29,31] Early biochemical fractionation studies[2] showed a small proportion of the emerin in rabbit brain in the microsomal ER fraction. This may represent emerin in the ER not yet trapped at the nuclear rim, though this is not proven. Lamins are not detected in this fraction and they, of course, are transported to the nucleus by a different mechanism (through nuclear pores by their nuclear localization signal sequences).

Immunolocalization of emerin in the heart at first suggested its additional presence at intercalated discs which might explain the conduction defects;[30] indeed, if emerin were also present at the myotendinous junction, a related structure, this might also provide the basis for contractures and muscle wasting. However, a subsequent study using fully-characterised antibodies showed that rabbit antisera can stain intercalated discs nonspecifically and that both affinity-purified rabbit antibodies and monoclonal anti-emerin antibodies stained only the nuclear membrane and not discs in the heart.[32] Interest in this hypothesis waned further with the subsequent identification of the lamin A/C gene as responsible for the autosomal dominant form of EDMD, since no-one has yet been bold enough to suggest that lamins have any function at intercalated discs. One study detected small amounts of over-expressed recombinant emerin in the plasma membrane of cultured cells[29] where it is less mobile than in the ER, but the significance of this has not yet been explored further.

What is responsible for trapping emerin at the nuclear rim as it diffuses through the ER? Colocalization of emerin and lamin A/C in cells and tissues[32,33] suggested that the nuclear lamina, below the inner nuclear membrane, might be involved and this was confirmed by the lamin A/C knockout mouse.[34] In the absence of lamin A/C, emerin remained to a large extent in the ER. In some knockout mouse tissues, such as heart and muscle, loss of emerin from the nucleus was incomplete, implying the presence of another trapping protein.[34] It seems unlikely that this protein is a B-type lamin, although there is some evidence that an interaction between emerin and lamin B may be possible in vitro.[35] The normal distribution of B-type lamins is altered in lamin A/C knockout mouse fibroblasts and LAP2, a known lamin B interactor, follows the abnormal distribution pattern, but emerin does not.[34] This is consistent with the suggestion[32] that an emerin-laminA system may be complementary to the LAP2-laminB system, perhaps even acting as a backup for some functions. The interaction of emerin with chromatin via BAF[36] remains a possibility for the additional nuclear trapping mechanism, though it is not clear why emerin should be retained in some knockout tissues and not in others.

The regions of emerin involved in nuclear targeting, lamin A/C interaction and chromatin binding have all been mapped[29,31,36] (Fig. 2). Removal of amino acids 107-175 (the Tsuchiya-Ostlund sequence) from emerin prevents targeting to the nuclear membrane; a C-terminal fragment containing this sequence (aa 107-254) targets normally.[29,31] The first 170 aa of emerin (which includes the Tsuchiya-Ostlund sequence) targeted correctly when an artificial transmembrane sequence was attached, but the Tsuchiya-Ostlund sequence alone (aa 117-170) did not.[29] This suggests that the LEM domain (aa 6-44) and its interaction with chromatin via BAF contributes to normal emerin localization. The lamin A binding region of emerin has recently been mapped to aa 70-164 minimum or 55-178 maximum, since mutations at amino-acids 70 and 164 (and points between) prevented binding to lamin A while mutations at amino-acids 54 and 179 did not[36]. This lamin A binding sequence overlaps with the Tsuchiya-Ostlund sequence for nuclear targeting, but there is a distinction in that aa 70-116 are evidently needed for lamin A binding but not for nuclear targeting. The logical extension of this data is that nuclear targeting can occur without lamin A binding and that a third protein, other than lamin A and BAF, may be implicated. Almost all potential phosphorylation sites in emerin lie within the BAF and lamin A binding regions. Phosphorylation of lamins and LAP2 by cdc2 kinase disrupts protein-protein interactions and is important for nuclear disassembly during mitosis.[37,38] Changes in emerin phosphorylation during the cell cycle have also been reported,[39] although emerin has no cdc2 kinase sites (Fig. 2). Clearly, the state of phosphorylation of emerin may have a bearing upon both its interaction with other proteins and its localization within the cell.

Although emerin is clearly concentrated at the nuclear rim in sections of skin or cardiac muscle, this is much less clear in cultured fibroblasts or COS cells where the internal nucleoplasmic staining can be almost as strong. The internal emerin has a finely-granular appearance with frequent brighter spots (Fig. 3A) and structures that resemble channels or invaginations of the nuclear membrane (Fig. 3B). Larger aggregates of internal emerin are sometimes observed (Fig 3C). Complete absence of antibody staining in emerin-negative EDMD fibroblast nuclei (Fig. 3D) shows that the internal nuclear stain is authentic emerin. It seems possible that some of the apparent invaginations in Fig. 3 are shrinkage artefacts during fixation, though channel-like structures were also observed in large cardiomyocyte nuclei in unfixed heart sections.[32] Internal nuclear staining observed by conventional microscopy of monolayer cultures might be explained by emerin in the upper surface of the nucleus, but confocal microscope sections clearly show the granular emerin staining throughout the nucleoplasm (Fig. 3). Lamins are also present in the nucleoplasm of interphase cultured cells and they have a very similar distribution to emerin.[33] This internal lamin staining presumably represents the nuclear matrix, of which lamins form a major part. Hozak et al[40] showed that some lamin antisera fail to detect internal nuclear lamins because of epitope masking and their presence was revealed only after removal of chromatin. A panel of mAbs against emerin that recognise six different epitope regions[32] revealed no evidence of masking of particular emerin regions; two mAbs showed brighter nuclear rim staining but this was not related to the position of their epitopes (unpublished data). We conclude that if masking of emerin does occur in fibroblast nuclei, whether at the rim or in the interior, it must be masking of the whole emerin molecule and not just some epitopes.

When the nuclear membrane breaks down during mitosis, emerin becomes dispersed throughout the mitotic cell cytoplasm, presumably by redistribution into the ER, which persists during mitosis.[35,41,42] It behaves, in this respect, like LAP2β. As the nucleus is re-assembled during telophase, emerin becomes associated briefly with chromosomes, an interaction mediated by the DNA-binding protein, BAF.[42] Here it associates with lamin A/C but not with LAP2 or B-type lamins.[41] Although emerin binds to the spindle attachment region of the chromosome, this binding is not affected by drugs that disrupt microtubules.[42] Emerin later begins to spread to cover the entire rim of the daughter nuclei.[42] However, when COS cell daughter nuclei separate after mitosis, a continued association is observed of both emerin and lamins with what appears to be remnant spindle mid-body connecting the daughter nuclei.[35]

Transfected emerin, detected with a "tag" antibody, shows a very similar distribution to endogenous emerin in about 50% of transfected cells. The other 50% of transfected cells, however, have emerin in larger aggregates in the nucleoplasm and half of these also have emerin aggregates in the cytoplasm.[43] This contrasts with transfected lamin A, which is similar to endogenous lamin A and emerin in 90% of transfected cells, the remaining 10% showing additional cytoplasmic staining.[44] This difference in behaviour between transfected emerin and lamin A may be a simple consequence of their different routes to the nucleus, ER in one case and nuclear pores in the other. The higher levels of nucleoplasmic emerin and lamins in cultured cells, compared for example with post-mitotic cardiomyocyte nuclei, may be connected with the suggested association of nucleoplasmic lamins with regions of DNA replication.[45]

Although emerin levels by western blotting are similar in all tissues,[2] many nuclei in tissue sections do not stain with emerin antibodies. Nagano and coworkers[1] saw no emerin staining in kidney tubule nuclei or nuclei from neurons, spleen or liver. Manilal and coworkers[32] found that nuclei in the heart that were strongly emerin-positive, such as cardiomyocytes, were also strongly stained by lamin A/C antibodies. Many nuclei with B-type lamin staining at the nuclear rim did not contain either emerin or lamin A/C. The reports described above of loss of emerin

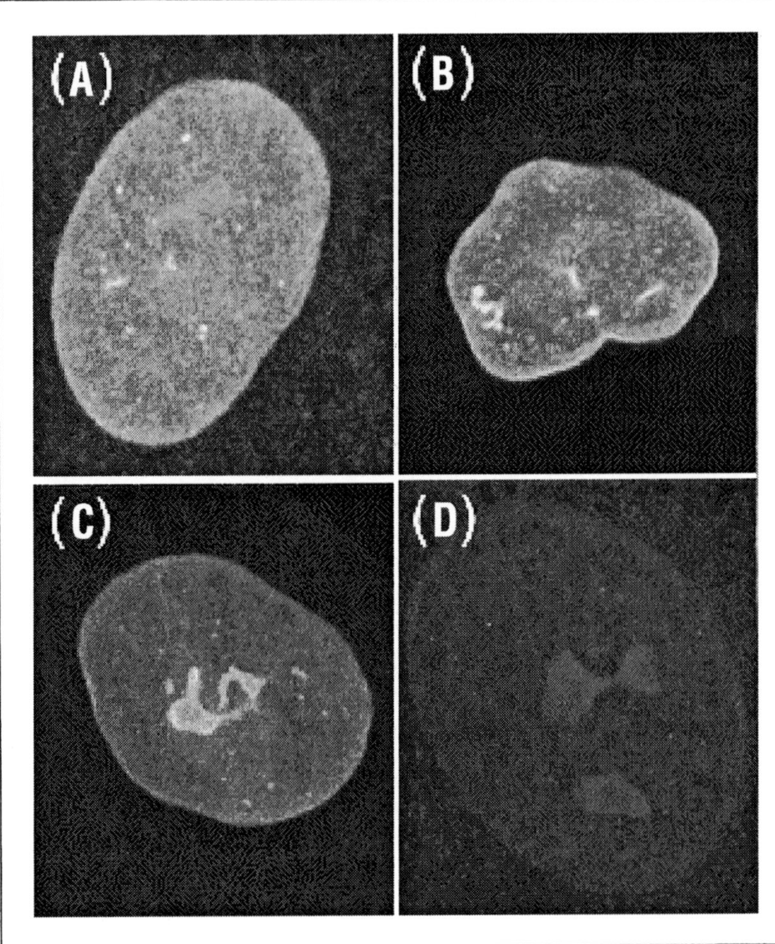

Figure 3. Emerin distribution in cultured human skin fibroblasts. Emerin (green) was detected using monoclonal antibody MANEM5 and FITC-labelled secondary antibody with a BioRad MicroRadiance confocal microscope. Nuclei were counterstained with ethidium bromide (red) and the more intensely-stained bodies are nucleoli. (A)-(C) show different patterns observed in normal nuclei and (D) shows absence of emerin in an X-EDMD nucleus. Cells were fixed with 50:50 acetone:methanol. A color version of this figure is included in the insert that precedes Chapter 1.

from the nuclear rim in cell lines lacking lamin A/C raise the interesting possibility that variable expression of lamin A/C in different cell types may also be responsible for differential emerin staining in adult tissues.

Distribution of Emerin and Lamins in X-Linked EDMD

In most cases of EDMD, emerin is completely absent, so the question of its distribution does not arise. The exceptions are the two or three missense mutations in emerin (Fig. 2), two small in-frame deletions and rare frameshifts in which some mutant emerin with an altered C-terminus is synthesized. One of the three missense mutations, g933t, causes either incorrect RNA splicing with a frameshift or, when normal splicing does occur, a Q133H amino-acid change. Emerin levels are greatly reduced, although some mutant emerin is synthesized.[46]

Although the mutation lies within the lamin A binding and nuclear targeting regions (Fig. 2), no direct effects on these processes were observed.[43] The reduction in emerin required to produce EDMD symptoms is not known. Carrier females with less than 5% of normal emerin levels have cardiac problems.[16] In such carriers, however, 5% of cells would have normal emerin levels and 95% would have no emerin at all, so comparison with EDMD patients is not valid. The other two mutations, S54F and P183H or T, have no effect on emerin levels, at least not in lymphoblastoid cells.[47,48] There is evidence from transfection studies, however, for reduced association with the nuclear rim and a greater tendency to aggregate in the ER/cytoplasm.[47,48] Since the disease is as severe in S54F (or nearly as severe in P183) as in emerin null mutants,[47,48] S54 and P183 presumably have an important role in emerin function, although they both lie outside any known functional region (Fig. 2) and neither of them affects interaction with lamin A in vitro.[49] One of the two in-frame deletions removes six amino-acids from the transmembrane sequence. This appears to prevent membrane integration and thus renders the mutant protein unstable, since mutant emerin is barely detectable on Western blots and is absent from the nuclear membrane.[16,47] The ΔYEESY deletion affects interaction with lamin A[49] and causes a greater reduction in nuclear rim association than the missense mutations.[47] Total levels of the ΔYEESY protein are also lower in patient lymphoblastoid cells.[49] Cartegni et al[30] described an interesting two-base frameshift deletion which by chance added a novel hydrophobic sequence to the first 169 amino-acids of emerin. This chimeric emerin, having both a nuclear targeting sequence and a hydrophobic C-terminal domain, located correctly to the nuclear rim (though in reduced amounts). The same effect was observed when this unusual mutant was transfected into COS cells.[29] A similar frameshift that added 101 amino-acids after the emerin transmembrane sequence resulted in no emerin production, presumably because normal insertion into the ER was prevented.[50]

Effects of absence of emerin in X-EDMD on the distribution of lamins and chromatin have been reported, similar to, though not as consistent as, the effects of absence of lamin A/C in the knockout mouse.[34] Ognibene and coworkers[51] studied skin fibroblasts from one X-EDMD patient and found patchy distribution at the nuclear rim of both A-type and B-type lamins in 18% of all nuclei. The areas of reduced lamin staining also showed reduced chromatin. In the electron microscope, areas of reduced peripheral heterochromatin were observed in 10% of skeletal muscle nuclei. Rather similar structural changes in muscle nuclei have been observed by electron microscopy in some, though not all, cases of autosomal dominant EDMD.[52] A resemblance to nuclear changes in apoptosis has been noted but it is unlikely that late stages of apoptosis would be occurring in such a high proportion of nuclei in muscle tissue.[53,54] It must be emphasized that EDMD patients do not have any skin disorders and the proportion of emerin-negative cells in EDMD carrier epidermis is usually around 50%, much higher than would be expected if such cells were less viable. Similarly, there is little obvious difference in lamin A/C distribution between emerin-positive and emerin-negative regions of carrier skin biopsies,[53,55] although this has not been studied by electron microscopy. The changes observed in cultured skin fibroblasts would seem to argue against these changes being secondary effects of the disease, so the precise significance of these interesting observations is not yet clear.

Acknowledgements

Work on EDMD in this laboratory is supported by grants from the British Heart Foundation (PG2000102) and the EU Fifth Framework (QLRT-1999-00870). We thank Kathy Wilson (Johns Hopkins University) for providing preprints of publications.

References

1. Nagano A, Koga R, Ogawa M et al. Emerin deficiency at the nuclear membrane in patients with Emery-Dreifuss muscular dystrophy. Nature Genet 1996; 12:254-259.
2. Manilal S, Nguyen thi Man, Sewry CA et al. The Emery-Dreifuss muscular dystrophy protein, emerin, is a nuclear membrane protein. Hum Mol Genet 1996; 5:801-808.
3. Bonne G, DiBarletta MR, Varnous S et al. Mutations in the gene encoding lamin A/C cause autosomal dominant Emery-Dreifuss muscular dystrophy. Nature Genet 1999; 21:285-288.

4. Clements L, Manilal S, Love DR et al. Direct interaction between emerin and lamin A. Biochem Biophys Res Commun 2000; 267:709-714.

5. Dechat T, Vlcek S, Foisner R. Review: Lamina-associated polypeptide 2 isoforms and related proteins in cell cycle-dependent nuclear structure dynamics. J Struct Biol 2000; 129:335-345.

6. Martin L, Crimaudo C, Gerace L. cDNA cloning and characterization of lamina-associated polypeptide 1C (LAP1C), an integral protein of the inner nuclear membrane. J Biol Chem 1995; 270:8822-8828.

7. Lin F, Blake DL, Callebaut I et al. MAN1, an inner nuclear membrane protein that shares the LEM domain with lamina-associated polypeptide 2 and emerin. J Biol Chem 2000; 275:4840-4847.

8. Ye Q, Worman HJ. Primary structure analysis and Lamin B and DNA binding of human LBR, an integral protein of the nuclear envelope inner membrane. J Biol Chem 1994; 269:11306-11311.

9. Rolls MM, Stein PA, Taylor SS et al. A visual screen of a GFP-fusion library identifies a new type of nuclear envelope membrane protein. J Cell Biol 1999; 146:29-44.

10. Gruenbaum Y, Wilson KL, Harel A et al. Review: nuclear lamins—Structural proteins with fundamental functions. J Struct Biol 2000; 129:313-323.

11. Emery AEH, Dreifuss FE. Unusual type of benign X-linked muscular dystrophy. J Neurol Neurosurg Psychiatry 1966; 29:338-342.

12. Emery AEH. Emery-Dreifuss muscular dystrophy—A 40 year retrospective. Neuromusc Disord 2000; 10:228-232.

13. Thomas NS, Williams H, Elsas LJ et al. Localisation of the gene for Emery-Dreifuss muscular dystrophy to the distal long arm of the X chromosome. J Med Genet 1986; 23:596-598.

14. Yates JR, Affara NA, Jamieson DM et al. Emery-Dreifuss muscular dystrophy: localisation to Xq27.3-qter confirmed by linkage to the factor VIII gene. J Med Genet 1986; 23:587-590.

15. Bione S, Maestrini E, Rivella S et al. Identification of a novel X-linked gene responsible for Emery-Dreifuss muscular dystrophy. Nature Genet 1994; 8:323-327.

16. Manilal S, Recan D, Sewry CA et al. Mutations in Emery-Dreifuss muscular dystrophy and their effects on emerin protein expression. Hum Mol Genet 1998; 7:855-864.

17. Wehnert M, Muntoni F. 60th ENMC International Workshop: Non X-linked Emery-Dreifuss Muscular Dystrophy. Neuromuscul Disord 1999; 9:115-121.

18. Fatkin D, MacRae C, Sasaki T et al. Missense mutations in the rod domain of the lamin A/C gene as causes of dilated cardiomyopathy and conduction-system disease. N Engl J Med 1999; 341:1715-1724.

19. Muchir A, Bonne G, van der Kooi AJ et al. Identification of mutations in the gene encoding lamins A/C in autosomal dominant limb girdle muscular dystrophy with atrioventricular conduction disturbances (LGMD1B). Hum Mol Genet 2000; 9:1453-1459.

20. Bonne G, Mercuri E, Muchir A et al. Clinical and molecular genetic spectrum of autosomal dominant Emery-Dreifuss muscular dystrophy due to mutations of the lamin A/C gene. Ann Neurol 2000; 48:170-180.

21. Raffaele Di Barletta M, Ricci E, Galluzzi G et al. Different mutations in the LMNA gene cause autosomal dominant and autosomal recessive Emery-Dreifuss muscular dystrophy. Am J Hum Genet 2000; 66:1407-1412.

22. Muntoni F, Lichtarowicz-Krynska EJ, Sewry CA et al. Early presentation of X-linked Emery-Dreifuss muscular dystrophy resembling limb-girdle muscular dystrophy. Neuromusc Disord 1998; 8:72-76.

23. Canki-Klain N, Recan D, Milicic D et al. Clinical variability and molecular diagnosis in a four-generation family with X-linked Emery-Dreifuss muscular dystrophy. Croat Med J 2000; 41:389-395.

24. Hegele RA. Molecular basis of partial lipodystrophy and prospects for therapy. Trends in Mol Med 2001; 7:121-126.

25. Furukawa K, Pant N, Aebi U et al. Cloning of a cDNA for lamina-associated polypeptide 2 (LAP2) and identification of regions that specify targeting to the nuclear envelope. EMBO J 1995; 14:1626-1636.

26. Harris CA, Andryuk PJ, Cline SW et al. Structure and mapping of the human thymopoietin (TMPO) gene and relationship of the human TMPO-beta to rat lamin-associated polypeptide-2. Genomics 1995; 28:198-205.

27. Yorifuji H, Tadano Y, Tsuchiya Y et al. Emerin, deficiency of which causes Emery-Dreifuss muscular dystrophy, is localized at the inner nuclear membrane. Neurogenet 1997; 1:135-140.

28. Squarzoni S, Sabatelli P, Ognibene A et al. Immunocytochemical detection of emerin within the nuclear matrix. Neuromusc Disord 1998; 5:338-344.

29. Östlund C, Ellenberg J, Hallberg E et al. Intracellular trafficking of emerin, the Emery-Dreifuss muscular dystrophy protein. J Cell Sci 1999; 112:1709-1719.

30. Cartegni L, di Barletta MR, Barresi R et al. Heart-specific localisation of emerin: new insights into Emery-Dreifuss muscular dystrophy. Hum Mol Genet 1997; 6:2257-2264.

31. Tsuchiya Y, Hase A, Ogawa M et al. Distinct regions specify the nuclear membrane targeting of emerin, the responsible protein for Emery-Dreifuss muscular dystrophy. Eur J Biochem 1999; 259:859-865.

32. Manilal S, Sewry CA, Pereboev A et al. Distribution of emerin and lamins in the heart and implications for Emery-Dreifuss muscular dystrophy. Hum Mol Genet 1999; 8:353-359.

33. Manilal S, Nguyen thi Man, Morris GE. Colocalization of emerin and lamins in interphase nuclei and changes during mitosis. Biochem Biophys Res Commun 1998; 249:643-647.

34. Sullivan T, Escalante-Alcalde D, Bhatt H et al. Loss of A-type lamin expression compromises nuclear envelope integrity leading to muscular dystrophy. J Cell Biol 1999; 147:913-920.

35. Vaughan A, Alvarez-Reyes M, Bridger JM et al. Both emerin and lamin C depend on lamin A for localization at the nuclear envelope. J Cell Sci 2001; 114:2577-2590.

36. Lee KK, Haraguchi T, Lee RS et al Distinct functional domains in emerin bind lamin A and DNA-bridging protein, BAF. J Cell Sci 2001; 114:4567-4573.

37. Heald R, McKeon F. Mutations of phosphorylation sites in lamin A that prevent nuclear lamina disassembly in mitosis. Cell 1990; 61:579-589.

38. Dechat T, Vlcek S, Foisner R. Review: Lamina-associated polypeptide 2 isoforms and related proteins in cell cycle-dependent nuclear structure dynamics. J Struct Biol 2000; 129:335-345.

39. Ellis JA, Craxton M, Yates JR et al. Aberrant intracellular targeting and cell cycle-dependent phosphorylation of emerin contribute to the Emery-Dreifuss muscular dystrophy phenotype. J Cell Sci 1998; 111:781-792.

40. Hozak P, Sasseville AM, Raymond Y et al. Lamin proteins form an internal nucleoskeleton as well as a peripheral lamina in human cells. J Cell Sci 1995; 108:635-644.

41. Dabauvalle MC, Muller E, Ewald A et al. Distribution of emerin during the cell cycle. Eur J Cell Biol 1999; 78:749-756.

42. T Haraguchi, T Koujin, M Segura-Totten et al. Emerin-BAF interactions are required for emerin assembly into the reforming nuclear envelope. J Cell Sci 2001; 114:4575-4585.

43. Holt I, Clements L, Manilal S et al. How does a g993t mutation in the rmerin gene cause Emery-Dreifuss muscular dystrophy? Biochem Biophys Res Commun 2001; 287:1129-1133.

44. Holt I, Clements L, Manilal S et al. The R482Q lamin A/C mutation that causes lipodystrophy does not prevent nuclear targeting of lamin A in adipocytes or its interaction with emerin. Eur J Hum Genet 2001; 9:204-208.

45. Moir RD, Montag-Lowy M, Goldman RD. Dynamic properties of nuclear lamins: Lamin B is associated with sites of DNA replication. J Cell Biol 1994; 125:1201-1212.

46. Mora M, Carregni L, Di Blasi C et al. X-linked Emery-Dreifuss muscular dystrophy can be diagnosed from skin biopsy or blood sample. Ann Neurol 1997; 42:249-253.

47. Fairley EAL, Kendrick-Jones J, Ellis JA. The Emery-Dreifuss muscular dystrophy phenotype arises from aberrant targeting and binding of emerin at the inner nuclear membrane. J Cell Sci 1999; 112:2571-2582.

48. Ellis JA, Yates JRW, Kendrick-Jones et al. Changes at P183 of emerin weaken its protein-protein interactions resulting in X-linked Emery-Dreifuss muscular dystrophy. Hum Genet 1999; 104:262-268.

49. Ellis JA, Craxton M, Yates J.R et al. Aberrant intracellular targeting and cell cycle-dependent phosphorylation of emerin contribute to the Emery-Dreifuss muscular dystrophy phenotype. J Cell Sci 1998; 111:781-792.

50. Ellis JA, Brown CA, Tilley LD et al. Two distal mutations in the gene encoding emerin have profoundly different effects on emerin production. Neuromusc Disord 2000; 10:24-30.

51. Ognibene A, Sabatelli P, Petrini S et al. Nuclear alterations in a skeletal muscle biopsy and in skin cultured from one patient affected by X-linked Emery-Dreifuss muscular dystrophy. Muscle & Nerve 1999; 22:864-869.

52. Sewry CA, Brown SC, Mercuri E et al. Skeletal muscle pathology in autosomal dominant Emery-Dreifuss muscular dystrophy with lamin A/C mutations. Neuropathol Appl Neurobiol 2001; 27:281-290.

53. Morris GE. Nuclear proteins and cell death in inherited neuromuscular disease. Neuromusc Disord 2000; 10:217-227.

54. Morris GE. The role of the nuclear envelope in Emery-Dreifuss muscular dystrophy. Trends Mol Med 2001; 7:572-577.

55. Manilal S, Sewry CA, Nguyen thi Man et al. Diagnosis of X-linked Emery-Dreifuss muscular dystrophy by protein analysis of leucocytes and skin with monoclonal antibodies. Neuromusc Disord 1997; 7:63-66.

CHAPTER 12

Laminopathies:
One Gene, Two Proteins, Five Diseases...

Corinne Vigouroux and Gisèle Bonne

Abstract

Lamins are the major components of the nuclear lamina, a network located between inner nuclear membrane and chromatin, which plays a fundamental role in the organization of the nuclear architecture in all human cells. Lamins A and C, which are alternatively spliced products of the A-type lamin gene (*LMNA*), are expressed in differentiated cells, whereas B-type lamins, arising from two different genes, are ubiquitous.

Recent familial genetic studies have shown, contrary to all expectations, that naturally occurring mutations in *LMNA* are responsible for two groups of diseases, apparently unrelated, affecting highly specialized tissues: dystrophies of skeletal and/or cardiac muscles, and partial lipodystrophies.

This review will first focus on the clinical aspects of these diseases and the phenotype-genotype correlations. We will summarize recent biological and experimental data on tissue and cellular alterations related to diverse molecular abnormalities in lamins A/C. This field of investigations provides informations of great interest for the understanding of the physiological role of lamins, and allows the exploration of new hypotheses on pathophysiological mechanisms leading to *LMNA*-linked diseases.

Introduction

Lamins are widely expressed proteins, but their precise physiological role remains still unknown. As other members of the intermediate filament protein family, their structure comprises a central α-helical coiled-coil rod domain flanked by globular N-terminal (head) and C-terminal (tail) domains.[1] Lamins A and C are the major alternative splice products of the *LMNA* gene. Their expression is developmentally regulated, increasing during differentiation. Two genes give rise to B-type lamins, which are constitutively expressed. Both types of lamins, dephosphorylated at the end of the mitosis, polymerize through coiled-coil interactions between their α-helical rod-domains, to form the nuclear lamina at the nucleoplasmic side of the inner nuclear membrane in interphase cells. The lamina interacts with integral proteins of the inner nuclear membrane like emerin, which binds A-type lamins. In addition, the lamina has narrow intrications with chromatin and nuclear pores complexes which mediate molecular trafficking between cytoplasm and the nucleus.[2]

The first evidence that *LMNA* was linked to a genetic disease was reported in 1999. By a positional approach, studies of familial genetic linkage in a large French pedigree with autosomal dominant Emery-Dreifuss muscular dystrophy (AD-EDMD) showed a strongly positive lod-score at an 8-cM locus on chromosome 1q21-q23. *LMNA*, which is located in this interval, was subsequently shown to be mutated in all affected subjects.[3] After this first publication, *LMNA* heterozygous alterations were shown to be responsible of two other diseases, which

Nuclear Envelope Dynamics in Embryos and Somatic Cells
edited by Philippe Collas. ©2002 Eurekah.com and Kluwer Academic/Plenum Publishers.

share with AD-EDMD several clinical symptoms: dilated cardiomyopathy with conduction defects (DCM-CD),[4,5] and limb-girdle muscular dystrophy with cardiac conduction disturbances (LGMD1B).[6]

A genetic linkage between the familial partial lipodystrophy of the Dunnigan type (FPLD) and a chromosome 1q21-22 locus has been known since 1998.[7-9] The discovery of *LMNA* mutations in this disease was surprising, since this disorder was clinically not at all suspected to be linked to muscular dystrophies.[10,11]

After a description of the clinical and genetic features of these two groups of diseases, we will report the recent findings on the nuclear alterations found in cells from patients and in experimental models (e.g., transgenic mice or genetically modified cell lineages). We will discuss rising pathophysiological hypotheses on these *LMNA*-linked diseases.

Disorders of Cardiac and/or Skeletal Muscles Linked to *LMNA* Alterations

Emery-Dreifuss Muscular Dystrophy (EDMD)

Benign forms of Duchenne muscular dystrophies were reported for the first time by Becker and Kiener in 1955.[12] In 1961, Dreifuss and Hogan described a Virginian family with an X-linked muscular dystrophy that was considered at that time as a benign form of Duchenne muscular dystrophy (DMD).[13] However, after a detailed clinical characterization of the family, Emery and Dreifuss suggested that this family presented a muscular dystrophy different from Duchenne and Becker muscular dystrophy.[14] This new clinical entity was later referred to as Emery-Dreifuss muscular dystrophy (EDMD).[15] Nevertheless, EDMD was most probably described for the very first time in 1902 by Cestan and Lejonne.[16] These authors reported a case of two brothers with a familial myopathy with severe and generalized contractures.

EDMD is a clinically and genetically heterogeneous condition. It is typically characterized by a triad of: 1) early contractures of the Achilles tendons, elbows and post- cervical muscles, often before there is any significant weakness, 2) slowly progressive muscle wasting and weakness with a distinctive humero-peroneal distribution early in the course of the disease and 3) by adult life, cardiomyopathy usually presenting as cardiac conduction defects (ranging from sinus bradycardia, prolongation of the PR interval on ECG to complete heart block requiring pacing). Thus affected individuals may die suddenly from heart block, or develop progressive heart failure. The latter may occur subsequent to the insertion of a pacemaker to correct an arrhythmia (Fig. 1, Table 1).[5,17,18] Skeletal muscle biopsies from patients with EDMD show dystrophic changes with a few necrotic and regenerating fibers, but this is not specific to this muscular disease. Usually fiber necrosis is less prominent than in DMD or Becker muscular dystrophy (BMD).[19] Skeletal muscles also show marked variations in fiber diameter and an increased number of hypertrophic fibers and internal nuclei.[20,21]

Two major modes of inheritance of EDMD exist, X-linked (XL-EDMD) and autosomal dominant (AD-EDMD). Rare cases of autosomal recessive transmission (AR-EDMD) have been also reported.[22-24] Defects in the emerin protein are responsible of XL-EDMD,[25] whereas mutations in lamin A/C gene cause AD and AR-EDMD.[3,24] The three forms are overall clinically identical,[18,26,27] even if some slight differences emerge between XL and AD forms of EDMD.[28] AD-EDMD exhibits wider clinical variability than XL-EDMD. AD-EDMD patients have more severe and progressive wasting of the biceps brachii compared to what is typically found in XL-EDMD.[29,30] Hypertrophy of the quadriceps and of the extensor digitorum brevis occurs in several AD-EDMD patients but not in XL-EDMD. Contractures, which are the first symptoms in XL-EDMD, might appear after weakness and difficulty in running in AD-EDMD patients. Loss of ambulation due to a combination of increasing joint stiffness and weakness is observed in AD-EDMD but is extremely rare in XL-EDMD.[18,28,31]

Limb-Girdle Muscular Dystrophy Associated with Atrioventricular Conduction Disturbances (LGMD1B)

Among the large family of muscular dystrophies, "limb-girdle muscular dystrophies" (LGMD) represent a genetically heterogeneous group of myogenic disorders with a limb girdle distribution of weakness.[32] The inheritance pattern in LGMD is heterogeneous. Four dominant (LGMD1) and eight recessive forms (LGMD2) have been identified to date. Van der Kooi et al[33] have described the LGMD1B form, inherited as an autosomal dominant trait. It is characterized by symmetrical weakness starting in the proximal lower limb muscles, and gradually proximal upper limb muscles also become affected. At variance with EDMD, early contractures of the spine are absent, and contractures of elbows or Achilles tendons are either minimal or late. Cardiac deficiency is not a constant feature in LGMD. However in LGMD1B, cardiological abnormalities are found in the majority of patients. These include dysrhythmias and atrioventricular conduction disturbances, such as bradycardia and syncopal attacks that require pacemaker implantation, and sudden cardiac death. In the three families described by van der Kooi et al,[33] there was a significant relation between the severity of atrioventricular conduction disturbances and age, and neuromuscular symptoms preceded cardiological involvement. These clinical features are very close to those of EDMD, however, the late appearance or absence of contractures led the authors to conclude that this disorder differed from EDMD. Muscle biopsies from patients with LGMD1B show non-specific myopathic changes similar to those observed in XL and AD-EDMD (Table 1).[33]

Because the locus of LGMD1B had been mapped to chromosome 1q11-21,[34] where *LMNA* is located, the *LMNA* gene became a good candidate for this muscular disease. Mutation analysis of *LMNA* in the three LGMD1B families described by van der Kooi et al identified three different *LMNA* mutations. This demonstrated that LGMD1B and AD-EDMD are allelic disorders.[6]

Dilated Cardiomyopathy and Conduction Defects (DCM-CD)

Cardiomyopathies are defined as diseases of the myocardium associated with cardiac dysfunction.[35] The most common forms are the dilated forms, responsible for approximately 60% of cases of cardiomyopathy with an annual incidence estimated to be 5-8 cases per 100,000 people.[36] Dilated cardiomyopathy (DCM) is characterized by dilatation and impaired contraction of the left ventricle or both ventricles. Histology is non-specific. Presentation is usually with heart failure, which is often progressive. Arrhythmias, thromboembolism and sudden death are common and may occur at any stage. Many causes of DCM have been described, but most commonly this disease is considered idiopathic. In the past years it has become increasingly clear that in at least 25% of DCM cases have a genetic basis.[37,38] Familial DCM is a heterogeneous disorder with different inheritance patterns. Autosomal dominant inheritance is the most common, but autosomal recessive, X-linked and mitochondrial inheritance have also been identified. In the last couple of years important progress has been made in unraveling familial DCM. Multiple genetic loci and several genes have been identified in this disease.

That *LMNA* could also be involved in DCM was completely unexpected. In the majority of affected members of one of the French families with AD-EDMD, the disease was confined exclusively to the heart and associated with arrhythmias, left ventricular dysfunction, dilated cardiomyopathy and a high incidence of sudden death.[3,5] These patients with exclusive cardiac involvement could easily have been diagnosed as DCM with conduction defects (DCM-CD, Table 1), suggesting that *LMNA* was one of the disease genes of familial DCM-CD. In agreement with this finding, mutations in *LMNA* were found in unrelated families with DCM-CD.[4] In addition, a *LMNA* mutation was identified in a family in which three phenotypes were described, DCM with EDMD-like skeletal muscle abnormalities, DCM with LGMD-like skeletal muscle abnormalities and pure DCM-CD.[39] So far, no sign of lipodystrophy has been reported in patients with striated muscle disorders due to *LMNA* mutations.

There is no specific treatment for these cardiac and/or skeletal laminopathies. However, great care should be given to proper diagnosis and follow-up of patients with EDMD, LGMD1B

Table 1. Clinical features of laminopathies affecting cardiac and/or skeletal muscles

	Emery-Dreifuss Muscular Dystrophy (AD-EDMD)	Limb-Girdle Muscular Dystrophy (LGMD1B)	Dilated Cardiomyopathy with Conduction Defects (DCM-CD)
Inheritable pattern	Autosomal dominant Autosomal recessive	Autosomal dominant	Autosomal dominant
Clinical features			
Neurological symptoms	- Slow progressive muscle wasting and weakness predominantly affecting humero-peroneal muscles - Early contractures of elbows, Achilles tendons and spine	- Slow progressive muscle wasting and weakness starting in the proximal lower limb muscles and gradually affecting proximal upper limb - Absence or minimal and late contractures of elbows, Achilles tendons and spine	No sign
Onset	Childhood (birth for severe cases)	Childhood to third decade	
Cardiological symptoms	Conductions defects (AV-block), and dysrhythmias requiring pacing - Dilated cardiomyopathy - Sudden death	Conductions defects (AV-block), and dysrhythmias requiring pacing - Dilated cardiomyopathy - Sudden death	Conductions defects (AV-block), and dysrhythmias requiring pacing - Dilated cardiomyopathy - Sudden death
Onset	Second to third decade	Second to fourth decade Significant relation between the severity of AV conduction disturbances and age.	Second to fourth decade
Biological features	CK level normal to mildly elevated	CK level normal to mildly elevated	CK level normal
Skeletal muscle histology	Mild dystrophic pattern, but not specific	Mild dystrophic pattern, but not specific	Normal

or DCM-CD. All patients should have a detailed cardiac examination and regular follow-up by a cardiologist since sudden death can occur, and early detection of arrhythmias can be lifesaving by defibrillator implantation. Also relatives of such patients should be cardiologically screened even if they show no subjective neuromuscular or cardiac symptoms. Apart from the patients with a typical presentation of weakness, contractures, and arrhythmia, especially in familial cases with a history of sudden death, *LMNA* mutations ought to be searched for.

Spectrum of LMNA *Mutations in Cardiac and/or Skeletal Muscle Disorders*

Lamins A and C (664 and 572 amino acids, respectively) are encoded by the *LMNA* gene through alternative splicing within exon 10. *LMNA* contains twelve exons, which are spread over approximately 24 kb of genomic DNA (Fig. 2).[40] Since the first description of *LMNA* mutations in AD-EDMD, a total of fifty-one different mutations were reported, forty- four in striated muscles disorders and seven in partial lipodystrophies. These mutations are: 2 nonsense mutations, 2 deletions with frameshift, 1 splice site mutation, 4 in-frame deletions and 42 missense mutations (Fig. 2). Two "hot spot" mutations are observed: the R453W missense mutation[24,28,41,42] in AD-EDMD and the missense mutation at R482. The R482 mutation exists in three different forms (R482W, R482Q and R482L) in FPLD.[10,11,43,44]

In cardiac and skeletal muscle disorders, the mutations are distributed along the gene between exons 1 and 10 in the region common to lamins A and C except for three missense mutations. One is located in exon 10 specific of lamin C and the other two in exon 11 specific of lamin A.[5,41,45] Thirty-five mutations were identified in 19 AD-EDMD families, 36 sporadic EDMD cases and 4 LGMD1B families.[3,6,24,28,41,42,46,47] The large proportion of EDMD sporadic cases reported underlines the high frequency of de novo mutations in the *LMNA* gene.[28] However, if mutations in *LMNA* gene are identified in 100% of EDMD familial cases, they are found in only 35% of sporadic cases with EDMD-like phenotype (Bonne G., personal communication). The clinical picture is often compatible with other muscular dystrophies, explaining the low efficiency of *LMNA* mutation detection in the EDMD-like isolated cases.

One mutation, H222Y, was identified at a homozygous state in a patient from a consanguineous family. The unaffected parents carried the mutation at a heterozygous state, demonstrating that *LMNA* mutations are also responsible of AR-EDMD.[24] The same amino acid, H222, was mutated to a proline (H222P) in another EDMD family with autosomal dominant transmission.[28] Thus, the pathogenic effect of mutations affecting the same amino acid is variable depending on the type of amino acid change (tyrosine versus proline). Another recessive case was recently described in which two heterozygous mutations were identified, one being specific of lamin A.[41] Finally, the R336Q mutation was identified in an EDMD family with an autosomal dominant transmission in which only two out of the four members carrying the mutation were affected.[24] Altogether, theses results demonstrate that *LMNA* mutations have variable pathogenic effects with a semi-dominant, dominant or recessive pattern of expression.[24,41]

Eleven mutations have been reported in patients with isolated DCM.[4,5,39,45,48] It was initially suggested that mutations in either the rod domain or the tail domain of lamin C might underlie the DMC-CD phenotype.[4] However, two reports demonstrated that within a family, the same *LMNA* mutation gives rise to various phenotypes ranging from an isolated DCM to LGMD1B or EDMD.[5,39]

Among all the reported cases with cardiac and/or skeletal disorders (EDMD, LGMD1B and DCM) due to *LMNA* mutation, there is no clear correlation between the phenotype and type or localization of the mutation in the gene.[5,6,24,28,39,47] Further studies are needed to identify the factors modifying striated muscle phenotypes among patients harboring mutations within lamin A/C. In contrast, 90% of mutations described in FPLD affect the same codon in exon 8 (arginine 482 is mutated in 28/32 reported FLPD cases). Besides this "hot spot", four other mutations were reported in FPLD families, two located in exon 8 and two in exon 11 that encodes only the tail domain of lamin A.[10,11,43,44]

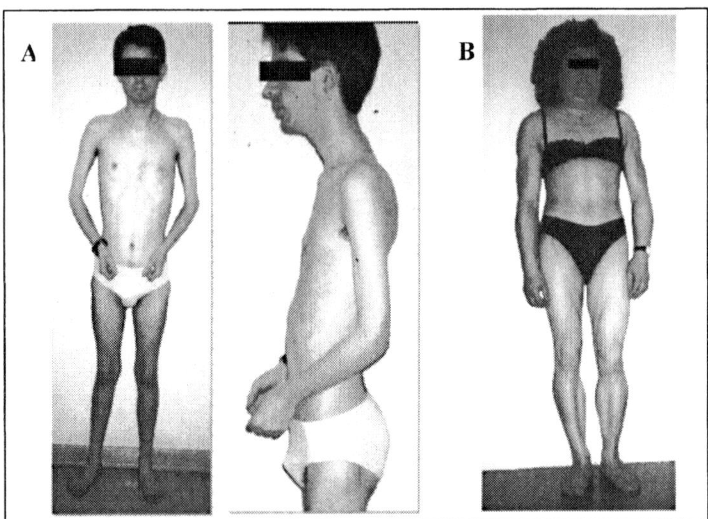

Figure 1. Clinical features of Emery-Dreifuss muscular dystrophy. (A) Clinical characteristics of a 18-year-old male patient with proven mutation in the *LMNA* gene (see Fig. 2). The phenotype is suggestive of EDMD with limitation of neck flexion and marked elbow and heel contractures. Due to complete atrioventricular block, an implantable defibrillator has been inserted at age 18. Courtesy of J.-A. Urtizberea, Service de Médecine Physique et Réadaptation de l'Enfant, Hôpital Raymond Poincaré, Garches, France. (B) Clinical features of partial lipodystrophy of the Dunnigan type in a 39-year-old woman. Note the accumulation of adipose tissue in the face and neck, constrasting with peripheral lipoatrophy with muscle prominence. This clinical aspect appeared gradually after puberty and the patient subsequently developed insulin-resistant diabetes and hypertriglyceridemia. Please see http://www.eurekah.com/chapter.php?chapid=716&bookid=56&catid=15 for color image.

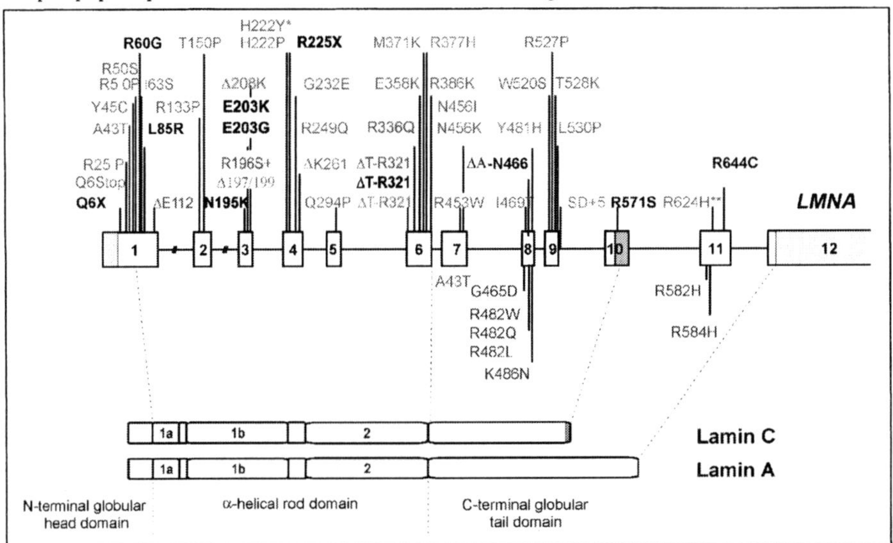

Figure 2. *LMNA* mutations identified in EDMD, LGMD1B, DCM-CD and FPLD. Mutations identified in AD-EDMD are shown in red.[3,24,28,39,41,46] The AR-EDMD mutation is depicted by an asterisk (*).[24] The double heterozygous mutations identified in an EDMD patient are depicted by a double asterisk (**).[41] LGMD 1B mutations are shown in green.[6,47] Mutations in black are responsible for DCM.[4,5,39,45,48] Mutations in blue are reported in FPLD.[10,11,43,44] Please see http://www.eurekah.com/chapter.php?chapid=716&bookid=56&catid=15 for color image.

Lipodystrophies and the Familial Partial Lipodystrophy of the Dunnigan Type (FPLD)

Lipodystrophies

Lipodystrophies represent a heterogeneous group of diseases characterized by generalized or partial alterations in body fat development or distribution, and insulin resistance. The other cardinal clinical signs of these syndromes are *acanthosis nigricans*, which is a skin disorder associated with insulin resistance, frequent hyperandrogenism in females, muscular hypertrophy and liver steatosis. Insulin resistance is associated with a progressive altered glucose tolerance leading to diabetes, and with hypertriglyceridemia.[49,50] When chronic hyperglycemia occurs, its treatment is often very difficult due to major insulin resistance, and the condition leads to early diabetic complications. Acute pancreatitis due to severe hypertriglyceridemia, and liver cirrhosis arising from the frequent nonalcoholic steatohepatitis are also responsible for the morbidity and mortality of these diseases. However, a broad pattern of severity is seen in lipodystrophies, ranging from the rare and serious congenital generalized form to the milder acquired partial one.

The main forms of lipodystrophies are classified according to their origin, either genetic or acquired, and to the clinical pattern of the lipoatrophy, either generalized or partial (Table 2). Etiologies of lipodystrophies are very heterogeneous. Two genetic loci, on chromosomes 9q34 and 11q13 are linked to the congenital generalized lipodystrophy (Berardinelli-Seip syndrome), which is transmitted as an autosomal recessive trait.[51,52] Recently, the 11q13 locus has been identified as the *BSCL2* gene, encoding seipin, a protein of unknown function mainly expressed in brain.[52]

LMNA is presently the only gene known to be involved in a genetic syndrome of partial lipodystrophy, i.e the familial partial lipodystrophy of the Dunnigan-type (FPLD). However, it was shown that dominant negative mutations in PPARγ, a transcription factor involved in adipogenesis, lead to severe insulin resistance, diabetes, and hypertension.[53] Recent studies suggest that affected subjects also present subcutaneous paucity of fat on limbs (S. O'Rahilly, personal communication).

Some forms of lipodystrophies could have an immunological basis: auto-immune diseases are sometimes associated with sporadic cases of generalized lipoatrophies (e.g., Lawrence syndrome). The C3 nephritic factor, an IgG antibody against complement components, can be detected in some cases of partial lipodystrophy of the Barraquer-Simons type (lipoatrophy of face and trunk, with excess accumulation of fat in the lower part of the body).[54] Finally, the probably most frequent form of lipodystrophy is the redistribution of fat that occurs in HIV-infected patients, mainly treated by antiretroviral medications. These patients frequently lose peripheral subcutaneous fat, accumulate visceral adipose tissue, and develop hypertriglyceridemia and insulin resistance.[55]

The pathophysiology of lipodystrophies is still unknown. However, murine models of lipoatrophic diabetes (aP2-nSREBP-1c and A-ZIP/F-1 mice) revealed that primary genetic alterations in fat development resulted in diabetes and dyslipidemia.[56,57] Diabetes could be reversed by fat transplantation in the A-ZIP/F-1 model.[58] Leptin deficiency, caused by the absence of adipose tissue, could be an important determinant of the metabolic abnormalities since exogenous administration or transgenic over-expression of leptin has been shown to markedly improve insulin sensitivity, glycemic control, dyslipidemia and hepatic steatosis in these mice.[59,60] Similarly, the defect in adiponectin, another fat-derived hormone, has recently been shown to be involved in insulin resistance.[61] Regarding the HIV-linked lipodystrophy syndrome, several authors recently pointed out the deleterious effects of some antiretroviral treatments on adipogenesis.[62-64] Altogether, these studies provide strong arguments for a primary role of disturbances in fat distribution or development, which lead secondarily to insulin resistance and metabolic complications.

Table 2. Classification of the main lipodystrophy syndromes

	Generalized Lipoatrophies		Partial Lipodystrophies		
	Berardinelli-Seip Syndrome	**Lawrence Syndrome**	**Familial Partial Lipodystrophy of the Dunnigan Type (FPLD)**	**Barraquer-Simons Syndrome**	**Lipodystrophy Linked to Antiretroviral Treatments of HIV Infection**
Clinical features	Congenital total lipoatrophy	Acquired total lipoatrophy	Lack of adipose tissue in the limbs, buttocks and trunk Fat accumulation in the neck and face	Lipoatrophy of face and trunk Fat excess in the lower part of the body More frequent in females More marked signs in females	Peripheral subcutaneous lipoatrophy (face and four limbs) and/or excess of central fat (visceral fat, or "buffalo neck")
Onset	At birth or in the firsts months of life	In infancy or early adulthood	At puberty	In infancy or adolescence	Most frequently a few months after antiretroviral therapy containing HIV-protease inhibitors
Inheritable pattern	Autosomal recessive	Acquired	Autosomal dominant	Unknown	-
Biological	Severe insulin resistance Diabetes occurring usually in late infancy or adolescence Severe hypertriglyceridemia		Severe insulin resistance Diabetes in adulthood Severe hypertriglyceridemia	Metabolic complications are uncommon (dyslipidemia, altered insulin sensitivity)	Mild insulin resistance Mild to severe hypertriglyceridemia

Continued on next page

Table 2. Cont.

	Generalized Lipoatrophies				Partial Lipodystrophies
	Berardinelli-Seip Syndrome	**Lawrence Syndrome**	**Familial Partial Lipodystrophy of the Dunnigan Type (FPLD)**	**Barraquer-Simons Syndrome**	**Lipodystrophy Linked to Antiretroviral Treatments of HIV Infection**
Clinical Etiology	- Mutations in the *BSCL2* gene encoding seipin	- Some forms could be of genetic origin but secondary expressed	Mutations in *LMNA* exon 8 (R482W/Q/L; K486N; G465D) or in exon 11 (R582H; R584H)	- Unknown cause	Inhibition of adipogenesis induced by HIV-protease inhibitors?
	- Forms linked to a chr 9q34 locus	- Some forms are associated with auto-immune diseases		- Presence of C3 nephritic auto-antibodies with glomerulonephritis in some cases	

Partial lipodystrophies with insulin resistance are pathophysiological models of great interest for diabetologists. Indeed, they represent stereotyped forms of the metabolic syndrome, largely prevalent in developed countries, which associates android repartition of fat, glucose intolerance, hypertension and dyslipidemia.[65] This frequent condition represents a strong risk factor for cardiovascular diseases. As abnormalities in the body distribution of fat are possible important primary etiologic factors for the development of type 2 diabetes,[66,67] partial lipodystrophies constitutes a new field of investigation in the pathophysiology of diabetes.

Familial Partial Lipodystrophy of the Dunnigan Type (FPLD)

Among lipodystrophic syndromes, the familial partial lipodystrophy of the Dunnigan-type (FPLD), dominantly inherited, is a rare disease characterized by the disappearance, after puberty, of adipose tissue in the limbs, buttocks and trunk. This progressive lipoatrophy spares the neck and face, where adipose tissue can accumulates, causing frequently a cushingoïd appearance.[68] However, in lean patients, this latter feature can be lacking, and differential diagnosis with total acquired lipoatrophy can be difficult if the familial dominant transmission of the disease is not evident. Prominence of muscles and superficial veins are partly due to the lipoatrophy, but, as in the other syndromes of insulin resistance, a genuine muscular hypertrophy is present.[69] The android aspect of patients is particularly striking in females, but does not systematically draw attention in males, in which this condition is frequently unrecognized. Garg et al performed magnetic resonance imaging studies in four affected patients, three females and one male.[70] They confirmed the clinically observed altered distribution of the subcutaneous adipose tissue, near-totally absent in areas from extremities and gluteal region, reduced in the truncal area, and increased in the neck and face. Intra-abdominal, intra-thoracic, and intermuscular fat is preserved, as well as mechanical adipose tissue (present in orbits, palm, sole, scalp and periarticular regions). In females, breasts have a markedly reduced subcutaneous fat whereas adipose tissue accumulates in the labia majora.[70] The partial adipose tissue loss in FPLD is associated with a reduced plasma leptin level, to about 40% of normal.[71]

Metabolic alterations associated with FPLD are responsible for the severity of the disease. Insulin resistance, usually attested by hyperinsulinemia with concomitant normal or elevated glycemia, has been confirmed by several tests, including euglycemic-hyperinsulinemic clamp.[72] Insulin-stimulated glucose transport and/or oxidation was found to be impaired in neck and abdomen adipocytes from affected patients, despite normal insulin binding, showing a post-receptor defect in the insulin action.[72] Clinically, acanthosis nigricans, brownish hyperkeratosic skin affection localized to the axillary and inguinal folds, is associated with insulin resistance. When hyperinsulinemia no longer compensates for insulin resistance, glucose intolerance, then diabetes, occur.

Dyslipidemia is frequent among FPLD patients. Hypertriglyceridemia, due to elevated very light density lipoprotein (VLDL) level, is the more prevalent feature. When severe, it can lead to a life-threatening acute pancreatitis. Other perturbations of the lipid profile can also be associated with FPLD, as decreased high-density lipoprotein (HDL)-cholesterol levels, with or without elevated total cholesterol.[73,74] As hypertension is also frequent in FPLD, these patients accumulate numerous cardiovascular risks, leading to early coronary heart disease.[75] Like in other lipodystrophic syndromes, a liver steatosis, with its risk of evolution towards cirrhosis, usually occurs in these patients.

Although this disease affects males and females, both clinical traits and metabolic complications are more severe in women.[44,76] Accordingly, this disease was first reported only in females, and a X-linked dominant transmission was initially evoked.[68] In addition, women affected by FPLD frequently complain of hirsutism, and a polycystic ovary syndrome with ovarian dysfunction, and hyperandrogenism is usual. So far, no specific cardiomyopathy or skeletal muscular dystrophy linked to *LMNA* mutations have been reported in FPLD patients.

Treatment of FPLD is difficult. Appropriate diet and physical training are important to minimize metabolic alterations. However, diabetes mellitus, which appears secondarily in the evolution of the disease, requires usually large doses of insulin. Insulin sensitizers, like metformin, could improve the control of glycemia. PPARγ agonists, such as thiazolidinediones, which promote both adipocyte differentiation and insulin sensitivity, seem promising in lipodystrophic syndromes.[77] A leptin treatment is currently evaluated in these patients.

Mutations causing FPLD cluster only in exons 8 and 11 of *LMNA*, coding for the globular C-terminal domain of type A-lamins (Fig. 2).[10,11,43,44,78,79] They all affect highly conserved residues among species. The most frequent mutation substitutes a basic amino acid at position 482 (arginine) for a neutral residue (tryptophan, glutamine, leucine). All patients are heterozygous for these mutations. The mutations occur in several different haplotypes in the families, suggesting that codon 482 is a site of recurrent mutation in unrelated pedigrees. The critical location at codon 482 of FPLD-linked *LMNA* alterations was recently confirmed by the observation of a patient with a LGMD1B phenotype due to a missense mutation at the adjacent codon 481.[47] The deamination of C to T at a CpG site is a likely mutational mechanism in the case of the R482W substitution. Other mutations also induce a complete or partial loss of a positive charge (from lysine for K486N substitution, or from arginine in the mutations R582H or R584H which affect exon 11) or the appearance of a negative charge (aspartate in the G465D mutation). Alterations in exon 11 are rare, concerning two reported families for the R584H alteration[44,78] and only one for the R582H substitution.[43] They affect specifically the A isoform of lamin. Two sisters affected by the R582H substitution have a less severe loss of subcutaneous adipose tissue and milder metabolic abnormalities.[79] However, we did not observe such an attenuate phenotype in the patient harboring the R584H mutation that we studied.[44] Environmental factors, such as diet and physical activity, are probably important determinants of the severity of metabolic complications.[78]

Studies of *LMNA* in other metabolic disorders have been performed. We excluded *LMNA* mutations as being responsible for generalized lipodystrophy.[44] No mutations in exon 8 of *LMNA* have been found in subjects with HIV therapy-associated lipodystrophy. Furthermore, lamins A/C and HIV-1 protease do not have any sequence homology, providing no evidence for the direct inhibition of lamins by HIV-1 protease inhibitors.[80] Hegele et al suggested that a common variation in *LMNA* could be associated with obesity-related phenotypes.[81,82] However, there is presently no evidence for an association of this variant with type 2 diabetes.[83]

Could Some Patients with *LMNA* Mutations be Affected by Both Skeletal or Cardiac Muscular Symptoms and Lipodystrophy?

The association of *LMNA* mutation-linked skeletal or cardiac muscular defects with lipodystrophy or metabolic abnormalities has not been reported so far. A muscular hypertrophy is usual in FPLD, as in other lipodystrophic syndromes. Calf hypertrophy could be reminiscent of what is observed in LGMD1B. Although some patients complain of cramps, their muscular strength is normal in most cases. However, patients with FPLD due to the *LMNA* R482W mutation and presenting muscular signs compatible with LGMD1B are being investigated.[84] A cardiac septal hypertrophy can be observed in FPLD patients, but is difficult to attribute to a specific genetic defect since it could be secondary to diabetes and hypertension. Further investigations of neuromuscular and cardiac phenotype in FPLD patients are needed. Likewise, an accurate evaluation of adipose tissue distribution, insulin sensitivity and lipid metabolism has not been reported in patients with AD-EDMD, LGMD1B or DCM-CD linked to *LMNA* mutations.

Experimental Models of Lamin A/C Alterations

lmna *Knock-Out Mice*

Sullivan et al reported the derivation of mice in which the lamins A/C have been eliminated by gene targeting (by deletion of a region extended from exon 8 to the middle of exon 11), to produce either homozygous or heterozygous offspring.[85] Homozygous mice, although normal at birth, rapidly exhibit a striking cardiac and skeletal muscular dystrophy with rigidity, more marked in proximal muscles, that resembles the EDMD phenotype in humans. In addition, postnatal growth is severely impaired and the mice exhibit premature mortality. Although a white fat atrophy has also been reported in this model, additional metabolic features of a lipodystrophic syndrome have not been described, so this appearance could be related to cachexia. Mice heterozygous for the *lmna* mutation are overtly normal at 6-10 months with minimal evidence of dystrophy.

Studies of cells from these mice clearly showed that loss of lamin A/C affects nuclear envelope integrity.[85] Indeed, nuclei from embryonic *lmna*[-/-] fibroblasts often exhibited an abnormal shape, with nuclear regions displaying disruption of heterochromatin, withdrawal of B-type lamins and other proteins of the inner nuclear envelope such as lamina-associated polypeptide (LAP)2β, irregular distribution of nuclear pores complexes, and partial mislocalization of emerin to the cytoplasm. Interestingly, this aberrant emerin distribution was rescued by over-expression of wild-type or R482W-mutated lamin A, contrary to the L85R, N195K or L530P mutated forms of the protein, which had no effect on emerin relocalization.[85,86]

Nuclear Alterations in Cells Harboring *LMNA* Mutations

From the studies in *lmna*[-/-] mice, it could be postulated that a loss of lamin A/C protein expression or function could underlie the muscular phenotypes. One the other hand, from studies of patients with XL-EDMD, it appears that the disease arises from either loss of emerin protein or mutations resulting in its subcellular mislocalization.[87-89] XL and AD-EDMD are almost clinically indistinguishable, thus a pathophysiological hypothesis would be that a defect in emerin distribution, secondary to *LMNA* mutations, is responsible for EDMD, as also suggested by studies of *lmna*[-/-] cells. However, heterozygous *lmna*[+/-] mice are healthy, and although their cells present frequent irregular nuclei, nuclear envelope proteins show a largely normal distribution.[85]

In human, immunocytochemical analysis of lamin A/C and emerin on skeletal muscle biopsies of AD-EDMD patients carrying *LMNA* mutation showed no detectable differences from control muscles, indicating that the mutations do not significantly alter the structure of the nuclear envelope.[21] The same analysis performed on a cardiac biopsy of an AD-EDMD patient with a non-sense mutation showed the same results.[3] The latter mutation leads potentially to the production of truncated lamins A/C with only five amino acids. Most probably such small peptides are degraded and only lamins A/C produced by the intact allele are expressed in the patients. This hypothesis was confirmed by the Western-blot analysis of the explanted myocardial tissue that showed a decreased expression of lamins A/C compared to that of a control heart tissue.[5]

In some of EDMD patients, an ultrastructural examination of the skeletal muscle biopsies showed a loss of heterochromatin from wide stretches along the nuclear envelope, in 10% of nuclei, with a rarefaction of nuclear pores complexes in these areas. Irregular shapes of some nuclei are also reported.[21,90] However, only a small proportion of muscle nuclei exhibit abnormalities, thus study of additional cases is required to draw any conclusion. Nevertheless, these alterations are reminiscent of features of *lmna*[-/-] mouse cells, albeit affecting a lower percentage of cells, and in skeletal muscle of XL-EDMD patients.[91,92] Further work is required using cells from patients with skeletal or cardiac muscular disease linked to *LMNA* alterations in order to evaluate the extent of nuclear disturbances.

We recently performed a study of skin fibroblasts of FPLD patients with R482W and R482Q *LMNA* mutations.[93] Protein expression of type A and type B lamins, LAP2β and emerin was normal in the whole population of fibroblasts. However, 5-25% of these cells had abnormal blebbing nuclei with A-type lamins forming a peripheral meshwork that was frequently disorganized (Fig. 3). Emerin strictly colocalized with this abnormal lamin A/C meshwork (Fig. 4), in agreement with the R482W mutated-lamin A dependent relocalization of emerin in *lmna*[-/-] fibroblasts,[86] and the preserved interactions between R482Q lamin A and emerin in vitro.[94] Cells from lipodystrophic patients often had other nuclear envelope defects, mainly herniations deficient in B-type lamins, nuclear pores and LAP2β. Furthermore, heterogeneous DNA staining

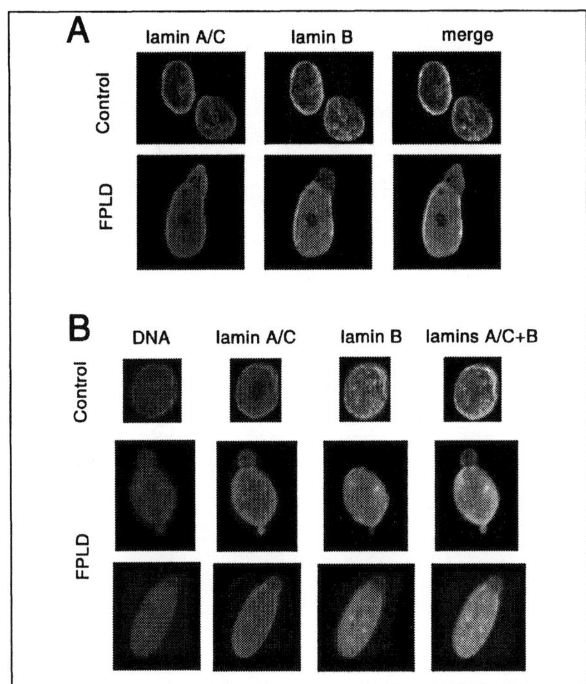

Figure 3. Fibroblasts from lipodystrophic patients with a *LMNA* R482W mutation show blebbing of nuclei with lamina disorganization and alterations in chromatin structure. Skin fibroblasts from control individuals and patients affected by FPLD due to a *LMNA* R482W mutation were fixed in methanol, then labeled with the DNA stain DAPI , and/or antibodies directed against lamins A/C (red) and B (green), as indicated, before analysis by confocal (A) or conventional (B) immunofluorescence microscopy. Five to 25% of fibroblasts from FPLD patients have abnormally shaped nuclei with buds containing a disorganized lamina with a honeycomb aspect or holes in the type A-lamin lattice and a weak or absent B type-lamin staining. The faint DAPI staining of DNA in the buds suggests that chromatin is decondensed in nuclear areas flanking disorganized nuclear envelope domains. See http://www.eurekah.com/chapter.php?chapid=716&bookid=56&catid=15 for color image.

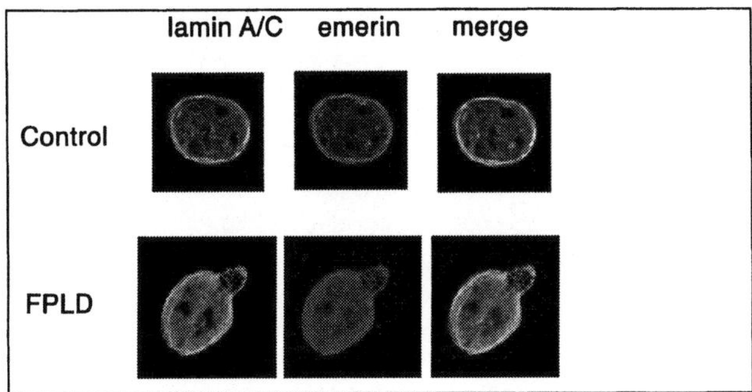

Figure 4. Emerin strictly colocalizes with the lamin A/C meshwork, even in nuclear herniations from FPLD fibroblasts. Fibroblasts from control individuals and patients affected by FPLD due to a *LMNA* R482W mutation were labeled for double immunofluorescence with antibodies directed against lamins A/C (green) and emerin (red), and observed with a confocal microscope. Emerin systematically colocalizes with lamin A/C in nuclei from both control subjects and patients fibroblasts, even in nuclear herniations with an abnormal type A-lamins meshwork, suggesting that the interactions between R482W mutated-laminA/C and emerin are preserved. See http://www.eurekah.com/chapter.php?chapid=716&bookid=56&catid=15 for color image.

by DAPI suggested that chromatin was decondensed in nuclear areas flanking disorganized nuclear envelope domains (Fig. 3). The mechanical properties of nuclear envelopes were altered, as judged from the extensive deformations observed in nuclei from heat-shocked cells, and from the low stringency of extraction of nuclear envelope proteins. These structural nuclear alterations were caused by the lamins A/C mutations, since the same changes were introduced in human control fibroblasts by expression of R482W mutated lamin A. However, despite these abnormalities, we showed that the fibroblasts from FPLD patients were euploid and able to cycle and divide.

Recently, transfection studies of mutant *LMNA* alleles have been performed in several cell types by three groups (Favreau C, Östlund C, Worman HJ, Courvalin JC, Buendia B, personal communication).[86,95] Data in HeLa cells and C2C12 myoblasts show heterogeneous defects in the shape of the nuclei, in the assembly of type A-lamins and/or in the distribution of emerin that concern some but not all EDMD or DCM-linked *LMNA* mutations. In human fibroblasts and in C2C12 myoblasts, overexpression of lamin A mutated in the carboxy-terminal domain generated an aberrant nuclear phenotype similar to that observed in cells from FPLD patients (Favreau C, Östlund C, Worman HJ, Courvalin JC, Buendia B, personal communication).[93]

The observation of heterogeneous, but very similar nuclear alterations in fibroblasts from FPLD patients or from XL-EDMD patients, from *lmna*[-/-] mice and from myocytes of AD-EDMD patients confirms that lamins A/C are major determinants of the nuclear architecture. However, these studies have shown that A-type lamins are not the only determinants of emerin localization. If altered interactions between FPLD-linked mutant forms of lamin A/C and emerin do not seem likely in FPLD, they probably neither represent the only pathophysiological mechanism in cardiac and skeletal muscular dystrophies linked to *LMNA* mutations.

Conclusion

Recent experiments evidenced the essential role of A-type lamins in nuclear architecture and integrity. However, the pathophysiology of the diseases linked to *LMNA* mutations remains unclear. A striking feature of these diseases is the tissue-specificity of the alterations, which is difficult to relate to the widespread expression of lamin A/C in differentiated cells.

Germline mutations of the RET protooncogene is another example of different phenotypes arising from mutations in the same gene. However, in the case of RET mutations, the phenotypes of either to congenital malformations or inherited cancer syndromes are due to the loss or gain of function of the mutated protein.[96] The divergent consequences of alterations in lamin A/C, a structural protein, seem more complex to understand. Knowledge about the other specific functions of lamin A/C is presently lacking to unravel the pathophysiological mechanisms of muscular dystrophies and lipodystrophies linked to *LMNA*. However, several hypotheses can be discussed.

The diverse types of *LMNA* mutations in patients with EDMD, LGMD1B and DCM-DC, some of them leading to a truncated form of the protein, suggest that a loss of function of lamin A/C could be responsible for the cardiac and/or skeletal muscle diseases. This loss of function could act via a secondary mislocalization of emerin, as suggested by the close clinical symptomatology induced by emerin mutations. However, from present cellular studies, emerin alterations were not systematically found when the *LMNA* gene was mutated, suggesting that this mechanism might not be unique. Further work on cells from patients will most likely provide more detailed information on this issue.

The compromised nuclear integrity could lead to nuclear fragility, and to mechanical damage during muscle contractions that could explain the striated muscle affections. However, the preferential alteration of the conduction pathway in cardiac tissue in *LMNA*-linked muscular dystrophies and cardiomyopathies and the nuclear fragility that we recently evidenced in cells from FPLD patients with *LMNA* R482W or R482Q mutations, are not in favor of this hypothesis.[93]

The mutational hot-spot found in FPLD-linked phenotype suggests a more specific alteration of a lamin A/C function that could only be expressed at the adipocyte level. The systematic alteration of the charge of an amino acid in the C terminal domain of the protein could evoke modifications of the binding of lamin A/C with others partners. The recent crystalization of the C-terminal end of lamin A showed that it is spatially organized as an immunoglobulin-like domain and that mutations in amino acids at positions 465, 482 and 486, found in FPLD, are localized at the external surface of the structure (S. Dhe-Paganon, E. Warner, S. Schoelson, personal communication). As immunoglobulin folds are known to mediate protein-protein, protein-lipid and protein-DNA interactions, this hypothesis could have a pathophysiological relevance, but data about of lamin A/C partners inducing a tissue-specific response are presently lacking.

A role for the inner nuclear membrane in the regulation of gene expression has been previously suggested.[97] In yeast, the inner nuclear membrane is involved in the spatial organization of chromatin and in the regulation of transcription.[98] In mammalian species, gene expression is also influenced by the spatial organization of the nucleus (for a review, see ref. 99). Disturbances in heterochromatin organization are a largely represented feature in cells affected by LMNA mutations. It is tempting to speculate that lamins could alter tissue-specific gene expression, and that the mutations could affect muscle or adipose tissue function, differentiation or survival. Indeed, several transcription factors, among which the retinoblastoma protein pRb, a key regulator of cell-cycle dependent transcription, are known to interact with A type-lamins (for a review, see refs. 100-102. Furthermore, the lamina provides an attachment site for apoptotic signaling machinery[100] and nuclear envelope proteins are early targets for caspase degradation.[103]

Finally, it has recently been shown in Drosophila that lamins, in addition to their role in nucleus integrity, are also required for cytoplasmic organization,[104] in accordance with the known mechanical interactions between nuclear scaffolding proteins and cytoskeletal filaments.[105] This aspect has now to be investigated in mammalian LMNA-mutated cells. Constant improvements in the description of lamin A/C-associated diseases and in experimental models for alterations in nuclear envelope proteins will provide important information for the understanding of nuclear physiology, adipose tissue and metabolic diseases, and skeletal and cardiac muscle dystrophies.

Addendum

During the editing process of this communication, a new phenotype also due to mutation of lamin A/C gene has been reported. This fifth laminopathy disease is an autosomal recessive form of axonal neuropathy (Charcot-Marie-Tooth disorder type 2, or CMT2B1).[106] The Charcot-Marie-Tooth (CMT) disorders comprise a group of clinically and genetically heterogeneous hereditary motor and sensory neuropathies, which are mainly characterized by muscle weakness and wasting, foot deformations, and electrophysiological as well as histological changes. A subtype, CMT2, is defined by a slight or absent reduction of nerve-conduction velocities together with the loss of large myelinated fibers and axonal degeneration. Homozygosity mapping in inbred Algerian families with autosomal recessive CMT2 (AR- CMT2) provided evidence of linkage to chromosome 1q21.2-q21.3 in two families. All patients shared a common homozygous ancestral haplotype that was suggestive of a founder mutation as the cause of the phenotype. A unique homozygous mutation in *LMNA* was identified in all affected members and in additional patients with CMT2 from a third, unrelated family. Ultrastructural exploration of sciatic nerves of *LMNA* null (i.e., *LMNA$^{-/-}$*) mice was performed and revealed a strong reduction of axon density, axonal enlargement, and the presence of non-myelinated axons, all of which were highly similar to the phenotypes of human peripheral axonopathies. The association of nerve abnormality with *LMNA* mutations enlarges the already broad range of phenotypes of laminopathies and shed further light on the important interactions between nerve and striated muscle.

Acknowledgments

Figure 1A is a courtesy of Prof J-A. Urtizberea, Service de Médecine Physique et Réadaptation de l'Enfant, Hôpital Raymond Poincaré, Garches, France. Figure 2 was adapted with permission from Roberts and Schwartz, 2000.[107] The authors thank Brigitte Buendia, Jean-Claude Courvalin and Jacqueline Capeau for critical reading of the manuscript.

References

1. Stuurman N, Heins S, Aebi U. Nuclear lamins: Their structure, assembly, and interactions. J Struct Biol 1998; 122:42-66.
2. Worman HJ, Courvalin JC. The inner nuclear membrane. J Membr Biol 2000; 177:1-11.
3. Bonne G, Di Barletta MR, Varnous S et al. Mutations in the gene encoding lamin A/C cause autosomal dominant Emery-Dreifuss muscular dystrophy. Nature Genet 1999; 21:285-288.
4. Fatkin D, MacRae C, Sasaki T et al. Missense mutations in the rod domain of the lamin A/C gene as causes of dilated cardiomyopathy and conduction-system disease. N Engl J Med 1999; 341:1715-1724.
5. Bécane H-M, Bonne G, Varnous S et al. High incidence of sudden death of conduction system and myocardial disease due to lamins A/C gene mutation. Pacing Clin Electrophysiol 2000; 23:1661-1666.
6. Muchir A, Bonne G, van der Kooi AJ et al. Identification of mutations in the gene encoding lamins A/C in autosomal dominant limb girdle muscular dystrophy with atrioventricular conduction disturbances (LGMD1B). Hum Mol Genet 2000; 9:1453-1459.
7. Peters J, Barnes R, Bennett L et al. Localization of the gene for familial lipodystrophy (Dunnigan variety) to chromosome 1q21-22. Nature Genet 1998; 18:292-295.
8. Jackson SN, Pinkney J, Bargiotta A et al. A defect in the regional deposition of adipose tissue (partial lipodystrophy) is encoded by a gene at chromosome 1q. Am J Hum Genet 1998; 63:534-540.
9. Anderson JL, Khan M, David WS et al. Confirmation of linkage of hereditary partial lipodystrophy to chromosome 1q21-22. Am J Med Genet 1999; 82:161-165.
10. Cao H, Hegele RA. Nuclear lamin A/C R482Q mutation in Canadian kindreds with Dunnigan-type familial partial lipodystrophy. Hum Mol Genet 2000; 9:109-112.
11. Shackleton S, Lloyd DJ, Jackson SN et al. *LMNA*, encoding lamin A/C, is mutated in partial lipodystrophy. Nature Genet 2000; 24:153-156.
12. Becker PE, Kiener F. Eine neue X-chromosomale muskeldystrophie. Arch Psychiatr Nervenkr 1955; 193:427.
13. Dreifuss FE, Hogan GR. Survival in X-chromosomal muscular dystrophy. Neurology 1961; 11:734-737.
14. Emery AEH, Dreifuss FE. Unusual type of benign X-linked muscular dystrophy. J Neurol Neurosurg Psychiat 1966; 29:338-342.
15. Rowland LP, Fetell M, Olarte M et al. Emery-Dreifuss muscular dystrophy. Ann Neurol 1979; 5:111-117.
16. Cestan R, LeJonne. Une myopathie avec rétractions familiales. Nouvelle iconographie de la Salpétrière 1902; 15:38-52.
17. Emery AEH. Emery-Dreifuss muscular dystrophy—A 40 year retrospective. Neuromusc Disord 2000; 10:228-232.
18. Wehnert M, Muntoni F. 60th ENMC International Workshop: Non X-linked Emery-Dreifuss Muscular Dystrophy, 5-7 June 1998. Neuromusc Disord 1999; 9:115-120.
19. Hopkins LC, Warren S. Emery-Dreifuss muscular dystrophy. In: Rowland LP, DiMauro S, eds. Handbook of Clinical Neurology: Myopathies. Vol 18. Amsterdam: Elsevier Science, 1992:145-160.
20. Hausmanowa-Petrusewicz I. The Emery-Dreifuss disease. Neuropat Pol 1988; 26:265-281.
21. Sewry CA, Brown SC, Mercuri E et al. Skeletal muscle pathology in autosomal dominant Emery-Dreifuss muscular dystrophy with lamin A/C mutations. Neuropathol Appl Neurobiol 2001; 27:281-290.
22. Takamoto K, Hirose K, Uono M et al. A genetic variant of Emery-Dreifuss disease. Muscular dystrophy with humeropelvic distribution, early joint contracture, and permanent atrial paralysis. Arch Neurol 1984; 41:1292-1293.
23. Taylor J, Sewry CA, Dubowitz V et al. Early onset, autosomal recessive muscular dystrophy with Emery-Dreifuss phenotype and normal emerin expression. Neurology 1998; 51:1116-1120.
24. di Barletta MR, Ricci E, Galluzzi G et al. Different mutations in the *LMNA* gene cause autosomal dominant and autosomal recessive Emery-Dreifuss muscular dystrophy. Am J Hum Genet 2000; 66:1407-1412.

25. Bione S, Maestrini E, Rivella S et al. Identification of a novel X-linked gene responsible for Emery-Dreifuss muscular dystrophy. Nature Genet 1994; 8:323-327.

26. Fenichel GM, Sul YC, Kilroy AW et al. An autosomal-dominant dystrophy with humeropelvic distribution and cardiomyopathy. Neurology 1982; 32:1399-1401.

27. Miller RG, Layzer RB, Mellenthin MA et al. Emery-Dreifuss muscular dystrophy with autosomal dominant transmission. Neurology 1985; 35:1230-1233.

28. Bonne G, Mercuri E, Muchir A et al. Clinical and molecular genetic spectrum of autosomal dominant Emery Dreifuss muscular dystrophy due to mutations of the lamin A/C gene. Ann Neurol 2000; 48:170-180.

29. Emery AEH. X-linked muscular dystrophy with early contractures and cardiomyopathy (Emery-Dreifuss type). Clin Genet 1987; 32:360-367.

30. Yates JR. 43rd ENMC International Workshop on Emery-Dreifuss Muscular Dystrophy, 22 June 1996, Naarden, The Netherlands. Neuromuscul Disord 1997; 7:67-69.

31. Hoeltzenbein M, Karow T, Zeller JA et al. Severe clinical expression in X-linked Emery-Dreifuss muscular dystrophy. Neuromuscul Disord 1999; 9:166-170.

32. Bushby KM. The limb-girdle muscular dystrophies-multiple genes, multiple mechanisms. Hum Mol Genet 1999; 8:1875-1882.

33. van der Kooi AJ, Ledderhof TM, de Voogt WG et al. A newly recognized autosomal dominant limb girdle muscular dystrophy with cardiac involvement. Ann Neurol 1996; 39:636-642.

34. van der Kooi AJ, van Meegen M, Ledderhof TM et al. Genetic localization of a newly recognized autosomal dominant limb-girdle muscular dystrophy with cardiac involvement (LGMD1B) to chromosome 1q11-21. Am J Hum Genet 1997; 60:891-895.

35. Richardson P, McKenna W, Bristow M et al. Report of the 1995 World Health Organisation/ International Society and Federation of Cardiology task force on the definition and classification of cardiomyopathies. Circulation 1995; 93:841-842.

36. Codd MB, Sugrue DD, Gersh BJ et al. Epidemiology of idiopathic dilated and hypertrohic cardiomyopathy. Circulation 1989; 80:564-572.

37. Michels VV, Moll PP, Miller FA et al. The frequency of familial dilated cardiomyopathy in a series of patients with idiopathic dilated cardiomyopathy. N Engl J Med 1992; 326:77-82.

38. Keeling PJ, Gang Y, Smith G et al. Familial dilated cardiomyopathy in the United Kingdom. Br Heart J 1995; 73:417-421.

39. Brodsky GL, Muntoni F, Miocic S et al. Lamin A/C gene mutation associated with dilated cardiomyopathy with variable skeletal muscle involvement. Circulation 2000; 101:473-476.

40. Lin F, Worman HJ. Structural organization of the human gene encoding nuclear lamin A and nuclear lamin C. J Biol Chem 1993; 268:16321-16326.

41. Brown CA, Lanning RW, McKinney KQ et al. Novel and recurrent mutations in lamin A/C in patients with Emery- Dreifuss muscular dystrophy. Am J Med Genet 2001; 102:359-367.

42. Colomer J, Iturriaga C, Bonne G et al. Autosomal dominant Emery-Dreifuss muscular dystrophy: : a new family with late diagnosis. Neuromusc Disord 2002;12:19-25.

43. Speckman RA, Garg A, Du F et al. Mutational and haplotype analyses of families with familial partial lipodystrophy (Dunnigan variety) reveal recurrent missense mutations in the globular C-terminal domain of lamin A/C. Hum Genet 2000; 66:1192-1198.

44. Vigouroux C, Magré J, Vantyghem MC et al. Lamin A/C gene: Sex-determined expression of mutations in Dunnigan-type familial partial lipodystrophy and absence of coding mutations in congenital and acquired generalized lipoatrophy. Diabetes 2000; 49:1958-1962.

45. Genschel J, Bochow B, Kuepferling S et al. A R644C mutation within lamin A extends the mutations causing dilated cardiomyopathy. Hum Mutat 2001; 17:154. ·

46. Felice KJ, Schwartz RC, Brown CA et al. Autosomal dominant Emery-Dreifuss dystrophy due to mutations in rod domain of the lamin A/C gene. Neurology 2000; 55:275-280.

47. Kitaguchi T, Matsubara S, Sato M et al. A missense mutation in the exon 8 of lamin A/C gene in a Japanese case of autosomal dominant limb-girdle muscular dystrophy and cardiac conduction block. Neuromusc Disord 2001; 11:542-546.

48. Jakobs PM, Hanson EL, Crispell KA et al. Novel lamin A/C mutations in two families with dilated cardiomyopathy and conduction system disease. J Card Fail 2001; 7:249-256.

49. Moller DE, O'Rahilly S. Syndromes of severe insulin resistance: clinical and patho-physiological features. In: Moller DE, ed. Insulin resistance. New York: Wiley and Sons, 1993:49-81.

50. Reitman ML, Arioglu E, Gavrilova O et al. Lipoatrophy revisited. Trends Endocrinol Metab 2000; 11:410-416.

51. Garg A, Wilson R, Barnes R et al. A gene for congenital generalized lipodystrophy maps to human chromosome 9q34. J Clin Endocrinol Metab 1999; 84:3390-3394.

52. Magré J, Delépine M, Khallouf E et al. Identification of the gene altered in Berardinelli-Seip congenital lipodystrophy on chromosome 11q13. Nature Genet 2001; 28:365-370.

53. Barroso I, Gurnell M, Crowley VE et al. Dominant negative mutations in human PPAR-gamma associated with severe insulin resistance, diabetes mellitus and hypertension. Nature 1999; 402:880-883.

54. Levy Y, George J, Yona E et al. Partial lipodystrophy, mesangiocapillary glomerulonephritis, and complement dysregulation. An autoimmune phenomenon. Immunol Res 1998; 18:55-60.

55. Carr A, Samaras K, Burton S et al. A syndrome of peripheral lipodystrophy, hyperlipidaemia and insulin resistance in patients receiving HIV protease inhibitors. AIDS 1998; 12:F51-F58.

56. Moitra J, Mason MM, Olive M et al. Life without white fat: a transgenic mouse. Genes Dev 1998; 12:3168-3181.

57. Shimomura I, Hammer RE, Richardson JA et al. Insulin resistance and diabetes mellitus in transgenic mice expressing nuclear SREBP-1c in adipose tissue: Model for congenital generalized lipodystrophy. Genes Dev 1998; 12:3182-3194.

58. Gavrilova O, Marcus-Samuels B, Graham D et al. Surgical implantation of adipose tissue reverses diabetes in lipoatrophic mice. J Clin Invest 2000; 105:271-278.

59. Shimomura I, Hammer RE, Ikemoto S et al. Leptin reverses insulin resistance and diabetes mellitus in mice with congenital lipodystrophy. Nature 1999; 401:73-76.

60. Ebihara K, Ogawa Y, Masuzaki H et al. Transgenic overexpression of leptin rescues insulin resistance and diabetes in a mouse model of lipoatrophic diabetes. Diabetes 2001; 50:1440-1448.

61. Yamauchi T, Kamon J, Waki H et al. The fat-derived hormone adiponectin reverses insulin resistance associated with both lipoatrophy and obesity. Nature Med 2001; 7:941-946.

62. Zhang B, MacNaul K, Szalkowski D et al. Inhibition of adipocyte differentiation by HIV protease inhibitors. J Clin Endocrinol Metab 1999; 84:4274-4277.

63. Dowell P, Flexner C, Kwiterovich PO et al. Suppression of preadipocyte differentiation and promotion of adipocyte death by HIV protease inhibitors. J Biol Chem 2000; 275:41325-41332.

64. Caron M, Auclair M, Vigouroux C et al. The HIV protease inhibitor indinavir impairs sterol regulatory element- binding protein-1 intranuclear localization, inhibits preadipocyte differentiation, and induces insulin resistance. Diabetes 2001; 50:1378-1388.

65. Reaven GM. Banting lecture 1988. Role of insulin resistance in human disease. Diabetes 1988; 37:1595-1607.

66. Danforth E, Jr. Failure of adipocyte differentiation causes type II diabetes mellitus? Nature Genet 2000; 26:13.

67. Joffe BI, Panz VR, Raal FJ. From lipodystrophy syndromes to diabetes mellitus. Lancet 2001; 357:1379-1381.

68. Köbberling J, Dunnigan MG. Familial partial lipodystrophy: two types of an X linked dominant syndrome, lethal in the hemizygous state. J Med Genet 1986; 23:120-127.

69. Wildermuth S, Spranger S, Spranger M et al. Köbberling-Dunnigan syndrome: A rare cause of generalized muscular hypertrophy. Muscle Nerve 1996; 19:843-847.

70. Garg A, Peshock RM, Fleckenstein JL. Adipose tissue distribution pattern in patients with familial partial lipodystrophy (Dunnigan variety). J Clin Endocrinol Metab 1999; 84:170-174.

71. Hegele RA, Cao H, Huff MW et al. *LMNA* R482Q mutation in partial lipodystrophy associated with reduced plasma leptin concentration. J Clin Endocrinol Metab 2000; 85:3089-3093.

72. Ursich MJ, Fukui RT, Galvao MS et al. Insulin resistance in limb and trunk partial lipodystrophy (type 2 Köbberling-Dunnigan syndrome). Metabolism 1997; 46:159-163.

73. Hegele RA, Anderson CM, Wang J et al. Association between nuclear lamin A/C R482Q mutation and partial lipodystrophy with hyperinsulinemia, dyslipidemia, hypertension, and diabetes. Genome Res 2000; 10:652-658.

74. Schmidt HH, Genschel J, Baier P et al. Dyslipemia in familial partial lipodystrophy caused by an R482W mutation in the *LMNA* gene. J Clin Endocrinol Metab 2001; 86:2289-2295.

75. Hegele RA. Premature atherosclerosis associated with monogenic insulin resistance. Circulation 2001; 103:2225-2229.

76. Garg A. Gender differences in the prevalence of metabolic complications in familial partial lipodystrophy (Dunnigan variety). J Clin Endocrinol Metab 2000; 85:1776-1782.

77. Arioglu E, Duncan-Morin J, Sebring N et al. Efficacy and safety of troglitazone in the treatment of lipodystrophy syndromes. Ann Intern Med 2000; 133:263-274.

78. Hegele RA, Cao H, Anderson CM et al. Heterogeneity of nuclear lamin A mutations in Dunnigan-type familial partial lipodystrophy. J Clin Endocrinol Metab 2000; 85:3431-3435.

79. Garg A, Vinaitheerthan M, Weatherall PT et al. Phenotypic heterogeneity in patients with familial partial lipodystrophy (Dunnigan variety) related to the site of missense mutations in lamin A/C gene. J Clin Endocrinol Metab 2001; 86:59-65.

80. Behrens GM, Lloyd D, Schmidt HH et al. Lessons from lipodystrophy: *LMNA*, encoding lamin A/C, in HIV therapy- associated lipodystrophy. AIDS 2000; 14:1854-1855.

81. Hegele RA, Cao H, Harris SB et al. Genetic variation in *LMNA* modulates plasma leptin and indices of obesity in aboriginal Canadians. Physiol Genomics 2000; 3:39-44.

82. Hegele RA, Huff MW, Young TK. Common genomic variation in *LMNA* modulates indexes of obesity in Inuit. J Clin Endocrinol Metab 2001; 86:2747-2751.

83. Wolford JK, Hanson RL, Bogardus C et al. Analysis of the lamin A/C gene as a candidate for type II diabetes susceptibility in Pima Indians. Diabetologia 2001; 44:779-782.

84. Vantyghem MC, Millaire A, Cuisset JM et al. Muscular and cardiac phenotype in patients affected by the familial partial lipodystrophy (FPLD) or Dunnigan syndrome. First International Workshop on Lipoatrophic diabetes and other syndromes of lipodystrophies, March 2001, Bethesda, USA, abstract P22.

85. Sullivan T, Escalante-Alcalde D, Bhatt H et al. Loss of A-type lamin expression compromises nuclear envelope integrity leading to muscular dystrophy. J Cell Biol 1999; 147:913-920.

86. Raharjo WH, Enarson P, Sullivan T. Nuclear envelope defects associated with *LMNA* mutations causing dilated cardiomyopathy and Emery-Dreifuss muscular dystrophy. J Cell Sci 2001; 114:4447-4457.

87. Nagano A, Koga R, Ogawa M et al. Emerin deficiency at the nuclear membrane in patients with Emery-Dreifuss muscular dystrophy. Nature Genet 1996; 12:254-259.

88. Ellis JA, Craxton M, Yates JR et al. Aberrant intracellular targeting and cell cycle-dependent phosphorylation of emerin contribute to the Emery-Dreifuss muscular dystrophy phenotype. J Cell Sci 1998; 111:781-792

89. Fairley EA, Kendrick-Jones J, Ellis JA. The Emery-Dreifuss muscular dystrophy phenotype arises from aberrant targeting and binding of emerin at the inner nuclear membrane. J Cell Sci 1999; 112:2571-2582.

90. Sabatelli P, Lattanzi G, Ognibene A et al. Nuclear alterations in autosomal-dominant Emery-Dreifuss muscular dystrophy. Muscle Nerve 2001; 24:826-829.

91. Fidzianska A, Toniolo D, Hausmanowa-Petrusewicz I. Ultrastructural abnormality of sarcolemmal nuclei in Emery-Dreifuss muscular dystrophy (EDMD). J Neurol Sci 1998; 159:88-93.

92. Ognibene A, Sabatelli P, Petrini S et al. Nuclear changes in a case of X-linked Emery-Dreifuss muscular dystrophy. Muscle Nerve 1999; 22:864-869.

93. Vigouroux C, Auclair M, Dubosclard E et al. Nuclear envelope disorganization in fibroblasts from lipodystrophic patients with heterozygous R482Q/W mutations in lamin A/C gene. J Cell Sci 2001; 114:4459-4468.

94. Holt I, Clements L, Manilal S et al. The R482Q lamin A/C mutation that causes lipodystrophy does not prevent nuclear targeting of lamin A in adipocytes or its interaction with emerin. Eur J Hum Genet 2001; 9:204-208.

95. Östlund C, Bonne G, Schwartz K et al. Properties of lamin A mutants found in Emery-Dreifuss muscular dystrophy, cardiomyopathy and Dunnigan-type partial lipodystrophy. J Cell Sci 2001; 114 :4435-4445.

96. Mak YF, Ponder BA. RET oncogene. Curr Opin Genet Dev 1996; 6:82-86.

97. Ye Q, Worman HJ. Interaction between an integral protein of the nuclear envelope inner membrane and human chromodomain proteins homologous to Drosophila HP1. J Biol Chem 1996; 271:14653-14656.

98. Galy V, Olivo-Marin JC, Scherthan H et al. Nuclear pore complexes in the organization of silent telomeric chromatin. Nature 2000; 403:108-112.

99. Gasser SM. Positions of potential: nuclear organization and gene expression. Cell 2001; 104:639-642.

100. Cohen M, Lee KK, Wilson KL et al. Transcriptional repression, apoptosis, human disease and the functional evolution of the nuclear lamina. Trends Biochem Sci 2001; 26:41-47.

101. Hutchison CJ, Alvarez-Reyes M, Vaughan OA. Lamins in disease: why do ubiquitously expressed nuclear envelope proteins give rise to tissue-specific disease phenotypes? J Cell Sci 2001; 114:9-19.

102. Wilson KL, Zastrow MS, Lee KK. Lamins and disease: insights into nuclear infrastructure. Cell 2001; 104:647-650.

103. Buendia B, Santa-Maria A, Courvalin JC. Caspase-dependent proteolysis of integral and peripheral proteins of nuclear membranes and nuclear pore complex proteins during apoptosis. J Cell Sci 1999; 112:1743-1753.

104. Guillemin K, Williams T, Krasnow MA. A nuclear lamin is required for cytoplasmic organization and egg polarity in Drosophila. Nature Cell Biol 2001; 3:848-851.

105. Maniotis AJ, Chen CS, Ingber DE. Demonstration of mechanical connections between integrins, cytoskeletal filaments, and nucleoplasm that stabilize nuclear structure. Proc Natl Acad Sci USA 1997; 94:849-854.

106. De Sandre-Giovannoli A, Chaouch M, Kozlov S, Vallat JM, Tazir M, kassouri N et al. Homozygous defects in LMNA, encoding lamin A/C nuclear-envelope proteins, cause autosomal recessive axonal neuropathy in human (Charcot-Marie-Tooth disorder type 2) and mouse. Am J Hum Genet 2002; 70:726-736.
107. Roberts R, Schwartz K. Myocardial diseases. Circulation 2000; 102:34-39.

Index